C. Robertson

Fundamentals of Petroleum

FOURTH EDITION

by Kate Van Dyke

Published by
PETROLEUM EXTENSION SERVICE
Division of Continuing Education
The University of Texas at Austin
Austin, Texas

in cooperation with
ASSOCIATION OF DESK AND DERRICK CLUBS
Tulsa, Oklahoma

1997

Library of Congress Cataloging-in-Publication Data

Van Dyke, Kate, 1951–
Fundamentals of petroleum / by Kate Van Dyke. — 4th ed.
p. cm.
ISBN 0-88698-162-X (pbk.)
1. Petroleum engineering. I. Title.
TN870.V28 1997 97-10098
665.5—dc21 CIP

First Edition published 1979. Fourth Edition 1997
Printed in the United States of America

Catalog No. 1.00040
ISBN 0-88698-162-X

Contents

Foreword

Our nation cannot live without energy. Oil and natural gas provide about 70 percent of the energy used in the United States. Because of its importance, people directly related to and affected by this industry should know more about it. The better people understand the petroleum industry, the greater the chances are for making the proper decisions toward solutions to our energy problems.

The petroleum industry is a leader in developing and applying advanced technology. The Association of Desk and Derrick Clubs recognizes the need for a better understanding of this complex and diversified technology and endorses *Fundamentals of Petroleum,* which is dedicated to individuals seeking knowledge of the petroleum industry. Designed for both the professional and layperson as a basic guide to the practical aspects of the petroleum industry, this book is intended to be used by the association as a primary text for training in house, in colleges, and through correspondence courses. It is also intended to serve as an educational tool for allied industries, as well as professional and governmental agencies. It provides a basic discussion of the petroleum industry from geology through exploration, drilling, production, transportation, refining and processing, and marketing.

Desk and Derrick is a unique organization of approximately ten thousand members employed in the petroleum and allied industries who are dedicated to the proposition that *greater knowledge* of the petroleum industry will result in *greater service* through job performance. Nonshareholding, noncommercial, nonprofit, nonpartisan, and nonbargaining in its policies, the organization has very positive concepts on the value of education for women.

Association of Desk and Derrick Clubs
4823 S. Sheridan, Suite 308-A
Tulsa, Oklahoma 74145

Preface

This fourth edition of *Fundamentals of Petroleum* retains the purpose of the first edition: namely, to give a nontechnical and overall view of the petroleum industry. Like its predecessors, this edition does not cover all the procedures and equipment used in the industry; nor does it give detailed descriptions. Instead, it presents general and useful information that should help lay persons understand the complex world of oil and gas. Technical advances and other changes in the industry made a new edition necessary.

While the book uses simple terms, readers may occasionally run across a word or phrase that they do not understand. In such cases, they may wish to refer to *A Dictionary for the Petroleum Industry* or the *Petroleum Fundamentals Glossary*, also published by PETEX.

PETEX produced the first edition with a great deal of help and support from the Association of Desk and Derrick Clubs (ADDC). Further, the ADDC Education Committee gave a considerable amount of input to PETEX's author and editors. Similarly, this fourth edition incorporates not only changes suggested by ADDC, but also by students, teachers, and readers of preceding editions.

PETEX sincerely hopes that this book will meet the needs of persons outside the industry who are interested in petroleum. We also hope it assists those in the industry, especially members of the Desk and Derrick Clubs everywhere, for it is these men and women who inspire PETEX to provide training materials.

1 Petroleum Geology

Geology is the science that deals with the origin, history, and physical structure of the earth and its life, as recorded in rocks. It is a science essential to the petroleum industry because most petroleum is found within rocks far underground. Anyone interested in the petroleum industry needs to be familiar with the basic principles of geology.

Geologists try to answer such questions as how old the earth is, where it came from, and what it is made of. To do this, they study the evidence of events that occurred millions of years ago, such as earthquakes, volcanoes, and drifting continents, and then relate these to the results of similar events happening today. For example, they try to discover where ancient oceans and mountain ranges were, and they trace the evolution of life through fossils. They also study the composition of the rocks in the earth's crust. In the course of their investigations, geologists rely on the knowledge derived from many other sciences, such as astronomy, chemistry, physics, and biology.

The petroleum geologist is concerned with rocks that contain oil and gas, particularly rocks that contain enough petroleum to be commercially valuable. The company that drills for oil wants a reasonable chance of making a profit on its eventual sale, considering the market price of oil and gas, the amount of recoverable petroleum, the expected production rate, and the cost of drilling and producing the well. So petroleum geologists have two jobs: first, they reconstruct the geologic history of an area to find likely locations for petroleum accumulations; then, when they find one of these locations, they evaluate it to determine whether it has enough petroleum to be commercially productive.

Before we go on, it is important to clear up a common misunderstanding about what an oil reservoir is. Many people think that a reservoir is a large, subterranean cave filled with oil or a buried river flowing with pure crude from bank to bank. Nothing could be further from the truth. Yet it is easy to understand how such notions come about. Even experienced oilfield workers often refer to a reservoir as an *oil pool*. And since many cities store their drinking water in ponds or lakes called reservoirs, this term adds to the confusion. In reality, a *petroleum reservoir* is a rock formation that holds oil and gas, somewhat like a sponge holds water.

And how big is a reservoir? In the oil business, a reservoir's size is determined by the amount of oil and gas it contains. Physically, however, a large reservoir may be broad and shallow, narrow and deep, or some shape in between. The East Texas field covers thousands of acres or hectares but is only 5 to 10 feet (1.5 to 3 metres) thick. On the other hand, the Gronigen field in Holland extends over only about 5 acres (2 hectares) but is some 85 feet (26 metres) thick.

BASIC CONCEPTS OF GEOLOGY

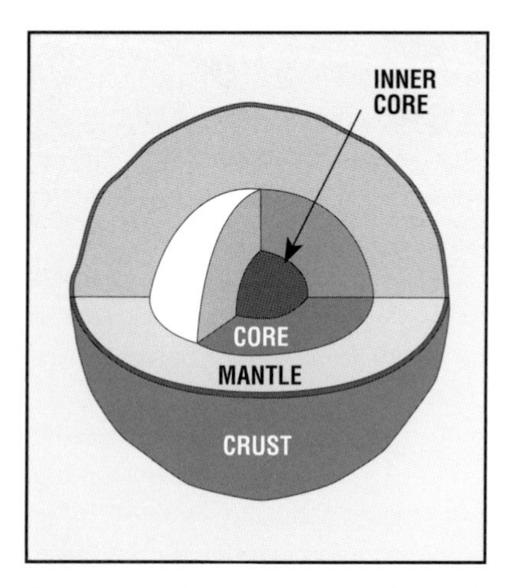

Figure 1.1 A cross section of the earth shows its inner and outer cores, the mantle, and the crust.

Astronomers and physicists think today that the earth was formed at least 4.55 billion years ago out of a cloud of cosmic dust. As gravity pulled the planet together, the heat of compression and of its radioactive elements caused it to become molten. The heaviest components, mostly iron and nickel, sank to the earth's center and became the *core*. Geologists believe that the core has two parts: an inner, solid core and an outer liquid core (fig. 1.1). Both are very hot, dense, and under tremendous pressure. Lighter minerals formed a thick, probably solid *mantle* around the outer core. Certain minerals rich in aluminum, silicon, magnesium, and other light elements solidified into a thin, rocky *crust* above the mantle.

Plate Tectonics

Geologists used to assume, quite naturally, that the continents always lay where they are now. However, resemblances between certain fossil plants in Europe and America, plus an apparent fit between the coastlines, led to the theory that the continents have moved over time (fig 1.2). Most geologists today believe that the crust is an assemblage of huge plates that fit together like a jigsaw puzzle. But unlike a jigsaw puzzle, pieces of the earth's

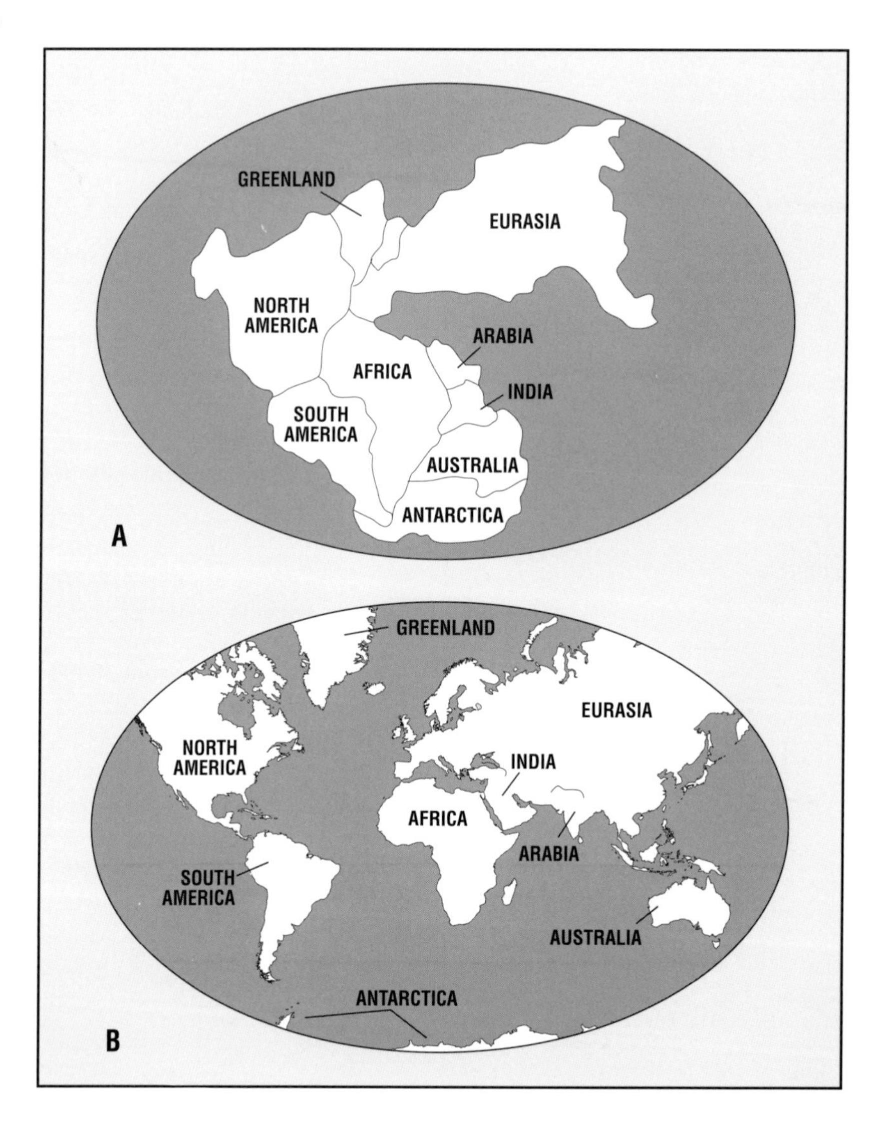

Figure 1.2 The relative positions of the continents as they looked 200 million years ago (A) and today (B).

crust continue to move and change shape. In some places they slide past one another; in others, they collide with or pull away from each other. The theory that explains these processes is called *plate tectonics.*

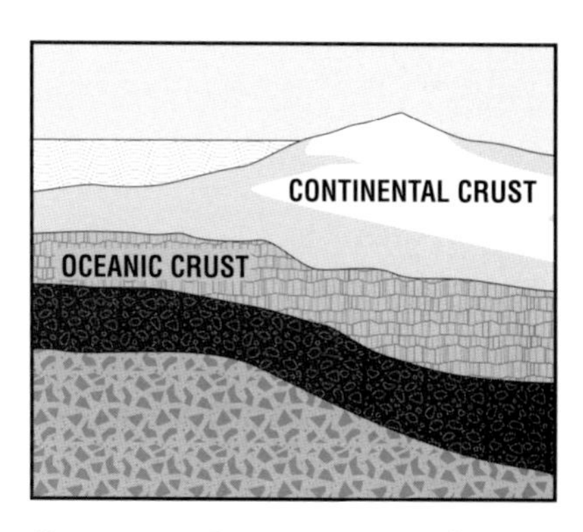

Figure 1.3 Oceanic crust is heavier than continental crust.

Crustal Plates

Geologists distinguish between *oceanic crust,* which lies under the oceans, and *continental crust,* of which the continents are made (fig. 1.3). Oceanic crust is thin—about 5 to 7 miles (8 to 11 kilometres)—and made up primarily of heavy rock that is formed when molten rock (*magma*) cools. The rock of continental crust, however, is thick—10 to 30 miles (16 to 48 kilometres)—and relatively light. Because of these differences, continents tend to float like icebergs in a "sea" of heavier rock, rising high above sea level where they are thickest—in the mountains.

Some of the best evidence for moving plates comes from the bottom of the sea. In the middle of the Atlantic Ocean is a mountain range 10,000 miles (16,100 kilometres) long that snakes from Iceland to the southern tip of Africa (fig. 1.4). It has a deep rift, or trench, along its crest. Investigation of this Mid-Atlantic Ridge has suggested that it is a place where two great plates are moving apart. Along the rift is a string of undersea volcanoes. Each time one erupts, the pressure of the lava pouring out pushes the sides of the rift farther apart. Lava then hardens into rock and becomes new crust between the two plates.

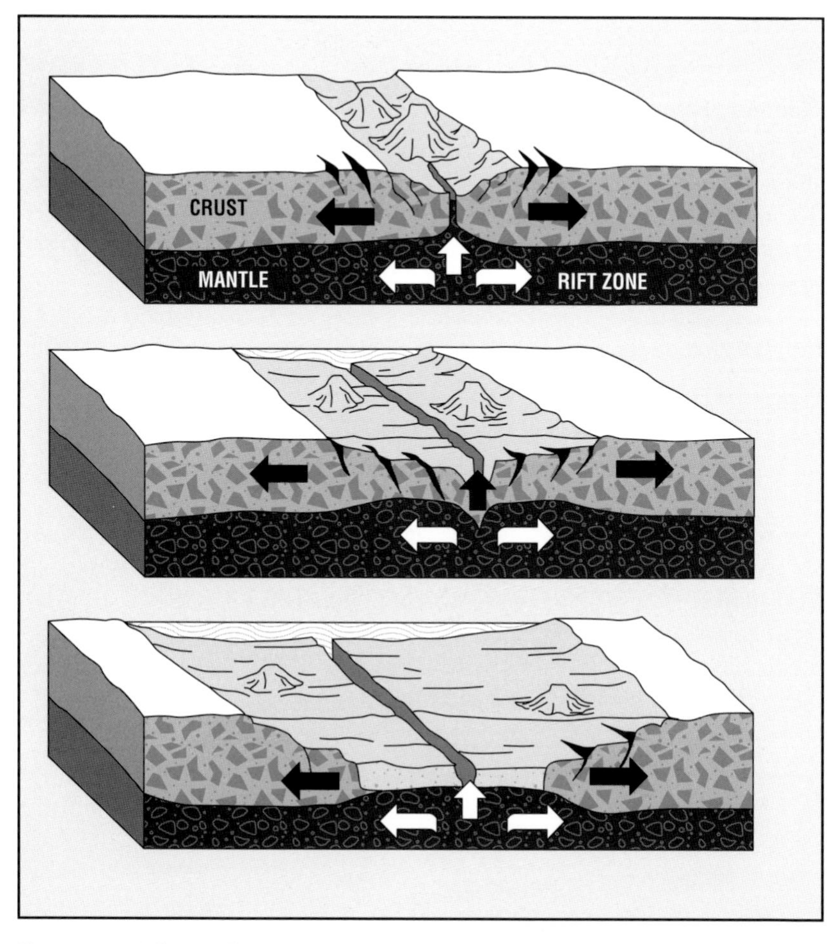

Figure 1.4 The Mid-Atlantic Ridge is an example of two plates moving apart, forming new oceanic crust as lava rising from beneath the plates hardens.

If the plates in the Atlantic are moving apart, then the Pacific Ocean basin is presumably becoming smaller. In fact, geologists believe that the westward movement of the North and South American continents (lighter continental crust) is forcing the Pacific plate (heavier oceanic crust) downward into the mantle (fig. 1.5). This collision of an oceanic and a continental plate accounts for the volcanoes and earthquakes common along this zone.

Geologists also have evidence of what happens when two continental plates collide. They believe that the tallest mountains in the world, the Himalayas, formed when India smashed into Asia. Like an incredibly slow head-on collision between two cars, the crustal collision buckled and folded the rocks along the edges of the two plates. In fact, the Himalayas are still rising by a measurable amount today.

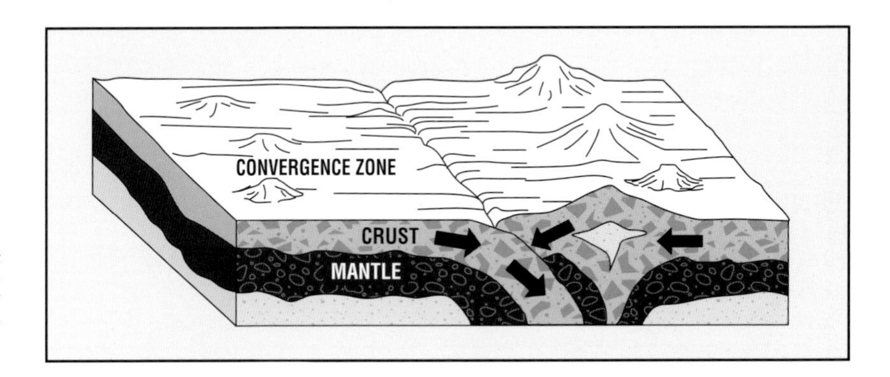

Figure 1.5 Along the Pacific coast, the North and South American continental plates are forcing the Pacific plate downward.

Geologic Structures

All this movement of the earth's crust over millions of years means that the shapes and locations of land masses and oceans have changed. Fossils of marine organisms found in some of the highest mountains and in the deepest oilwells prove that the rocks there were formed in ancient seas and then rose or fell to their present positions.

Geologists describe three basic structures that occur when rocks deform, or change shape, due to tectonic movement: warps, folds, and faults.

Warps and Folds

Near the surface of the crust, at atmospheric temperatures and pressures, rocks tend to break when subjected to great stresses such as earthquakes. However, deeper into the crust, heat rising from the mantle raises the rocks' temperature, and the pressure of overlying rocks compresses them. At these higher temperatures and pressures, the rocks become somewhat flexible. Instead of breaking, they tend to warp or fold when stressed.

Warps occur when broad areas of the crust rise or drop without fracturing. The rock strata in these areas appear to be horizontal but, on closer inspection, are actually slightly tilted, or dipping (fig. 1.6).

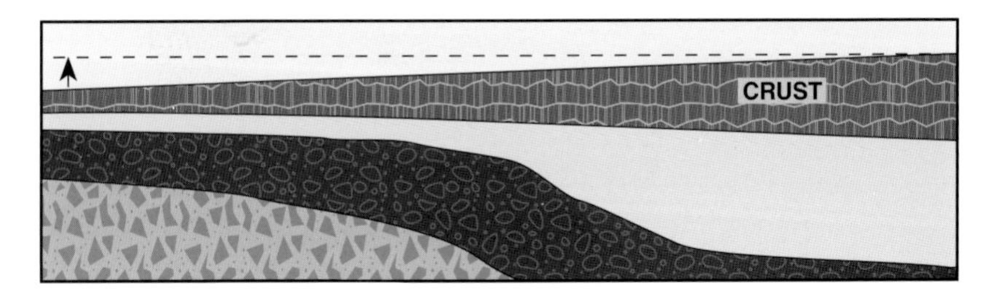

Figure 1.6 A warp is a gently tilted area of horizontal crust.

Figure 1.7 This photo shows deformation of the earth's crust by the buckling of layers into folds.

Rock strata that have crumpled and buckled into wavelike structures are called *folds* (fig. 1.7). Folds are the most common structures in mountain chains, ranging in size from wrinkles of less than an inch to great arches and troughs many miles across. The upwarps or arches are called *anticlines;* the downwarps or troughs are *synclines* (fig. 1.8).

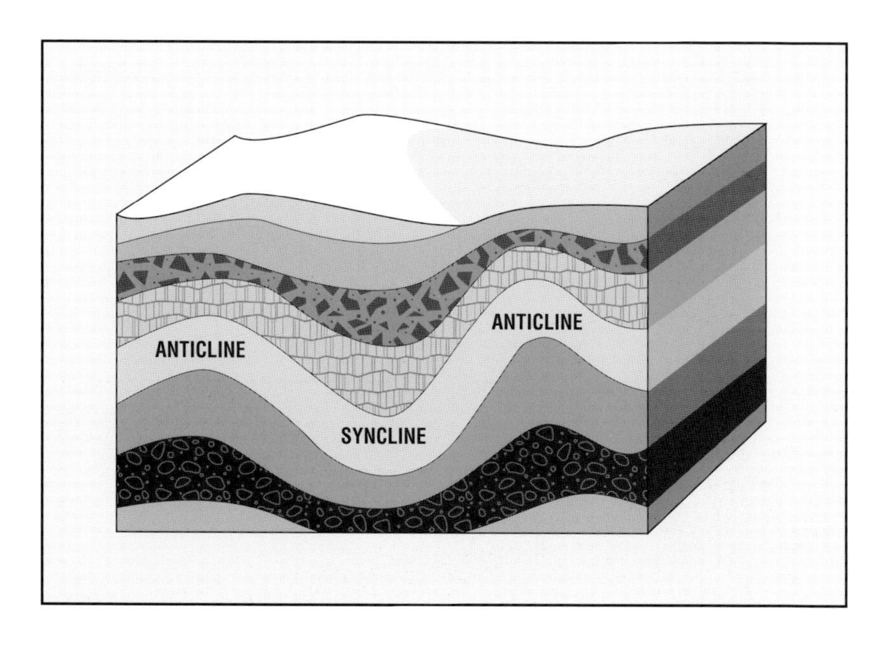

Figure 1.8 Geologists group folds into anticlines and synclines.

Geologists further divide anticlines and synclines according to how the folds tilt. A short anticline with its crest plunging downward in all directions from a high point is called a *dome* (fig. 1.9). Many domes are almost perfectly circular. Some of them have a core of one type of rock that has pushed up into the surrounding rock and lifted it, such as the salt domes along the U.S. Gulf Coast. A syncline that dips down toward a common center is called a *basin* (fig. 1.10). Anticlines and synclines are important to petroleum geologists because they often contain petroleum.

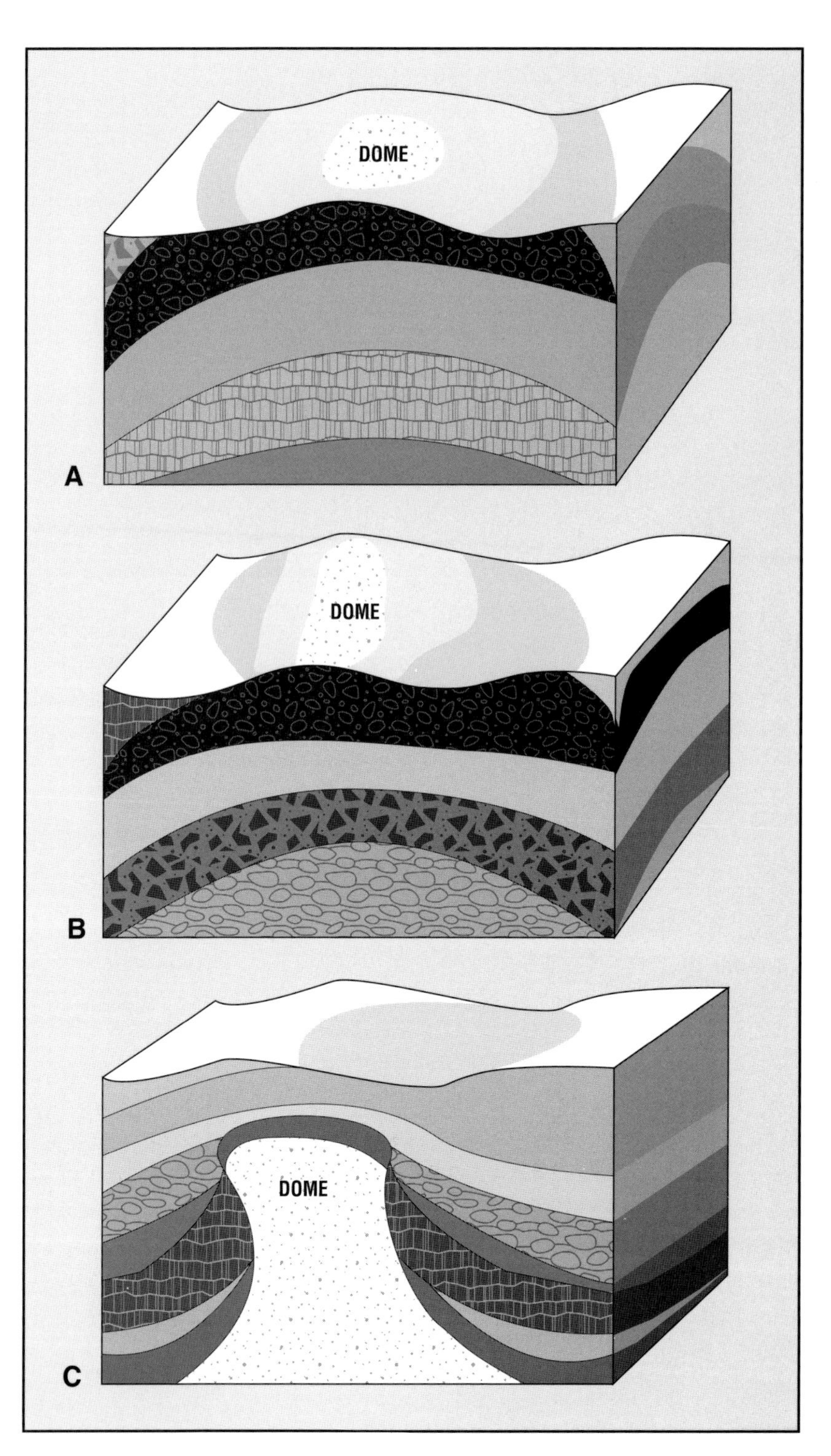

Figure 1.9 A dome may be nearly circular (A) or elongated (B). Some have an intrusive core of salt or other type of rock that pushes up the surrounding rock (C).

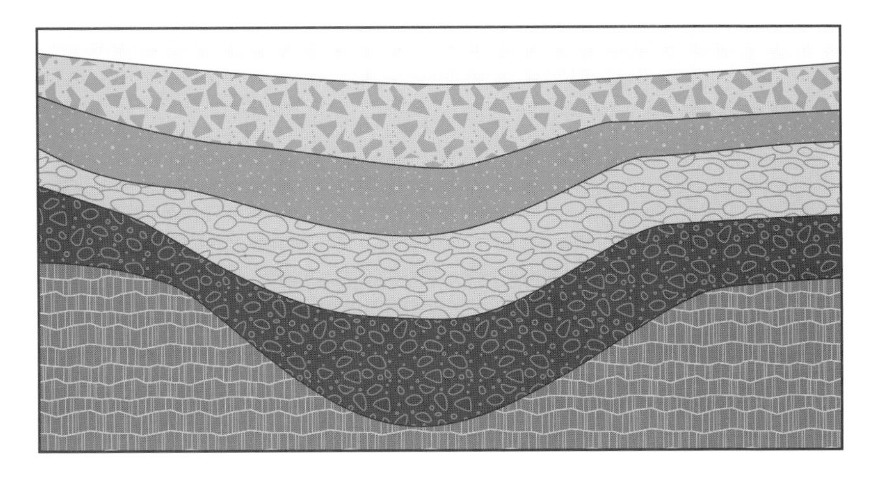

Figure 1.10 A basin is the opposite of a dome.

Faults

When rocks near the surface break, or fracture, the two halves may move in relation to each other. If they do, the fracture is called a *fault*. The two halves along a fault may move apart only a few millimetres or many yards or metres, as along the San Andreas Fault in California (fig. 1.11). Remember that the west coast of the Americas is the boundary between two of the earth's largest crustal plates. The ground next to parts of the San Andreas fault moved horizontally 21 feet (6.4 metres) during the great San Francisco earthquake of 1906.

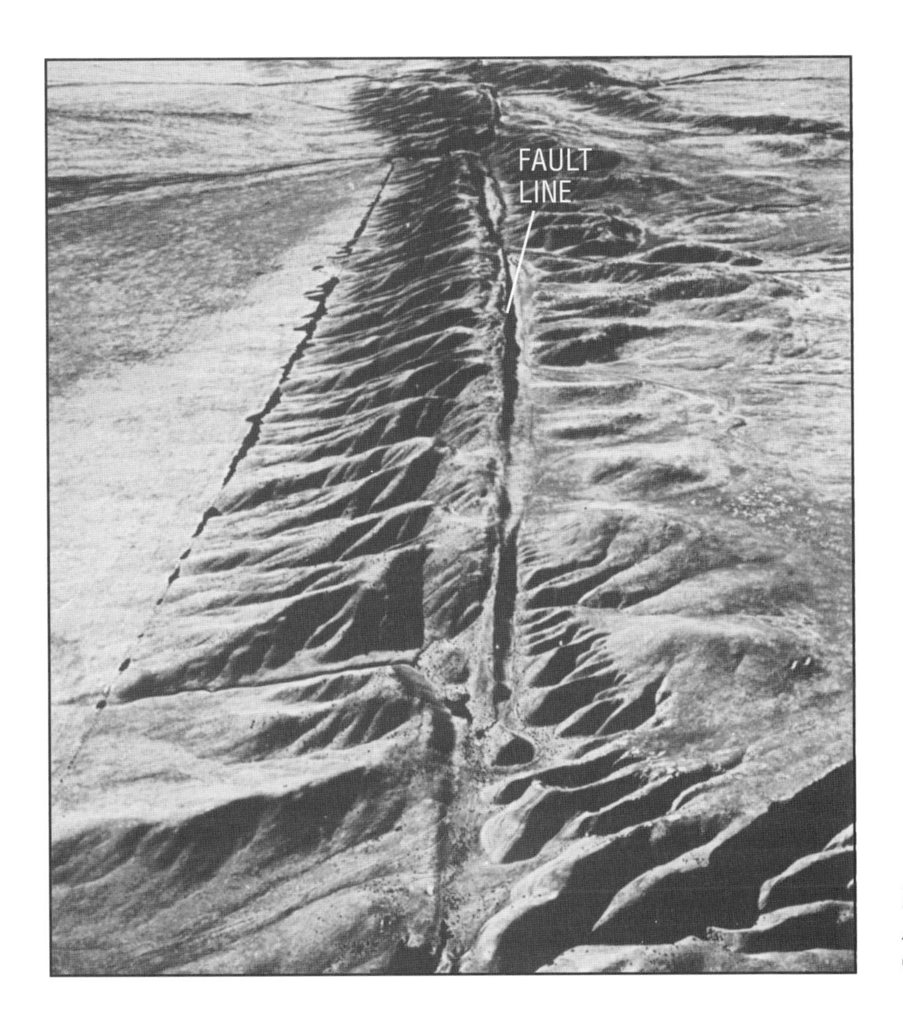

Figure 1.11 This view of the San Andreas Fault in the Carrizo Plain in California shows a distinct fault line.

Geologists classify faults mainly by the direction of the movement. Movement is mostly vertical in *normal* and *reverse faults* but horizontal in *overthrust* and *lateral faults* (fig. 1.12). Combinations of vertical and horizontal movement are also possible, as in *growth faults*. Faults are important to the petroleum geologist because they affect the location of oil and gas accumulations. For example, if a fault runs through a bed of rock containing oil, the geologist can predict where in the same area another part of the original oil-containing rock might have moved.

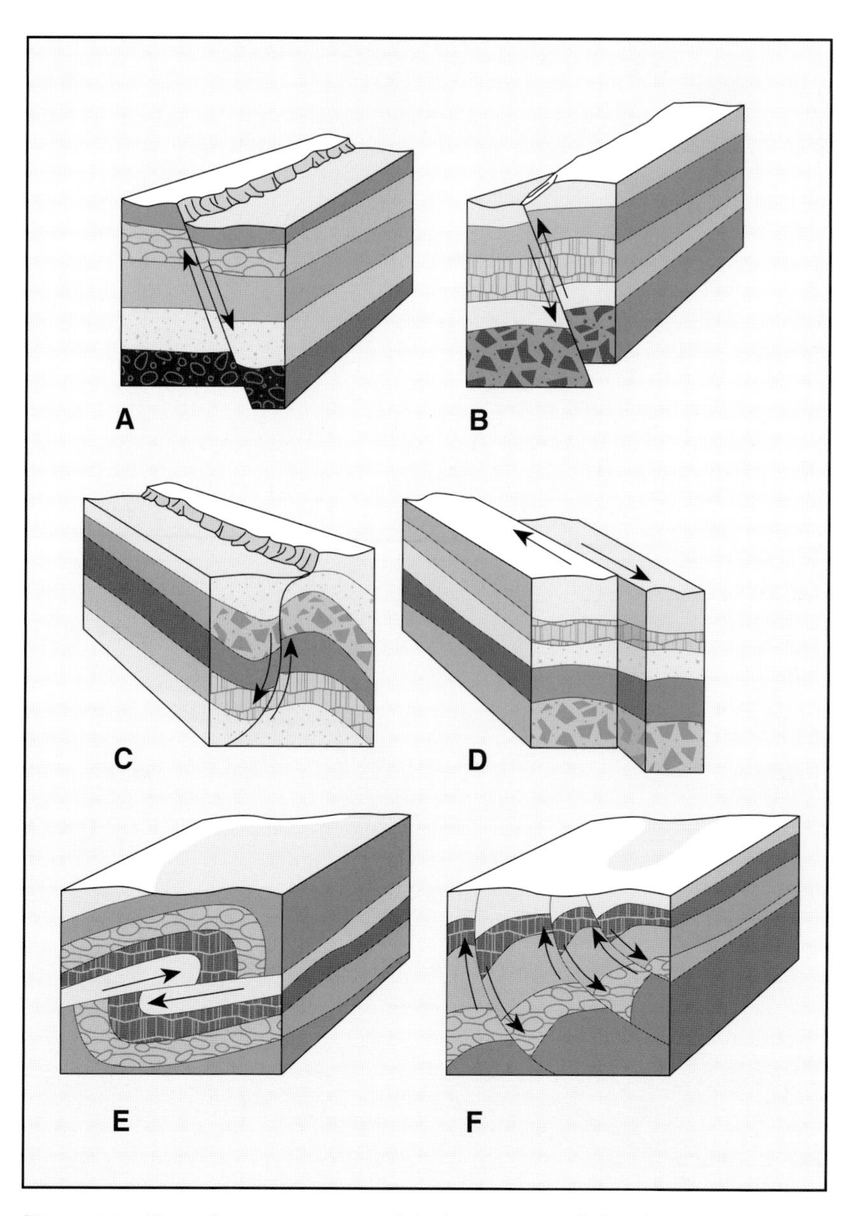

Figure 1.12 Several common types of faults are normal dip slip (A), reverse or thrust dip slip (B and C), lateral (D), overthrust (E), and growth (F) faults.

Sometimes faulting can produce certain recognizable surface features (fig. 1.13). A *graben* is a long, narrow block of crust between two faults that has sunk relative to the surrounding crust. A *horst*, on the other hand, is a similar block that has risen. In the North Sea, oil has accumulated in sediment-filled grabens beneath the ocean floor.

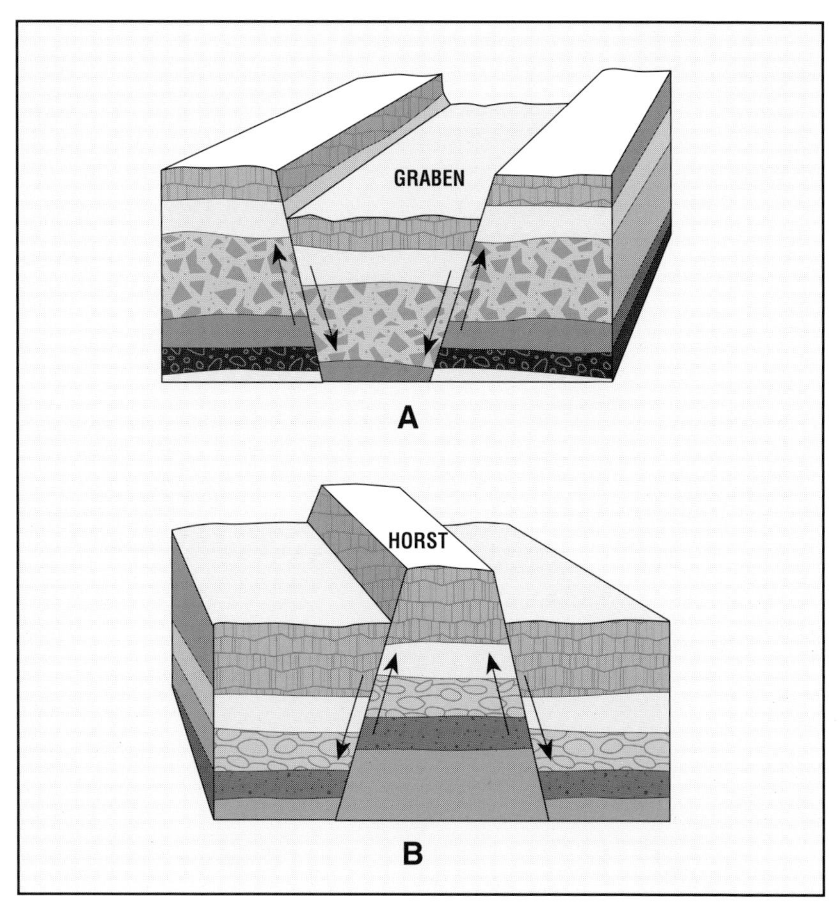

Figure 1.13 Two landscape features formed by faults are the graben (A) and the horst (B).

Life on Earth

About 1 to 1.5 billion years after the earth formed, geologists believe, simple living organisms appeared in the oceans. However, more complex forms (fig. 1.14) did not appear in abundance until about 2.5 billion years later—at the beginning of the Cambrian period, only 550 million years ago. Not until the Devonian period, about 350 million years ago, did vegetation become widespread on the land. Land animals were scarce until even later.

Figure 1.14 Abundant sea life helped form petroleum beneath the ocean floor.

Because life has evolved continuously from Precambrian times, the fossil remains of animals and plants succeed one another in a definite, known order. Geologists have classified rocks in groups based upon this succession, as shown in table 1.1. Geologists have estimated the durations of the eras, periods, and epochs from studies of radioactive minerals. The presence of life may be essential to the petroleum story because, according to the prevailing theory, organic matter is necessary for the formation of oil.

Table 1.1

GEOLOGIC TIME SCALE

Era	Period	Epoch	Duration (millions of years)	Dates (millions of years)
				0.00
		Recent	0.01	
	Quaternary			0.01
		Pleistocene	1	
				1
		Pliocene	10	
Cenozoic				11
		Miocene	14	
				25
	Tertiary	Oligocene	15	
				40
		Eocene	20	
				60
		Paleocene	10	
				70 ± 2
	Cretaceous		65	
				135 ± 5
	Jurassic		30	
Mesozoic				165 ± 10
	Triassic		35	
				200 ± 20
	Permian		35	
				235 ± 30
	Pennsylvanian		30	
				265 ± 35
	Mississippian		35	
Paleozoic				300 ± 40
	Devonian		50	
				350 ± 40
	Silurian		40	
				380 ± 40
	Ordovician		70	
				460 ± 40
	Cambrian		90	
				550 ± 50
Precambrian			4,500 ±	

(After R.M. Sneider)

Categorizing Rocks

Up to now, we have grouped all rock together as the material that the crustal plates are made of. But, of course, it is not all the same. Different kinds of rock contain different minerals, have different physical properties, and were formed in different ways. Geologists group the rocks of the earth's crust into three types according to how they were formed: igneous, sedimentary, and metamorphic.

Types of Rock

Deep in the earth's crust, temperatures are high enough to melt rock into magma. Magma sometimes erupts to the surface as lava, or it may force its way into other solid rock underground. In either case, when magma cools, it solidifies, forming *igneous* rocks, such as granite and basalt.

Sedimentary rocks are rocks formed in horizontal layers, or *strata*, from sediments. A sediment can consist of eroded particles of older rock (igneous, sedimentary, or metamorphic) that wash downhill to lakes or to the oceans. Sediment may also consist of minerals that precipitate out of water. In any case, water is a crucial ingredient in forming sedimentary rocks. Over tens of thousands of years, the layers become thick, and the weight of the overlying sediments compacts the earlier deposits (fig. 1.15). Minerals in the water cement these deposits together into sedimentary rocks. Limestone, sandstone, and clay are typical sedimentary rocks.

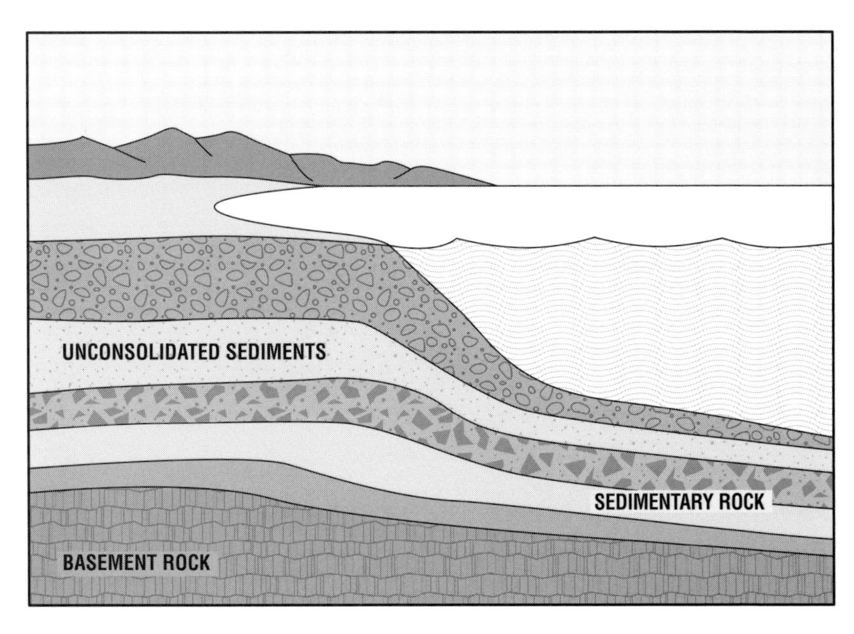

Figure 1.15 The weight of overlying sediments, combined with minerals in water, compacts sediments into rock.

Metamorphic rocks are rocks—either igneous, sedimentary, or other metamorphic rocks—that have been buried deep in the earth where they were subjected to high temperatures and pressures. The term comes from the Greek *meta*, to change, and *morphe*, form or shape. During the metamorphic process, the original rock undergoes physical and chemical changes that may greatly alter its composition and appearance. So, for example, limestone can be metamorphosed into marble, and sandstone into quartzite.

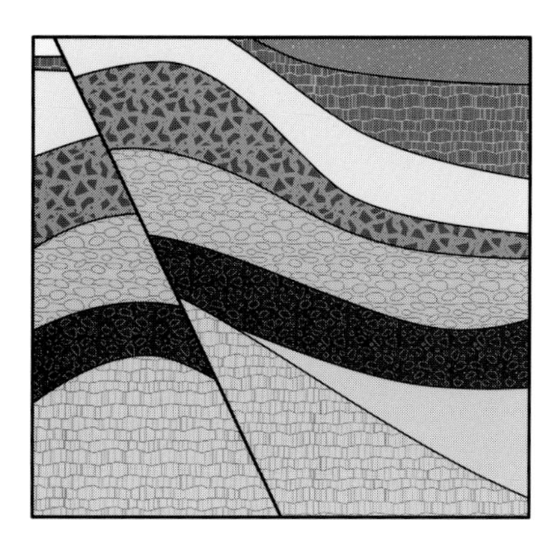

Figure 1.16 Several formations stacked on top of one another are deformed.

A belt of one type of rock is known as a *formation*. Formations are stacked on top of one another and then deformed by folding, warping, and faulting (fig. 1.16).

The Rock Cycle

As you can see, over time, igneous, sedimentary, and metamorphic rocks are all changed into one another. Wind, water, and moving ice erode all types of rock, carry the particles to the ocean or lakes, and create new sedimentary rock. The movement of magma into rock not only creates new igneous rock when it cools, but also metamorphoses the existing rock with its heat. Tectonic movement raises buried rock to the surface, where it erodes, or pushes it deeper into the earth, where it may metamorphose or become magma. Erosion, movement of crustal plates, and movement of molten rock continuously create new types of rock from the old (fig. 1.17).

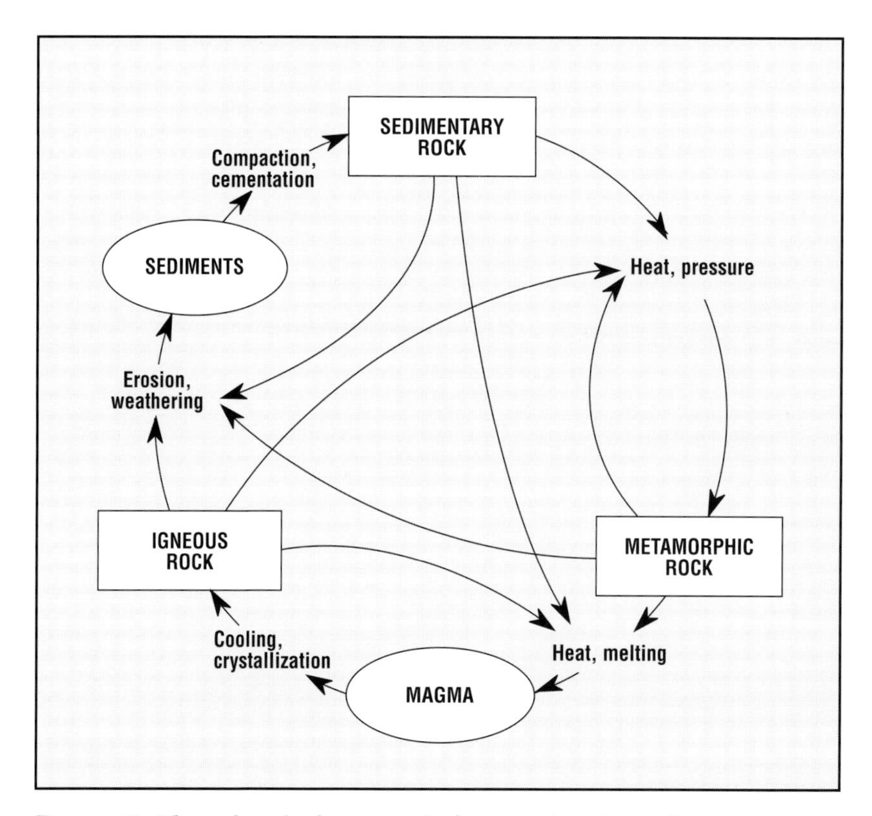

Figure 1.17 The rock cycle changes rocks from one type to another.

Petroleum-Bearing Rocks

Sedimentary rocks are the most interesting to petroleum geologists because most oil and gas accumulations occur in them; igneous and metamorphic rocks rarely contain oil or gas. Furthermore, most of the world's oil lies in sedimentary rock formed from marine sediments deposited on the edges of continents. This is why many of the largest deposits lie along seacoasts, such as along the Gulf of Mexico and the Persian Gulf, for example.

ACCUMULATIONS OF PETROLEUM

Geologists are not sure how petroleum came to exist in sedimentary rocks. But they do know what characteristics petroleum-filled rocks must have in order to remove the petroleum. And they know what sorts of geologic structures and processes create conditions favorable to forming reservoirs large enough to exploit.

Origin of Petroleum

Many geologists believe that petroleum is the result of the breakdown of organic matter—plants and animals—by some unknown process (the organic theory). Some, however, think that living things had nothing to do with it (the inorganic theory).

Organic Theory

The *organic theory* holds that oil and gas formed from the remains of plants and animals. (The term organic refers to living things, or organisms.) Most geologists think that the plants and animals that gave rise to oil and gas were very small, even microscopic in size. In theory, these tiny organisms lived in ancient rivers and seas, as they do today. The rivers carried the plants and animals to the sea, along with silts and muds. The same kinds of organisms lived in shallow seas and the marginal waters of the warmer oceans, where they fell in a slow, steady rain to the bottom as they died. Most were eaten or oxidized before reaching bottom, but some escaped destruction and were entombed in the ooze and mud on the seafloor. As a result, a rich mixture of sediment (grains of silt, sand, and mud) and organic material formed that was cut off from any oxygen dissolved in the water. Without oxygen, the organic material could not decay normally.

In time, after thousands and thousands of years, a thick body of sediment and organic remains built up on the bottom of the sea. Eventually, more sediments were deposited on top of the organic mixture until the great weight of the overlying sediment pushed the lower sediments deep into the earth, where the bottom beds became rock (fig. 1.18). Geologists believe that high heat and pressure, bacteria, chemical reactions, and other forces worked on the organic remains and transformed them into oil and gas.

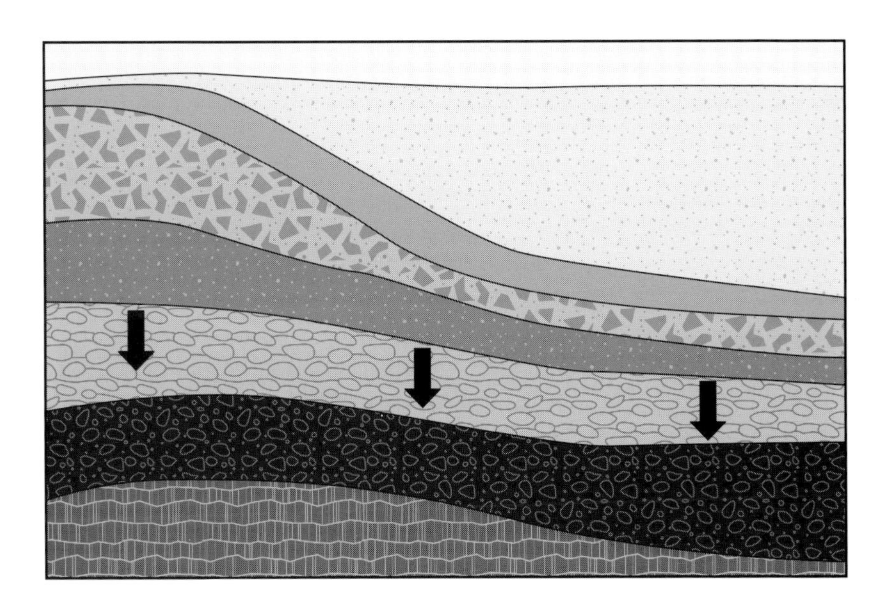

Figure 1.18 According to the organic theory of petroleum formation, beds of silt, mud, and sand were buried deep beneath the earth. The deepest layers were turned into rock by the weight of the overlying beds.

Inorganic Theory

The *inorganic theory,* first put forth in the early 1800s, holds that petroleum is either left over from the formation of the solar system or was formed later deep within the earth. This theory has the advantage of explaining why petroleum deposits are often very deep, in patterns that relate more to large-scale structural features of the crust than to smaller-scale sedimentary rocks. Geologists that support the inorganic theory also believe that it explains why petroleum taken from a large area is often chemically similar even though the formations where it was found are made of different types of rocks of different geological ages.

The Chemistry of Hydrocarbons

The deeper in the crust a rock is, the higher its pressure and temperature. When the temperature reaches about 150°F (66°C), carbon and hydrogen in rocks begin to combine chemically to form hundreds of different kinds of hydrocarbon molecules. *Hydrocarbons* are chains of carbon atoms with hydrogen atoms attached (fig. 1.19). Oil and natural gas are mixtures of different types of hydrocarbons. The chemical process continues to a maximum temperature of 225° to 350°F (107° to 177°C). Above this temperature, the heavier long-chain molecules break into smaller, lighter hydrocarbons, such as methane gas. However, above 500°F (260°C), organic material is destroyed as a source for petroleum. The organic theory holds that organic sediment buried too deeply produces no hydrocarbons because of extreme temperatures.

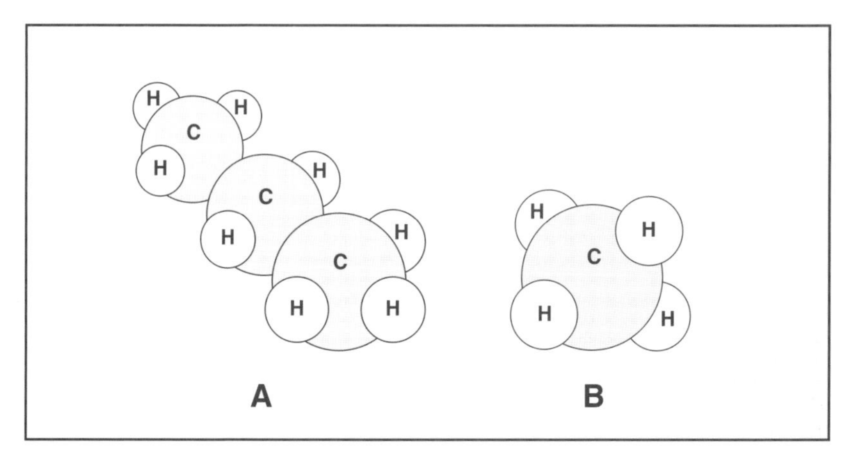

Figure 1.19 Hydrocarbon molecules contain hydrogen and carbon atoms. Heavier propane (A) has a chain of carbon atoms with hydrogen atoms attached. Methane (B) with only one carbon atom, is a smaller, lighter molecule.

Figure 1.20 When reservoir rock is magnified, its porosity can be seen.

Porosity and Permeability of Oil-Bearing Rocks

Although rock appears solid to the naked eye, some rocks have tiny openings, called *pores,* that can be seen under a strong magnifying glass (fig. 1.20). A rock that has pores is *porous,* and its *porosity* can, with difficulty, be measured. Any oil or gas that exists in the rock is in these pores (remember that a reservoir is something like a sponge).

The greater the porosity of a formation, the more petroleum it is able to hold. Porosity may vary from less than 5 percent in a tightly cemented sandstone or carbonate to more than 30 percent in unconsolidated sands.

To be commercially valuable, reservoir rock must have a porosity of 10 percent or more—that is, at least 10 percent of the rock must be pore space, capable of containing petroleum.

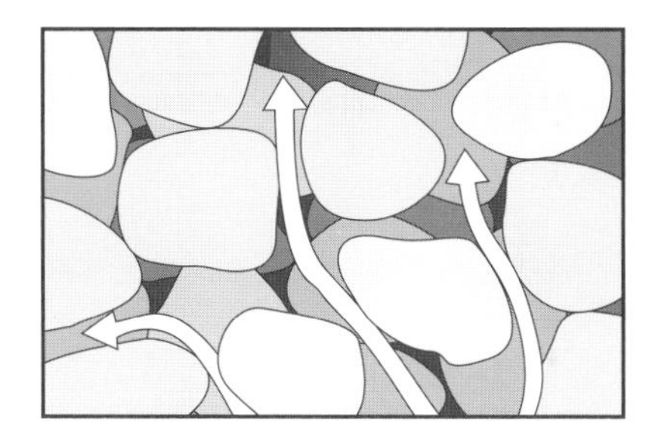

Figure 1.21 A rock is *permeable* when the pores are connected.

Porosity is of no use if the rock is not also *permeable*. A rock is permeable when its pores are connected—that is, oil, gas, and water can flow through it by moving from one pore to another (fig. 1.21). Hydrocarbons in rock with low permeability have difficulty moving about in the rock, and therefore they cannot flow out of the rock and into a well. The unit of measurement of permeability is the *darcy*. Most petroleum reservoirs have permeabilities so small that they are measured in thousandths of a darcy, or *millidarcys*. A porous formation is not necessarily permeable, but highly porous formations are often highly permeable as well. Sandstones and carbonates (such as limestone and dolomite) are generally the most porous and permeable rocks and are also the most common reservoir rocks.

Rock that is porous and contains hydrocarbons but has practically no permeability is called *shale*, and the oil in it is *shale oil*. The western U.S., for example, has vast shale deposits whose petroleum will not flow into a well. Advanced mining techniques and other innovative technology have to be used to get hydrocarbons from this type of rock.

Tar sands are another source of hydrocarbons. These are formations made up of sand cemented together by tar or asphalt, very sticky types of oil. Like shale, tar sands require environmental, economic, and technological innovations to recover the hydrocarbons. Canada, Venezuela, and Russia have large tar sand deposits.

Migration of Petroleum

In the organic theory, petroleum did not form in large concentrations; initially, it was as dispersed as the organic matter it originated from. Once formed, however, it migrated through permeable rock and accumulated in relatively large amounts. As tectonic forces moved the petroleum-bearing rocks out of their birthplace, great pressures from overlying formations squeezed the petroleum out of the relatively impermeable shale into cracks and into more permeable formations such as sandstone. It oozed constantly, following a circuitous path upward, always seeking the surface. Oil and gas tend to seek shallower levels. Unless they are trapped underground by geological formations, they will continue to move upward until they escape at the surface. In fact, sometimes they do exactly this. Since ancient times people have found oil in places known as *seeps*.

Traps

If the reservoir rock is porous and permeable enough, then the petroleum will migrate. But if it is to accumulate, something must stop the migration.

A *trap* is an arrangement of rock layers that contains an accumulation of hydrocarbons, yet prevents them from rising to the surface. The trap consists of an impermeable layer of rock above a porous, permeable layer containing the hydrocarbons. Traps come in all shapes, sizes, and types. Perhaps the easiest way to group them is by looking at the geologic features that caused them to form. The basic kinds of traps are those that formed due to folding, faulting, unconformities, domes or plugs, changes in permeability within a formation, or combinations of these.

Geologists group traps into two basic types: *structural* and *stratigraphic* (fig. 1.22). Structural traps occur when the reservoir formation deforms. Stratigraphic traps are those where porosity or permeability has changed within a formation.

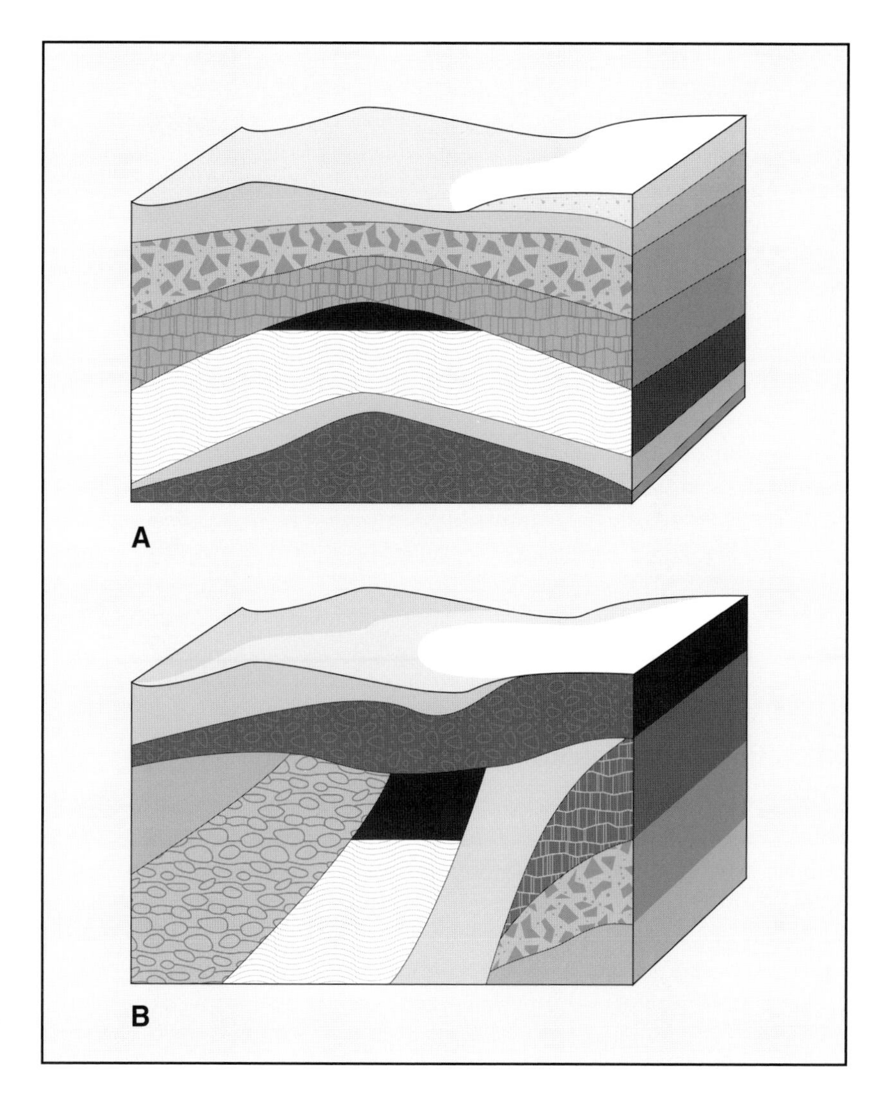

Figure 1.22 Basic types of hydrocarbon traps are structural (A) and stratigraphic (B).

Structural Traps

Structural traps come in many sizes and shapes. Most are formed by the folding or faulting of reservoir rock (fig. 1.23). Some of the more common structural traps are anticlinal traps, fault traps, and dome plug traps.

Anticlinal Traps

Reservoirs formed by folding usually have the shape of anticlines or domes (fig. 1.24). In an *anticlinal trap*, the rock layers that were originally laid down horizontally were folded upward into an arch or dome. Later, hydrocarbons migrated from below into one of the porous and permeable beds in the anticline or dome and accumulated in the top of the folded porous layer. Further upward movement was stopped by the shape of the structure and by a seal or *caprock*—a layer of impermeable rock above the reservoir. Two examples of oilfields with anticlinal traps are the Santa Fe Springs field in California and the Agha Jari field in Iran.

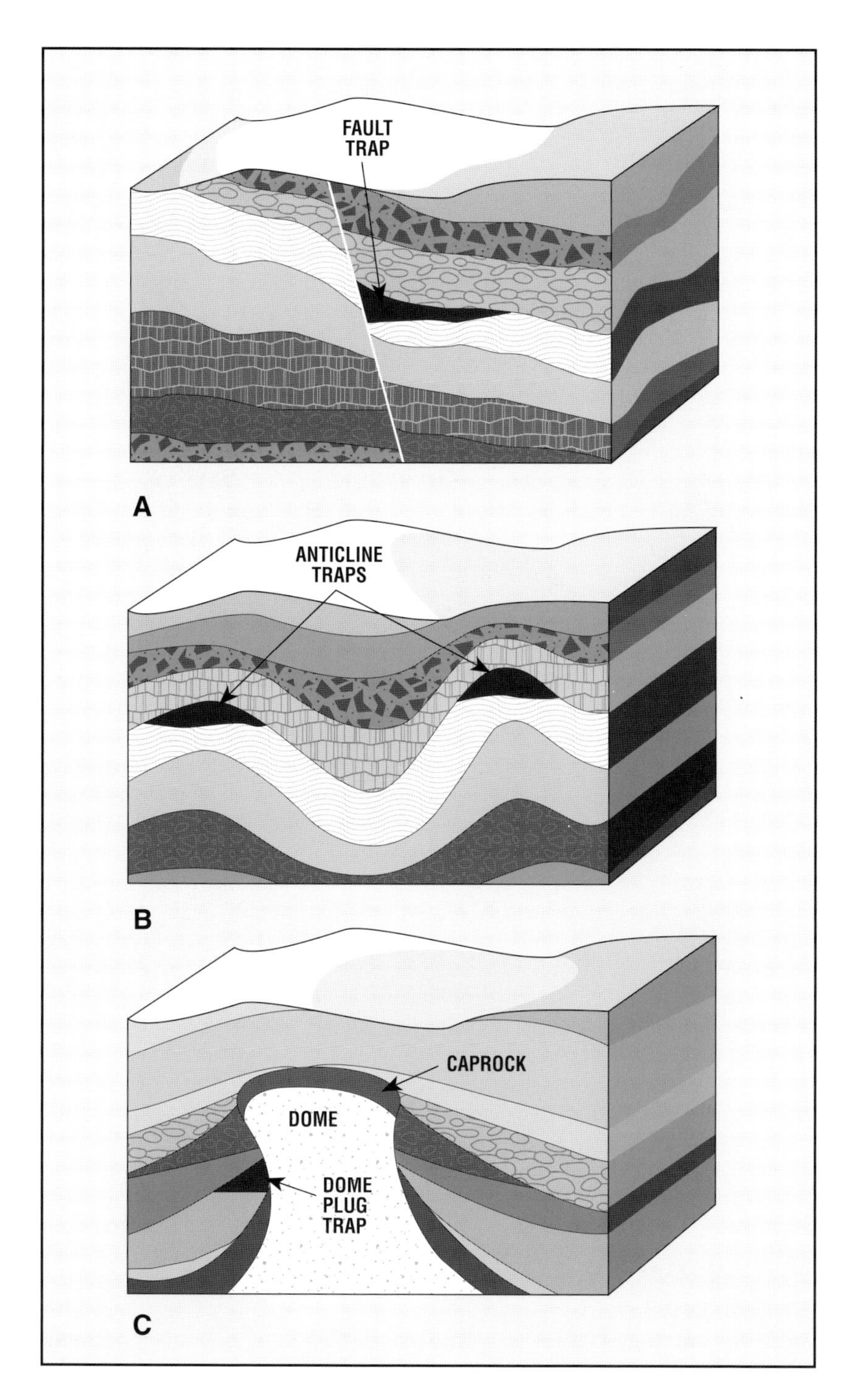

Figure 1.23 Common types of structural traps (shown as dark areas) are fault, anticlinal, and dome plug traps.

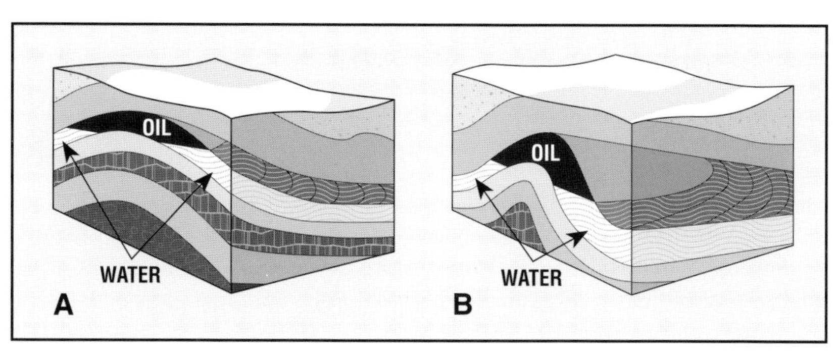

Figure 1.24 Oil accumulates in a dome-shaped structure (A) and an anticlinal type of fold structure (B). An anticline is generally long and narrow, while the dome is circular in outline.

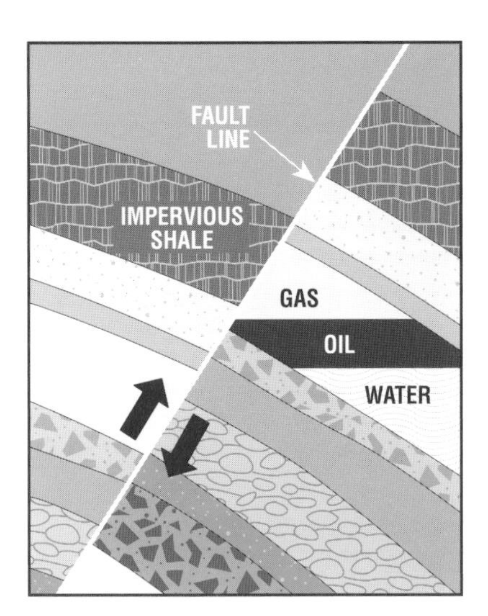

Figure 1.25 Gas and oil are trapped in a fault trap—a reservoir resulting from normal faulting or offsetting of strata. The section to the right of the fault line has moved up, leaving impervious shale opposite the hydrocarbon-bearing formation.

Fault Traps

A *fault trap,* as the name suggests, was formed by the movement of rock along a fault line. The reservoir rock is on only one side of the fault. On the other side, in one type of fault trap, is an impermeable layer that moved opposite the reservoir and prevents hydrocarbons from migrating further (fig. 1.25). In another type of fault trap, the impermeable material is a rock called *gouge* within the fault zone itself. A fault trap depends on the effectiveness of the seal that the gouge or impermeable layer provides.

A simple fault trap stops hydrocarbon migration with a single fault. However, it is possible for two or even three faults to form a trap (fig. 1.26). Petroleum in fault traps tends to accumulate lengthwise, parallel to the fault trend. For example, the accumulations in the many oilfields along the Mexia-Talco fault zone extend from central to northeastern Texas.

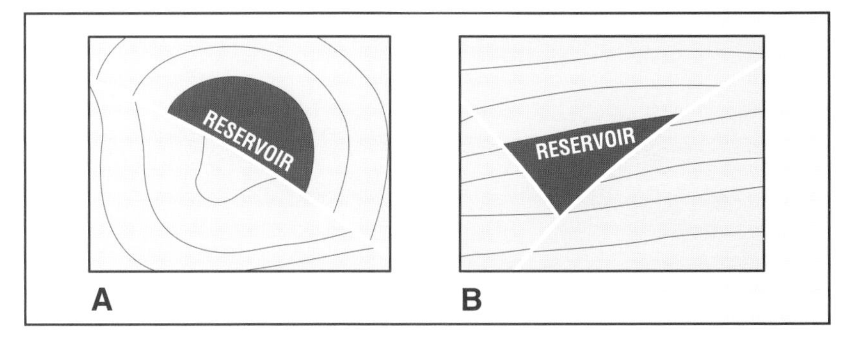

Figure 1.26 Structure contour maps show simple (A) and compound (B) faulting.

Plug Traps

Oil and gas are also found associated with domes. A dome that has a core of rock, called a *plug,* that has pushed into the other formations may create a plug trap. Usually the plug is made of nonporous salt that has pierced, deformed, or lifted the overlying strata (fig. 1.27). Hydrocarbons migrated into any porous and permeable beds on both sides of the column of salt and were trapped there, since they could not flow into the salt. The plug may be nearly circular, as in a typical salt dome oilfield on the U.S. Gulf Coast or in Germany, or it can be long and narrow, as in the Romanian oilfields.

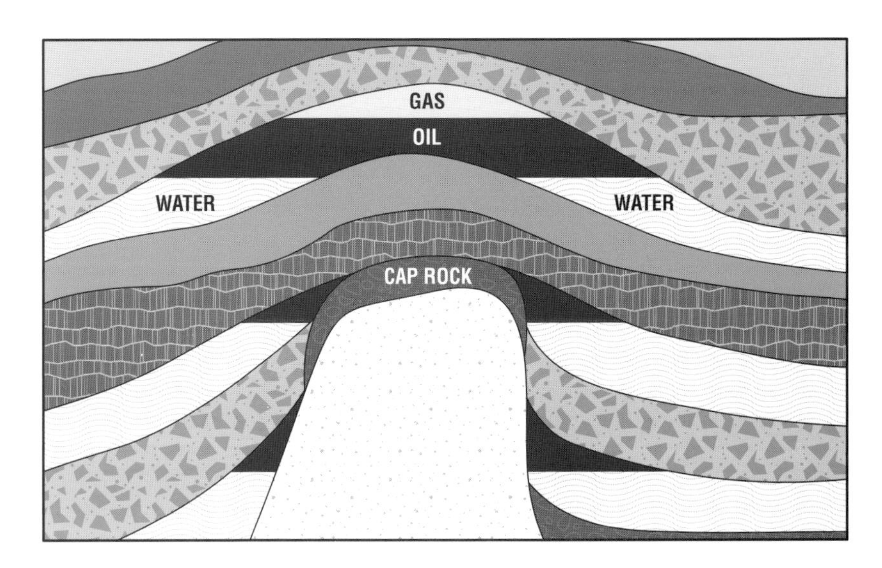

Figure 1.27 A nonporous salt mass has formed dome-shaped traps in overlying and surrounding porous rocks.

Hydrocarbon accumulations in the traps around the outside of a salt plug are usually not continuous but broken into separate segments by faulting (fig. 1.28). This discontinuity can make plug traps difficult to drill successfully. The geologist knows the traps are there but cannot predict their precise locations accurately. As a result, the oil company may drill many dry holes in an attempt to tap the reservoirs. Recent advances in interpretation of exploration data, however, have improved the success rate.

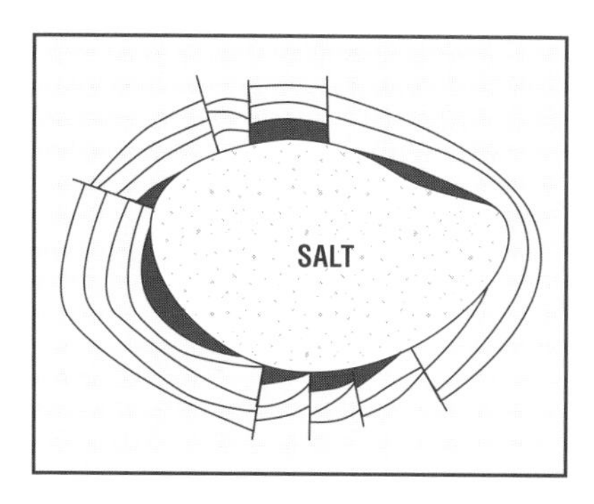

Figure 1.28 Discontinuous peripheral traps form around a piercement salt dome.

Stratigraphic Traps

A stratigraphic trap is caused either by a nonporous formation sealing off the top edge of a reservoir bed or by a change of porosity and permeability within the reservoir bed itself (fig. 1.29).

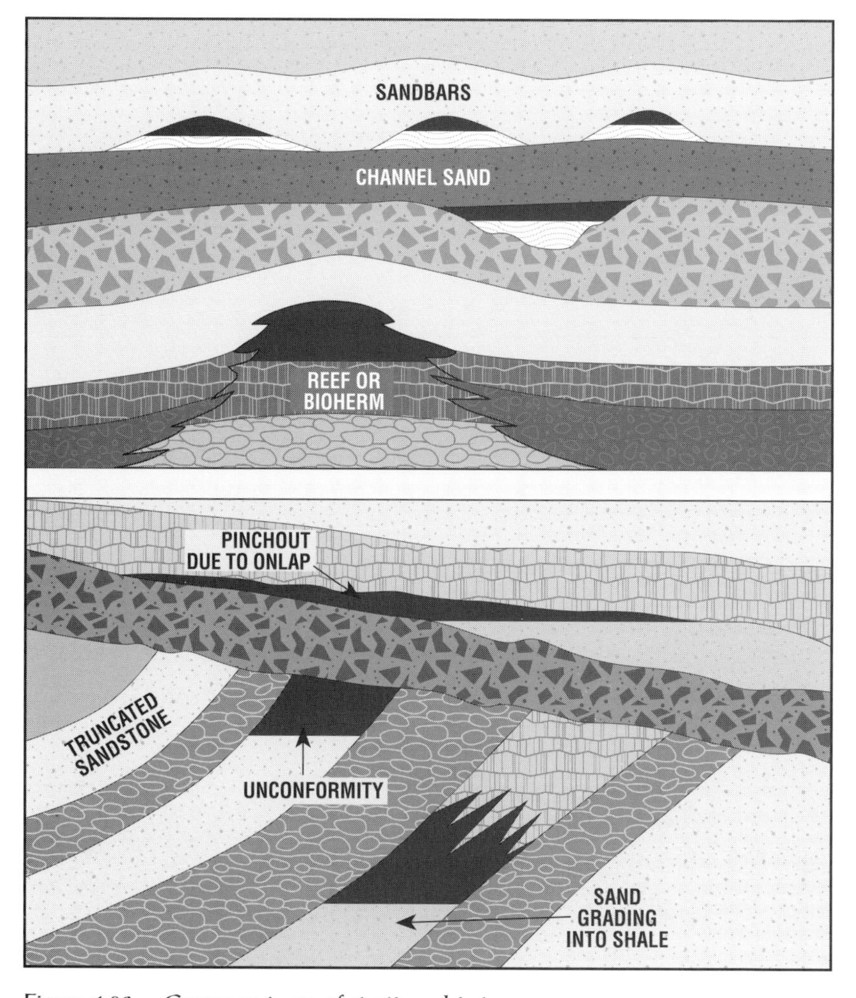

Figure 1.29 Common types of stratigraphic traps

Unconformity

In some places in the earth's crust, a layer of rock has formed and then eroded, and new sediment has been deposited on top of it to form a younger layer of rock. Such a time gap in the geologic record is an *unconformity*. Geologists distinguish among three types of unconformities: the nonconformity, the disconformity, and the angular unconformity.

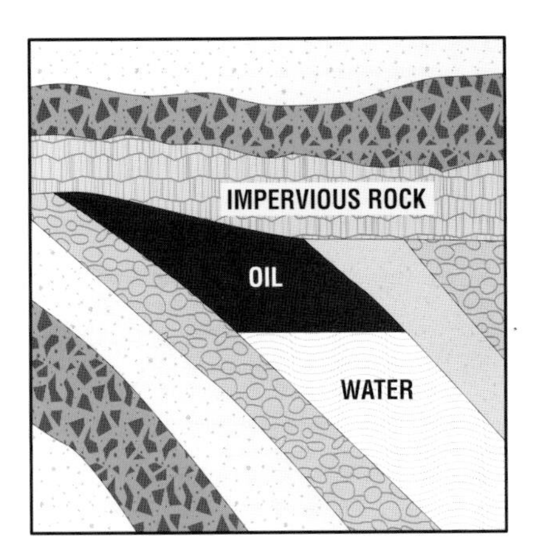

Figure 1.30 Oil is trapped under an unconformity.

A *nonconformity* is an unconformity whose older, eroded layer of rock is igneous and whose newer layer is sedimentary (fig. 1.30). A *disconformity* is an unconformity where the old and new rock layers are parallel to each other. In an *angular unconformity*, the older rock below the unconformity has deformed before the overlying rocks were deposited. This causes the two strata to be tilted relative to each other.

An unconformity can form a trap if part of a porous bed eroded and was then overlaid with impermeable caprock. An example of a reservoir formed by an angular unconformity is the East Texas field.

Lenticular Trap

A change of permeability within a rock layer causes a *lenticular trap*. Abrupt changes in the amount of connected pore space seal off the hydrocarbons in the more permeable part of the bed. These changes may have been caused by an uneven distribution of sand and clay as the sediment was being deposited—for example, in river delta sandbars (fig. 1.31).

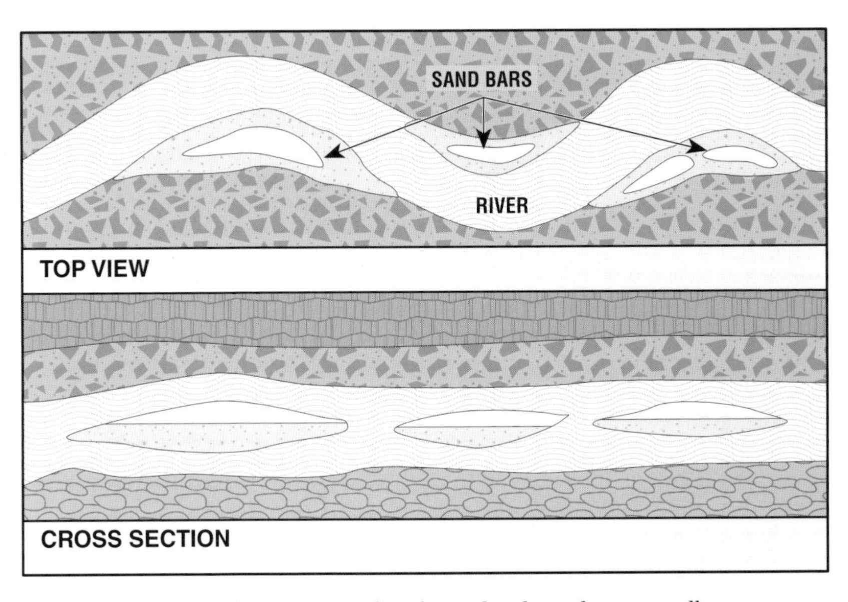

Figure 1.31 Lenticular traps are often formed in buried river sandbars.

Combination Traps

Probably the most common type of reservoir is one that was formed by a combination of folding, faulting, permeability changes, and other conditions. For example, a reservoir could occur in an anticline that is faulted and associated with an unconformity (fig. 1.32). The Seeligson field in Southwest Texas and parts of the East Texas field are reservoirs with combination traps.

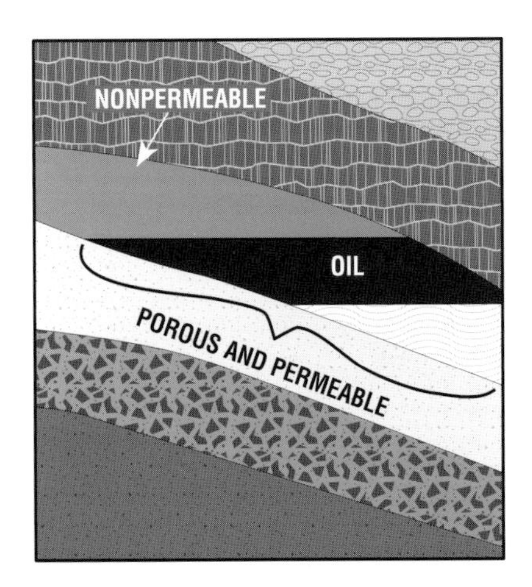

Figure 1.32 A change of permeability within a rock layer can form a trap. The lower part of the layer is porous and permeable, but the upper part is not.

RESERVOIR FLUIDS

A fluid is any substance that will flow. Reservoir rock usually contains three fluids: in addition to oil and gas, it contains salt water. Oil and water are liquids as well as fluids. Natural gas is a fluid but not a liquid in its natural state, although it can be liquefied by artificial means.

In a reservoir, the three fluids can be mixed together or layered or both. When they are layered, the lightest (gas) is on top, the oil is in the middle, and the heaviest (water) is on the bottom, like the contents of a bottle of Italian salad dressing (fig. 1.33). Petroleum companies prefer that a reservoir contain all three fluids in layers because the pressure of gas and water can often drive oil out of the rock to the surface, making pumping unnecessary.

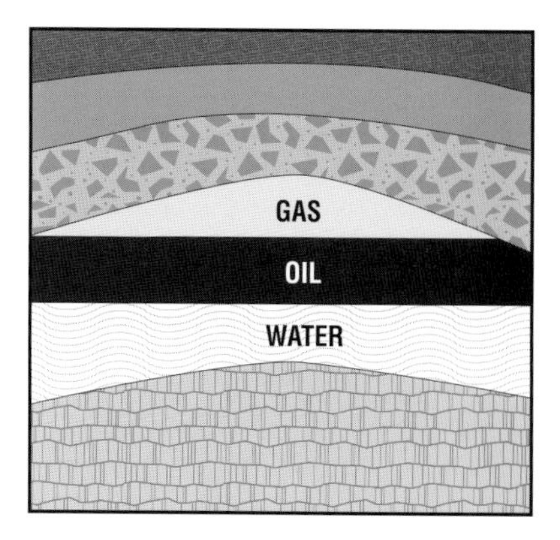

Figure 1.33 When hydrocarbons are layered in a reservoir, the water is on the bottom and the gas on the top.

Water

Most oil reservoirs are sedimentary formations that were deposited in or near the sea. These sedimentary beds were originally saturated with salt water. The forming petroleum displaced part of this water, but some remained. The salt water that remains in the formation is called *connate interstitial water*—connate from the Latin meaning "born with" and interstitial because the water is found in the interstices, or pores, of the formation. By common usage this term has been shortened to *connate water* and always means the water that stayed in the formation when the reservoir was being formed.

Connate water is distributed throughout the reservoir. However, nearly all petroleum reservoirs have additional water that accumulated along with the petroleum. It is this "free" water that supplies the energy for a water drive. *Bottom water* occurs beneath the oil accumulation; *edgewater* is found at the edge of the oil zone (fig. 1.34).

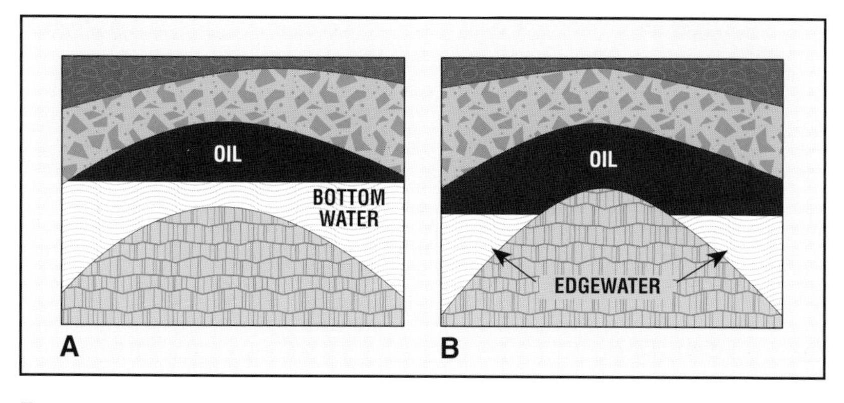

Figure 1.34 Bottom water is below the petroleum in a reservoir (A), and edgewater is at the edge of the oil zone (B).

Oil

Oil, which is lighter than water and will not readily mix with it, makes room for itself in the pores of reservoir rock by pushing the water downward. However, oil will not displace all the water. A film of water sticks to, or is *adsorbed* by, the solid rock material surrounding the pore spaces (fig. 1.35). This film is called *wetting water*. In other words, water is not only in the reservoir below the oil accumulation but also within the pores along with the oil. The rare exceptions are oil-wet reservoirs, which have no film of water lining the pores but which may have an oil saturation of 100 percent of the available porosity.

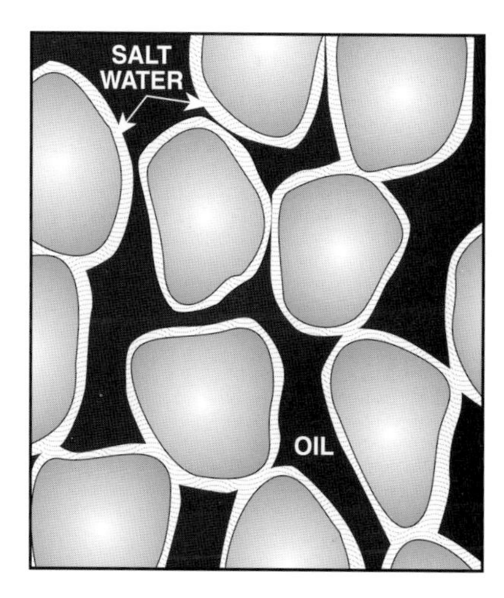

Figure 1.35 Wetting water usually coats the grains of the reservoir rock.

Natural Gas

Reservoirs usually contain natural gas along with oil. The energy supplied by gas under pressure is probably the most valuable drive for forcing oil out of reservoirs. The industry has come a long way since the day it was general practice to "blow" the gas into the atmosphere. Gas occurs with oil and water in reservoirs in two principal ways—as *solution gas* and as *free gas* in gas caps.

Given proper conditions, such as high pressure and low temperature, natural gas will stay in solution in the oil while in the reservoir (fig. 1.36). When the oil comes to the surface and pressure is relieved, the gas comes out of solution, much as a bottle of soda water fizzes when you remove the cap. Gas in solution occupies space in a reservoir, and geologists allow for this space when calculating how much oil is in the reservoir.

Free gas—gas that is not dissolved in oil—tends to accumulate in the highest structural part of a reservoir, where it forms a *gas cap* (fig. 1.37). As long as there is free gas in a reservoir gas cap, the oil in the reservoir will remain saturated with gas in solution. Dissolved gas lowers the *viscosity* of the oil (its resistance to flow), making the oil easier to move to the wellbore.

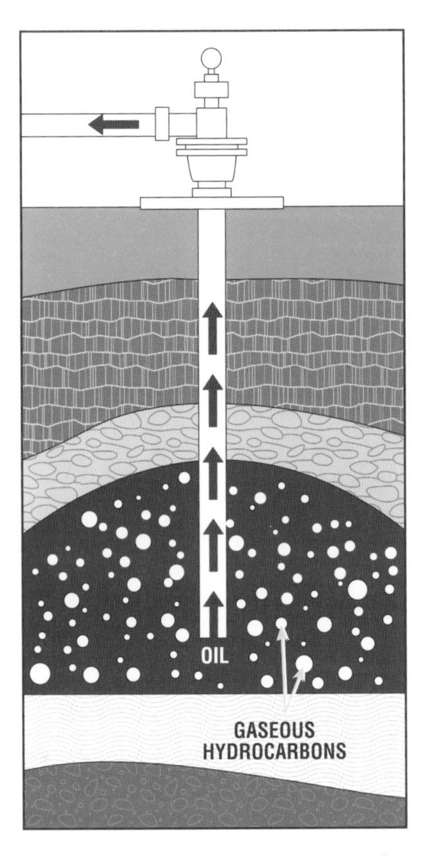

Figure 1.36 Solution gas stays in solution until a well is drilled into the reservoir.

Distribution of the Fluids

Geologists want to find the level in the reservoir where layers of oil and water touch, the *oil-water contact line*. This line is important because, to get the maximum amount of oil from the reservoir, the oil company does not want to pump up water with the oil. Practically all reservoirs have water in the lowest portions of the formation, with oil just above it. However, the oil-water contact line is usually neither sharp nor horizontal throughout a reservoir. Instead, it is a zone of part water and part oil several feet or metres thick. Much the same holds true for the gas-oil contact; but oil, being much heavier than gas, does not tend to rise as high into the gas zone as water does into the oil zone.

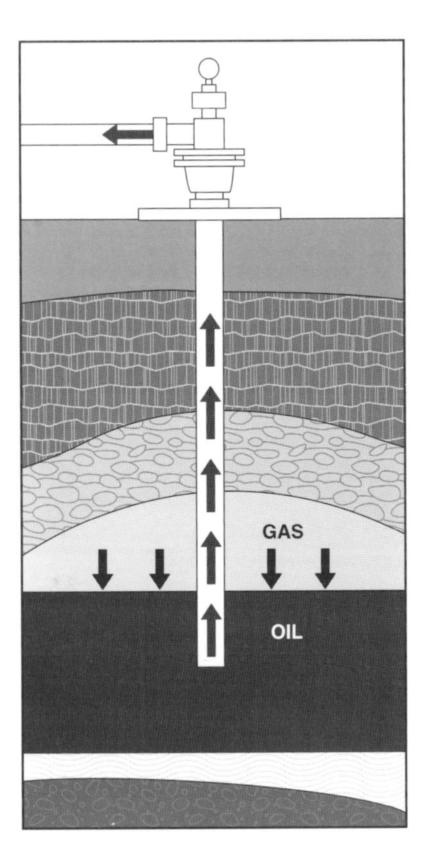

Figure 1.37 Free gas forms a gas cap.

RESERVOIR PRESSURE

All reservoir fluids are under pressure. The weight of the fluid itself creates a normal pressure. Abnormal pressure occurs when the weight of the formations on top of the reservoir is added to the fluid pressure.

Normal Pressure

Fluid pressure exists in a reservoir for the same reason that pressure exists at the bottom of the ocean. Imagine a swimmer in a large swimming pool who decides to see whether he or she can touch bottom. Everything is going well except that the swimmer's ears begin to hurt. The deeper the dive, the more the ears hurt. The reason for the pain is that the pressure of the water is pressing against the eardrums. The deeper the swimmer goes, the greater the pressure.

Just as water creates pressure in a swimming pool, fluids in a reservoir create pressure. When the reservoir has a connection to the surface (fig. 1.38), usually the only pressure in it is the pressure caused by fluid in and above it. As long as this connection to the surface exists, rocks that overlie a reservoir do not create any extra pressure in the reservoir. Even though their weight bears down on the formation, fluids can rise to the surface and escape. Imagine again the swimming pool full of water. Dump a huge load of rocks into it. The rocks do not increase the water pressure; instead the water sloshes over the sides.

The same thing happens in a reservoir. Unlike a swimming pool, however, a reservoir's connection to the surface is usually circuitous. It may outcrop at the surface many miles away, or it may be connected to the surface through other porous beds that overlie it. In most cases, though, as long as the reservoir has some outlet to the surface, the pressure in it is caused only by the fluids and is considered to be normal pressure.

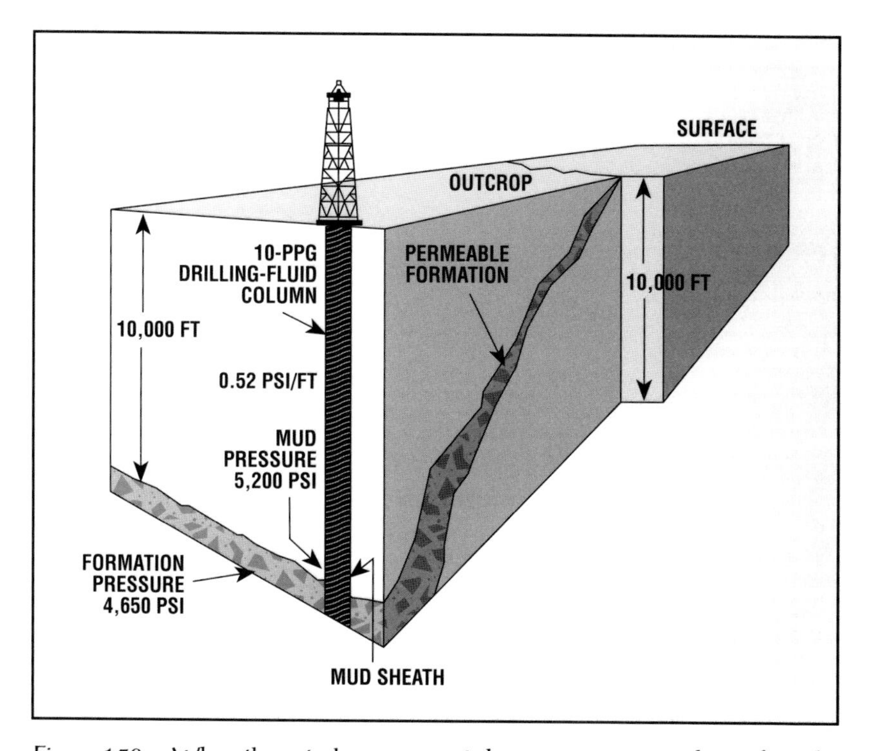

Figure 1.38 When the petroleum reservoir has a connection to the surface, the pressure is considered normal.

Abnormal Pressure

Reservoirs that do not have a connection with the surface are usually surrounded by impermeable formations. In these cases, the overlying rock formations do add to the reservoir pressure. What happens in this case is that the heavy weight of the overlying beds presses down and squeezes the reservoir. Since fluids in the reservoir cannot escape to the surface, the reservoir pressure builds up to abnormally high levels.

If our swimming pool had an airtight lid right on the surface of the water and you dumped that huge load of rocks on the lid, the weight of the rocks would press down on the water, and the water would have no place to escape. The pressure on the water would build and build, as more weight pressed on it.

Abnormally high pressure can also build up due to an artesian effect (fig. 1.39). In this case, formations surrounding the reservoir trap the oil and gas, but allow the water below the oil to reach up toward the surface some distance away. Since water seeks its own level, when the well provides an outlet for the reservoir, the water under it pushes the hydrocarbons up forcefully.

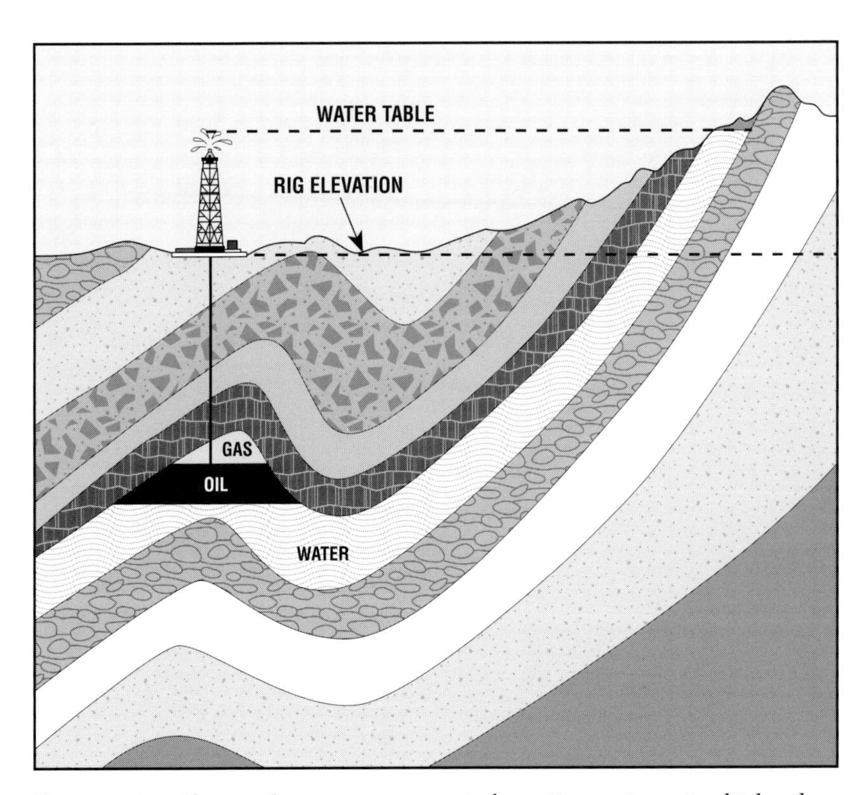

Figure 1.39 Abnormal pressure can occur in formations outcropping higher than the rig elevation.

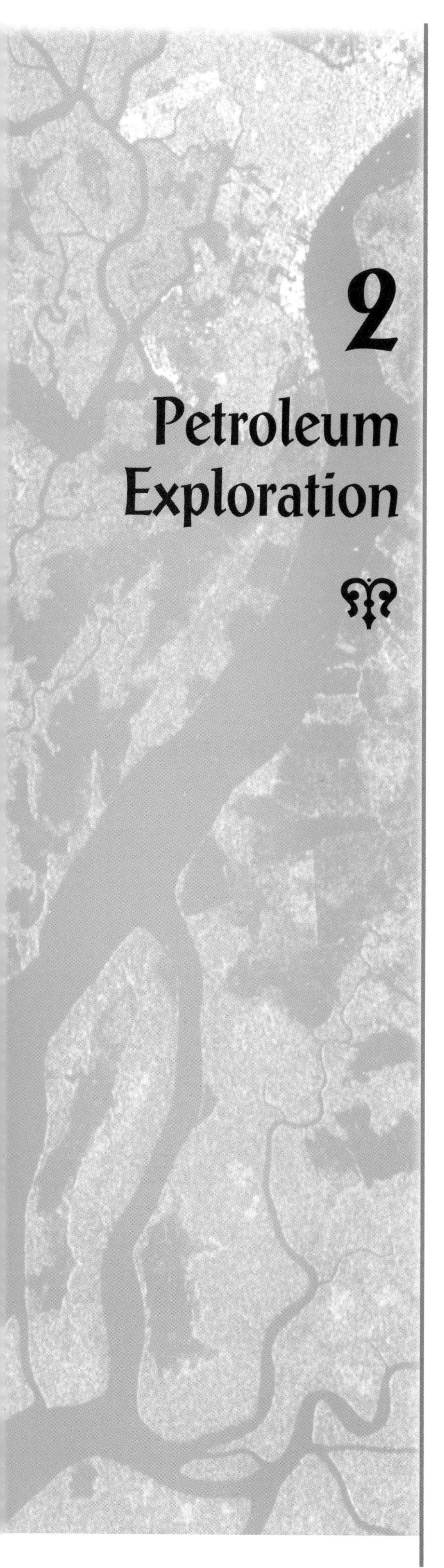

Exploring for petroleum was once a matter of good luck and guess work; now petroleum explorationists use many technical and scientific principles to guide them. The most successful oil-finding method in the early days of oil exploration was to drill near oil seeps, where oil was actually present on the surface. Today, surface and subsurface geological studies drive the discovery of oil and gas. Aerial photographs, satellite images, and various geophysical instruments provide information that helps determine where to drill an exploratory well. Then specialists examine the rock fragments and core samples brought up while drilling the exploratory well and run special tools into the hole to get more information about the formations underground. Examining, correlating, and interpreting this information make it possible for the petroleum prospector to accurately locate subsurface structures that may contain hydrocarbon accumulations worth exploiting.

SURFACE GEOGRAPHICAL STUDIES

In a relatively unexplored area, explorationists first study the *topography*, the natural and man-made features on the surface of the land. Sometimes they can deduce the character of underground formations and structures largely from what appears on the surface.

Aerial Photographs and Satellite Images

Before choosing a site for study, geologists must contend with an unexplored area that may cover tens of thousands of square miles. In order to narrow this vast territory down to regions small enough for detailed surface and subsurface analyses, they may use a combination of aerial and satellite imaging. A series of landscape features that seem unrelated or insignificant to the ground observer may line up quite distinctly when viewed from the air.

In the past, aerial photography was the only way to look at the land from the air. It had some serious disadvantages, however. Besides the expense of flying, aerial exploration requires that the photographer take a large number of photographs from varying camera angles and distances. Because of the unequal quality of these pictures, large-scale geological interpretation is difficult.

Remote sensing has largely replaced aerial photography. *Remote sensing* refers to using infrared or other means to map an area. Both satellites and airplanes may carry remote sensing equipment.

Landsat

Landsat satellites have mapped all of the earth's landmasses many times since the United States launched the first one in 1972. The purpose of the Landsat program has been primarily to map vegetation and observe long-term changes to the earth's surface. Landsat 7, to be launched in 1998, will carry on board ASTER, a sensor especially built for geological applications.

It will provide visible, thermal, and infrared images of all land-masses and coastal areas on the earth.

Geologists use Landsat data to detect the presence of clays often associated with mineral deposits. After sensors on the satellite scan large areas of the earth, they send images of each scene back to earth as electronic pulses. Ground stations throughout the world receive these signals (fig. 2.1). When computers process the Landsat images, earth characteristics as small as 100 feet (30 metres) in diameter are enhanced (fig. 2.2). Companies using Landsat data, however, still need traditional exploration information to pinpoint the location of commercial deposits.

Explorationists can buy satellite data from the government relatively cheaply or from companies that enhance and interpret the satellite images. To analyze the raw data can be expensive, although it costs less than seismic surveys.

Figure 2.1 Landsat photos such as this are received by remote sensing systems and processed by computers. *(Courtesy of the United States Geological Survey EROS Data Center)*

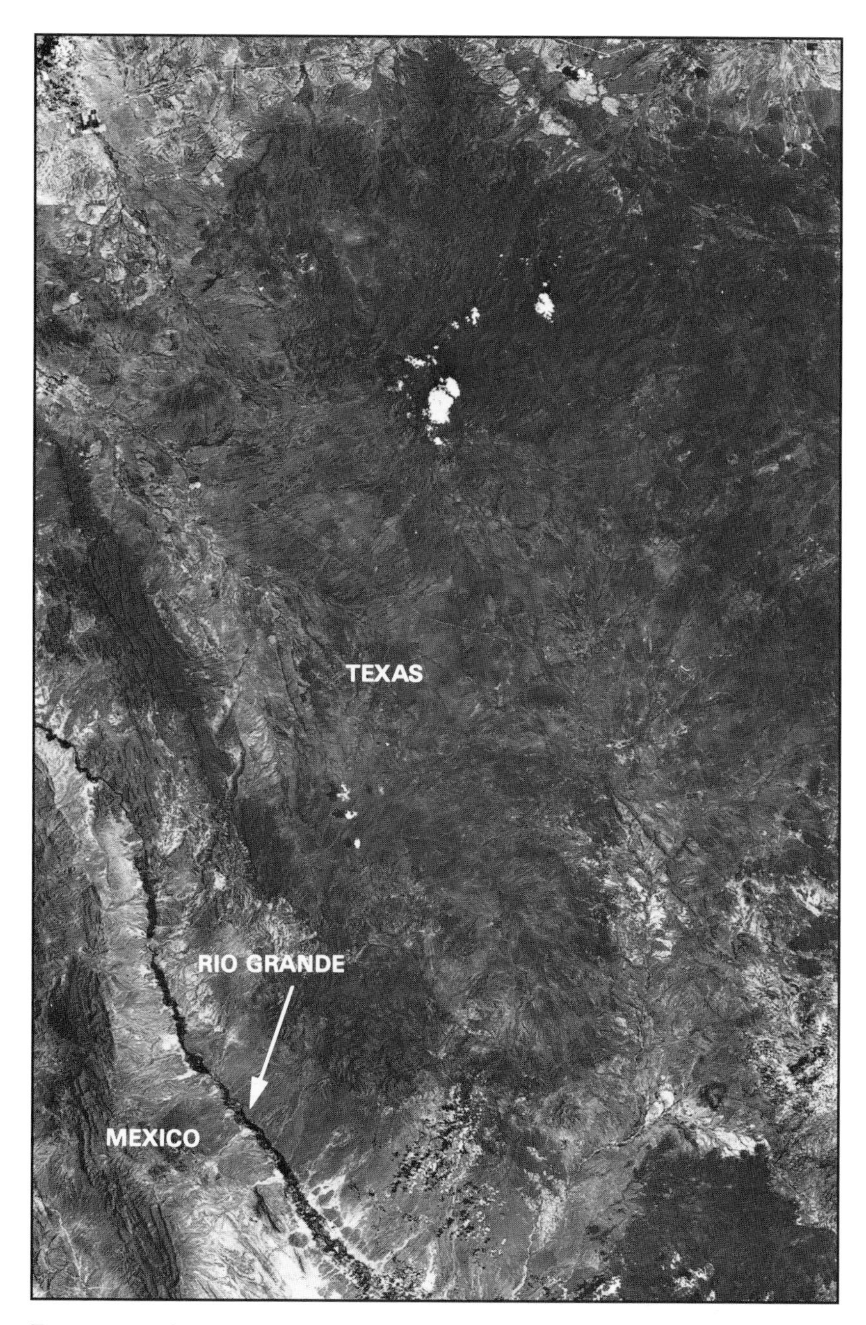

Figure 2.2 This Landsat image of the West Texas area near Presidio shows the Rio Grande bordering Texas and Mexico at lower left. The original photograph is a false-color composite; thus, dark areas are not discernible on this black and white copy. *(Image provided by the Center for Remote Sensing and Energy Research at TCU and printed by the Environmental Research Institute of Michigan)*

Radar

Another type of remote sensing uses radar. Radar devices bounce high-frequency radio waves off land features to a satellite or an airplane. Return echoes are recorded to form a low-resolution relief map, which is useful in searching unexplored areas for potential oil-trapping structures and for discerning large-scale terrain features at a glance. Imaging radar used in airplanes is called *side-looking airborne radar (SLAR).*

Oil and Gas Seeps

Oil and gas seeps are obvious signs of a subsurface petroleum source. Often, however, seepage is so slow that it is not easy to detect, since bacteria and weathering may decompose it as soon as it surfaces. Chemically testing soil or water may reveal traces of hydrocarbons. Even plumes of gas rising from seeps on the ocean floor have led to offshore exploration.

Many of the great oilfields of the world were discovered, in part at least, because of the presence of oil seeps at the surface. Oil seeps occur either along fractures that pierce the reservoir or at spots where formations dip up to the surface (fig. 2.3). Sometimes the seeps at the outcrop of a reservoir are active—oil or gas is still flowing out slowly, as at Mene Grande, Venezuela. In other cases (for example, at Coalinga Field, California), later sediments have buried the part of the reservoir near the surface and completely sealed it, so that the seep is no longer active. The Athabasca tar sands in Canada appear to be a similar seep of the Cretaceous age. Seepage from fractures and faults is common and may be oil, gas, or mud (for example, the mud volcanoes of Trinidad and Russia).

As early as 1842, observers noticed that oil seeps often occur at anticlinal crests. However, not until after the drilling of the famous Drake well in Pennsylvania in 1859 did they note that newly discovered wells were on anticlines. Explorationists made little practical use of this information until 1885, when I. C. White applied it to search for gas in Pennsylvania and nearby states. During the latter part of the nineteenth century, geologists searched for oil in the East Indies and Mexico. By 1897, some U.S. oil companies had established geology departments.

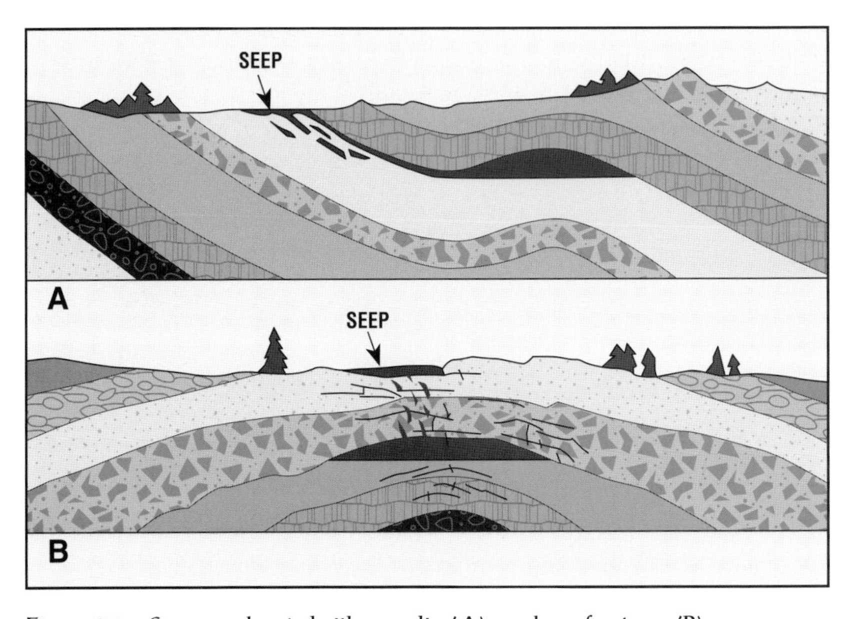

Figure 2.3 Seeps are located either updip (A) or along fractures (B).

COLLECTING DATA

Much of the information that explorationists work with comes from the files, libraries, and databases of the company they represent, government agencies, or privately operated service companies. Some of the information is proprietary and accessible only to those collecting it, while other information is available for a fee or is in the public domain.

Private Company Libraries

A private company or geological society may maintain a large collection of drilling and production data, maps, or well logs (data taken from existing wells). Members pay dues for access to the information. The libraries get information both from public agencies and from members who contribute their own data. An oil company usually has a data library including all information collected on the areas that it has already explored or developed. The company's geologist would use this information.

Public Agency Records

Agencies in the United States that regulate the oil and gas industry collect and file all types of data relating to drilling and production. These files are usually open to the public and accessible to anyone who knows the name of a well, the site of the well, who drilled it, and when it was drilled. Geologists may also get information from a well log or a core sample library, often operated by a public university.

Databases

Both public and private organizations have established computer databases that offer access to a variety of information, generally regional and classified by field and reservoir. Most major oil companies subscribe to petroleum databases, and nonsubscribers may request service on a per-search basis. Many government agencies and universities have made or are in the process of making data available on the Internet for anyone to use.

GEOPHYSICAL SURVEYS

By 1920, explorationists knew of other geological factors besides anticlinal folding that control oil and gas accumulations. Surface mapping alone could not help in searching for these other factors. Fortunately, geophysical methods of exploration came into existence about this time.

Geophysics is the study of the physics of the earth, its oceans, and its atmosphere. Petroleum geologists are most interested in the earth's magnetism, gravity, and especially seismic vibrations, or vibrations of the earth. Sensitive instruments can measure variations in one of these physical qualities that may be related to conditions under the surface. These conditions, in turn, may point to probable oil- or gas-bearing formations. Searching for oil and gas through geophysical means does not guarantee a successful find, but the combination of geophysical information and geological know-how reduces the chances of drilling a dry hole.

Magnetic and Electromagnetic Surveys

Geologists searching for petroleum have made good use of the fact that the earth has a strong magnetic field. Applying the principle that like rocks have like magnetic fields enables the geologist to compare slight differences in the magnetism generated by the minerals in the rocks. These differences suggest the locations of different formations. For example, igneous rock, frequently found as the *basement rock* under sedimentary layers, often contains minerals that are magnetic. Such rock seldom contains hydrocarbons, but, as described in chapter 1, it sometimes intrudes into the overlying sedimentary rock, creating arches and folds that may serve as hydrocarbon traps. Through magnetic surveys, geophysicists can get a fairly good picture of the configuration of the geological formations. A formation out of place with its surroundings may indicate the presence of a fault, which may be associated with hydrocarbon deposits.

Magnetometer Surveys

The *magnetometer* detects slight variations in the earth's magnetic field. These variations mainly show how deep magnetized rocks are buried. A magnetometer can be towed behind a ship or an airplane to cover large areas. It transmits data to a device on board which records the information onto paper or magnetic tape.

A development of airborne magnetics is the *micromagnetic* technique for oil exploration. An airplane tows a micromagnetometer from a low altitude, normally about 300 feet (91 metres) above the ground. It detects micromagnetic *anomalies*, or deviations from the norm. A computer then processes the magnetic data tapes from the aircraft, and geologists use these analyses to predict fractures in the basement and the characteristics of the overlying sediments. The micromagnetic technique is an efficient exploration method that narrows down relatively small areas for seismic surveys (see later section in this chapter) or exploratory drilling.

Magnetotellurics

Magnetotellurics operates on the theory that rocks of differing composition have different electrical properties. Geologists record and measure the naturally occurring flow of electricity between rocks or across salt water, for example, and then analyze and interpret the information to reveal subsurface structures.

Company geologists use magnetotellurics primarily in reconnaissance (exploratory) surveying, although improved data processing techniques have made it increasingly useful for development surveys. The amount of detail that the method yields depends on the distance between survey sites. Closely spaced sites result in a detailed survey, which is helpful in deciding where to drill new wells near an area of proven production. Regional surveys provide geologic cross sections, or cutaway views of the formations.

Although techniques continue to improve, magnetic and electromagnetic surveying does not assure the detection of all traps that contain hydrocarbons. Such surveys are useful, however, for giving the geologist a general idea of where oil-bearing rocks are most likely to be found.

Gravity Surveys

Geophysicists also make use of slight variations in the earth's gravitational field caused by the varying weight of rocks. Some rocks are denser than others; that is, a square yard (or metre) of dense rock weighs more than a square yard (or metre) of less dense rock, in the same way that a lead ball is denser than a cotton ball.

Very dense rocks close to the surface exert a gravitational force more powerful than that of a layer of very light rocks. Geophysicists applied this knowledge particularly in the early days of prospecting off the Gulf Coast (fig. 2.4). They could often locate salt domes by gravitational exploration because ordinary domes and anticlines are associated with maximum gravity, whereas salt domes are usually associated with minimum gravity.

The *torsion balance,* first marketed commercially in 1922, was one of the earliest gravitational instruments invented. It, as well as another early instrument—the pendulum—was rather difficult to use. Today, the most common instrument is the *gravimeter* or gravity meter, a sensitive weighing instrument for measuring variations in the gravitational field of the earth.

Although the basic principle of the gravitational method remains the same, new technology and instruments continue to improve data collection. A small, portable, highly accurate gravimeter is now available for land work.

Figure 2.4 Seismic exploration leads geophysicists from the deserts of Saudi Arabia (A), to Utah's canyon lands (B), and into the freezing plains of Alaska (C). *(Courtesy of American Petroleum Institute)*

Also, gravity data can be collected onboard a ship with a great deal of accuracy, and from the air as well, but with less resolution. Interpreters now use computers to help analyze how gravity variations relate to geology. Gravity maps and models help the geologist examine large areas of development and provide guidelines for planning a seismic exploration program.

Seismic Surveys

A seismic survey is usually the last exploration step before an oil company actually drills a prospective site. Unlike gravity, magnetic, and electromagnetic surveys that provide general information, seismic surveys give the explorationist more precise details on the formations beneath the surface.

Seismology

Seismology works because the earth's crust has many layers with different thicknesses and densities. When energy from the surface, such as an explosion, strikes the layers, part of it travels through the layers and part of it is reflected back to the surface. It is like bouncing a rubber ball. If you drop the ball on the sidewalk, its bounce will be quite different than if you drop it onto a pile of sand. In a similar way, each different layer in the earth "bounces" seismic energy back to the surface with its own particular characteristics.

Seismic surveys start with small artificially produced "earthquakes." Sensors called *geophones* pick up the reflected seismic waves and send them through cables to a recorder. The recorder, a *seismograph*, amplifies and records their characteristics to produce a *seismogram* (fig. 2.5). Seismograms generate a *seismic section*, which is a two-dimensional slice from the surface of the earth downward. The information from a seismic survey indicates the types of rock, their relative depth, and whether a trap is present.

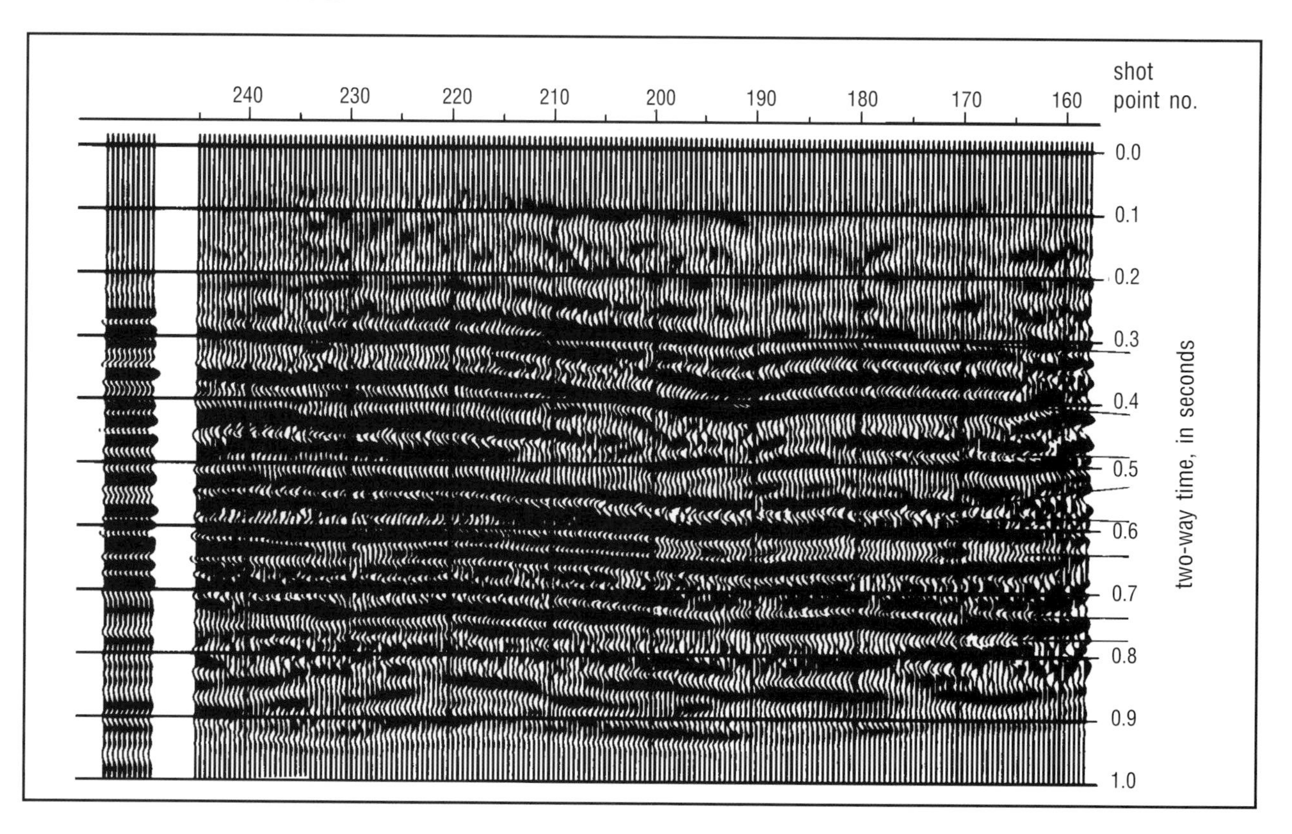

Figure 2.5 A seismic section indicates boundaries between formations.

3D Seismic

The type of seismic section described above is now known in the industry as a 2D (two-dimensional) seismic section because a new technique has developed called 3D (three-dimensional) seismic surveying, or just *3D seismic* for short. In this technique a company runs many seismic surveys close together to create a series of seismic sections of an area perhaps 2 or 3 miles (3 to 5 kilometres) square. Computer programs "paste" these sections together to form a cubic picture of the area. The advantage of 3D seismic is that, with the help of computers, an explorationist can slice the cube in any direction—north-south, east-west, horizontally, or on any other plane (fig. 2.6). Focusing on an area in this way provides much more reliable information about the geologic structures it contains.

3D seismic is becoming popular offshore, particularly in the Gulf of Mexico, where the areas surveyed tend to be larger than on land. An exploration company uses it where it already has a good idea that a large enough oil accumulation exists to justify the expense.

You may also hear the term *4D seismic,* which refers to repeated 3D surveys (through time—the fourth dimension) to monitor changes in the formations, primarily changes in fluid levels.

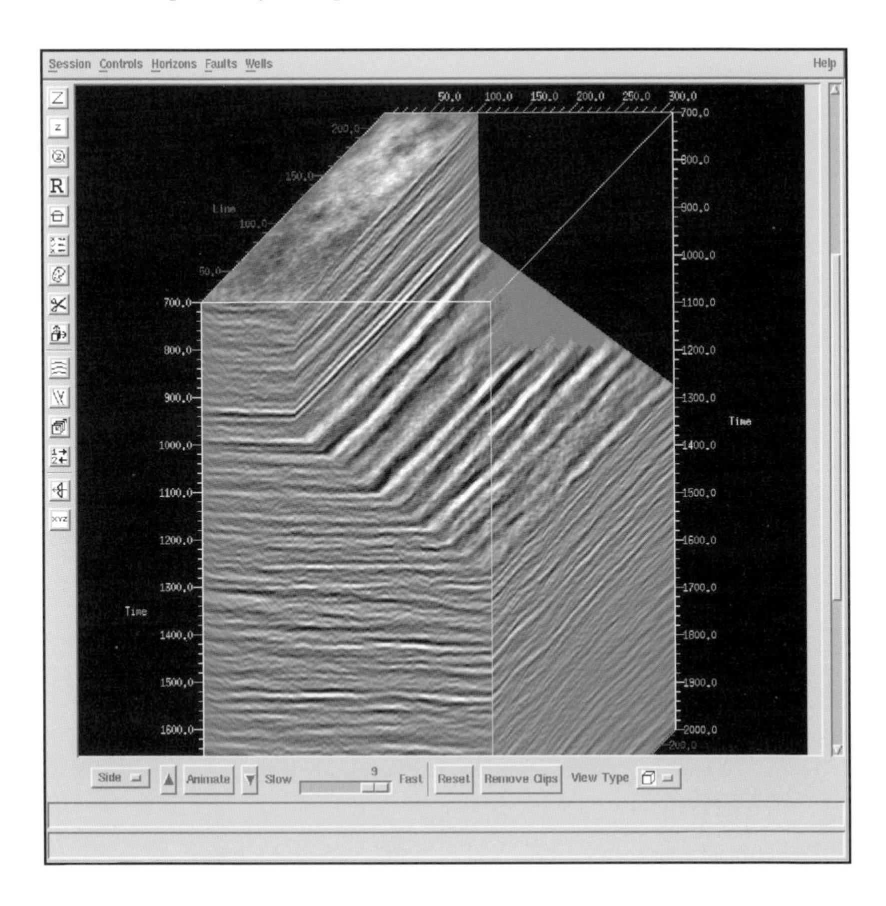

Figure 2.6 A 3D seismic image is a cube that shows the types of rock, their depths, and whether a trap with hydrocarbons is present. *(Courtesy of the Bureau of Economic Geology, The University of Texas at Austin)*

Early Methods

The first seismometer, as inventor David Milne called it, was used in 1841 to measure and record the vibrations of the ground during earthquakes. A few years later, Italian L. Palmieri set up a similar instrument, which he called a seismograph, on Mount Vesuvius. From these simple beginnings seismic exploration evolved.

Dr. L. Mintrop, a German scientist, developed one of the first practical uses of seismic data during World War I. He invented a portable seismograph machine for the German armies to use in locating the positions of the Allied guns. Mintrop set up three seismographs on the battlefield opposite Allied artillery, and when a gun fired, he calculated the precise location so accurately that often the Germans could wipe out that position with one try.

While fighting the war, the Germans realized that miscalculations in distance were due to the variation of seismic vibrations caused by geological formations through which shock waves passed. They then applied basic geological concepts to correct their computations of the distance.

After the war, Mintrop reversed the process, setting off an explosion a known distance from the seismograph and measuring how long it took for the subsurface shock wave reflections to return. Using this information, he estimated the depths of formations. Finding that field exploration confirmed his theories, he put them into practice, forming the first seismic exploration company, Seismos. Soon after Mintrop started the company, the Gulf Production Company hired one of the Seismos crews and brought them to the Gulf Coast of Texas. With their arrival, news of seismic exploration spread, and soon rival companies were opening all over the state.

Two enterprising young brothers—Dabney E. and O. Scott Petty—decided to improve on Dr. Mintrop's methods. They resigned their jobs and spent a year developing a machine much more sensitive than Mintrop's. Their new seismograph used a vacuum tube sensitive enough to register the vibration of a "fly landing on a bar of steel," as O. Scott Petty explained it. They established the Petty Geophysical Engineering Company, which became one of the early leaders in the field.

Explosive Methods

In the past, the most common method of creating seismic vibrations was to place dynamite (the *shot*) into a shallow hole in the ground and explode it (fig. 2.7). Geophones embedded in the earth at various locations around the dynamite picked up the reflected waves. Under certain conditions, dynamite is still used to create seismic waves.

A seismic, or *doodlebug,* crew consists of the *party chief*, geologists or geophysicists to plan the locations of the equipment, surveyors to mark the

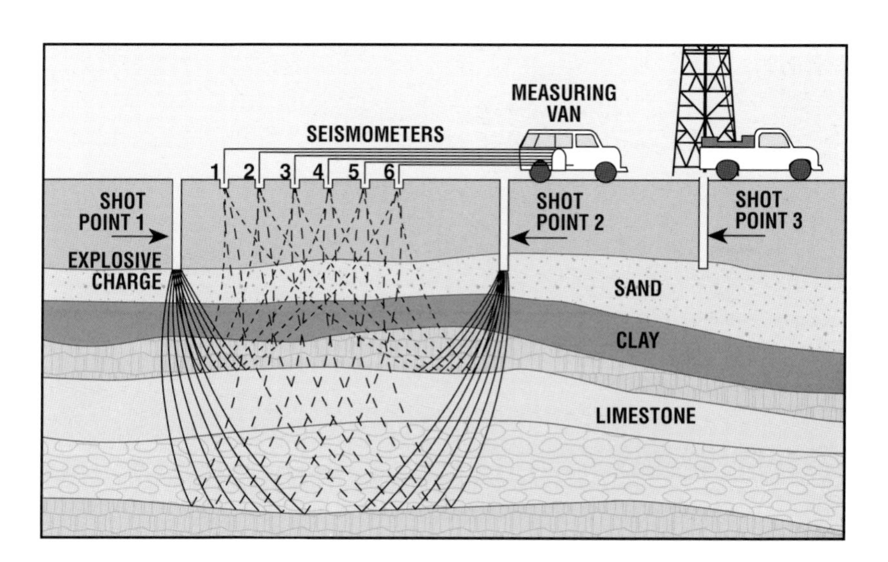

Figure 2.7 In seismic surveying, an explosion at shot point 1 creates shock waves. The subsurface formations reflect the waves to seismometers, and equipment in the measuring van records them.

locations, drillers to drill the holes for the shot, loaders to make up and load the shot, shooters to set it off, and *jug hustlers* to attach the geophones to the recording equipment by means of cables. *Jug* is another name for a geophone.

Modern Land Methods

For the most part, newer methods of creating vibrations in the earth have replaced dynamite. Engineers have developed various mechanical impactors and vibrators to create seismic waves on the earth's surface that penetrate down into the rock layers.

One of the first nondynamite sources of surface energy was the Thumper, developed by Petty-Ray Geophysical (fig. 2.8). This impactor drops a heavy steel slab from as high as 9 feet (2.7 metres) onto the ground to create shock waves. Later, Sinclair Oil and Gas Company developed the Dinoseis, which uses a mixture of propane and oxygen in an expandable chamber to create an explosion. The explosion chamber is mounted under a truck and is lowered to the ground for use. The most popular seismic device used today for land exploration is the Vibroseis, developed by Conoco (fig. 2.9).

Figure 2.8 The Thumper drops a 6,000-pound steel slab (surrounded by safety chains to warn personnel) 9 feet to strike the earth and create shock waves.

Figure 2.9 The Vibroseis truck has a vibrator mounted underneath it that creates low-frequency sound waves. Geophones pick up the sonic reflections for recording *(inset). (Courtesy of Trapp and Eskew Geophysical Consultants)*

The Vibroseis generates continuous low-frequency sound waves whose reflections are picked up and changed into electrical impulses by the geophones. These recorded impulses are sent to a computer for analysis, then printed in the form of a *seismic reflection profile*. Geologists then analyze the profile to determine subsurface structures.

Although explorationists usually use geophones on the surface of the earth to gather seismic data, they may also use them in the borehole of an existing well. In this method, technicians run geophones into a well and attach them to the wall of the hole at intervals of 20 to 100 feet (6 to 30 metres). With the receivers deep in the earth, surface noise is less likely to distort the signal from the seismic source. Another advantage of this method is that it can gather information about the geological structures in the immediate vicinity of the hole.

Marine Seismic Methods

Seismic exploration offshore uses similar equipment as exploration on land but uses it from a ship (fig. 2.10). A sound source sends sound waves through the water, and formations beneath the seafloor reflect the seismic waves to *hydrophones*, the marine version of geophones. Most commonly, hundreds

Figure 2.10 The *R/V Hollis Hedberg*, a geophysical vessel, has logged over 100,000 miles of geophysical data gathering since its christening in 1974. *(Courtesy of Gulf Oil Exploration and Production Co.)*

of hydrophones trail behind the ship on steel cables (fig. 2.11). Another method of positioning the hydrophones is a *vertical-cable survey*. In this type of seismic survey, the ship's crew places cables with hydrophones attached in a specific pattern in the ocean. Each cable has an anchor on one end and strong buoy on the other. Then the ship sets off the source of seismic waves in a certain pattern around the cables.

A seismic ship can stay at sea several months if necessary. It has a double crew, one for ship operations and one for seismic operations.

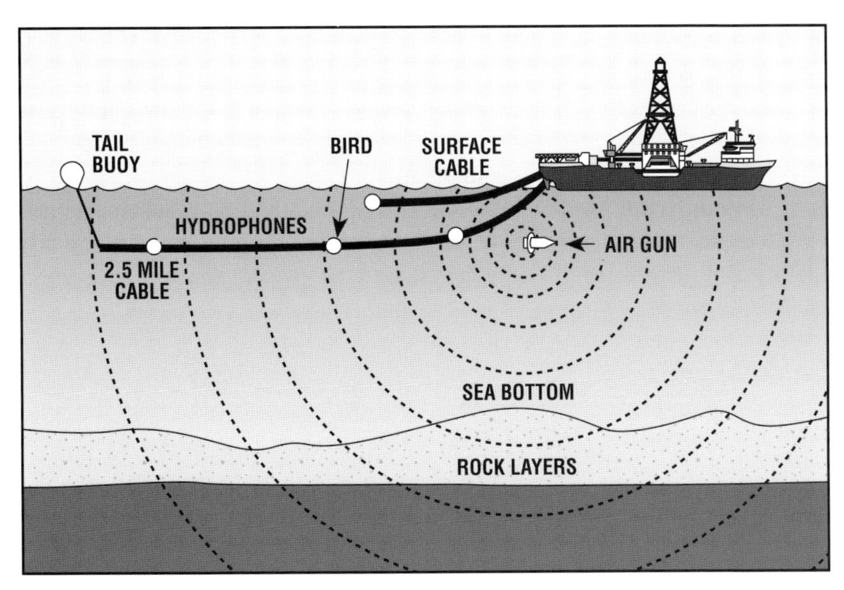

Figure 2.11 Seismic operations at sea use sound to create shock waves whose reflections are picked up by hydrophones.

Sound Sources

In the early days of marine seismic exploration, explosive charges suspended from floats generated the sound waves. Although used for many years, this method is now banned in many parts of the world because of environmental considerations.

Today an air gun is one of the most common sound sources on a seismic ship. Air guns contain chambers of compressed gas. As they release the gas under water, it makes a loud "pop," and the rock layers below the ocean floor reflect this sound. Both the "pops" and their reflected echoes are recorded on magnetic tape.

Another sound source used in offshore exploration is steam. An on-deck boiler produces the steam, which is suddenly released from a special chamber into the ocean. The steam hits the cold sea and condenses rapidly, generating the necessary sound. As with the air guns, hydrophones trailing behind the ship pick up the reflected echoes and record them on magnetic tape for future analysis.

Analyzing the Data

Some geophysical ships have computers on board and geophysicists who can process the seismic information at sea (fig. 2.12). On other ships, the information is held until reaching shore, where it is sent to a supercomputer for processing. Data collected this way can take months to collect and interpret, but supercomputers on shore can handle larger volumes of data than the PC-type workstations on a ship. Sometimes, a helicopter may transport the information directly to the operating company's offices to speed up processing.

The most recent advance is a satellite system developed by the U.S. government and the oil industry in 1996. The satellite can receive the data directly from the ship and beam it back to a receiver on land, where it is sent immediately to a supercomputer. The supercomputer translates the data into a map that is sent to exploration company offices, where geologists can see it in minutes and can direct the ship's captain to adjust course to focus on particularly promising formations.

Figure 2.12 Onboard computers allow geologists to analyze the seismic data as it comes in. *(Courtesy of David M. Stephens, photographer, The University of Texas at Austin)*

RESERVOIR DEVELOPMENT TOOLS

If all the surface and subsurface information indicates a strong possibility of hydrocarbons, the oil company may drill an exploratory well or wells. As drilling progresses, the geologist keeps a watchful eye on the underground rock by means of core samples, well logs, and test results. Specialists combine all of the geological information gathered and recorded as the well is drilled with the surface data and the subsurface findings of the geophysicists in order to try to predict whether any reservoir found has enough oil or gas to justify completing the exploratory well or drilling additional wells in the reservoir.

Well Logs

A *log* is a record of information about the formations through which a well has been drilled. Exploration companies use several different kinds of logs, and each supplies them with specific information about the particular well and the formations encountered in it.

Driller's Log

One of the most common well logs is the *driller's log*, which provides basic information to the geologist. The driller, the person in charge of drilling operations, keeps a record of the kinds of rocks and fluids encountered at different depths and anything else of interest. Particularly when formations are alternating from soft to hard rock, a driller's log may be a very useful tool to geologists. It keeps track of exactly how long it took to drill through a particular formation, and the drillers can correlate that time record with wells drilled later. An astute driller learns to recognize key formations and reports them on the log.

Wireline Logs

Geologists also have several methods of getting indirect information about the formations down the hole, using *wireline logs*. A *wireline* is a metal line that can be run into the hole with a tool attached to its end. A wireline that can carry electricity to the tool is a *conductor line*, but in oilfield slang both are often referred to as wireline.

Wireline logging often involves complex calculations and interpretation of the information the tool relays to the surface. The logging specialist uses a computer to compare, or correlate, data from various surveys, print it out in the form of charts or graphs, keep track of the depth of the logging tool, and warn of malfunctions.

Wireline logs gather data in many different ways under many different conditions; however, the procedure for each log is basically the same. An instrument called a *sonde* is lowered into the wellbore on conductor line. A highly sophisticated electronic instrument, the sonde measures and records electrical, radioactive, or acoustic properties of formations. Then it transmits the signals up the conductor line as it is being raised to the surface at a predetermined speed. At the surface, the wireline unit has equipment—frequently, computers—that translates these signals into graphs that geologists and engineers can interpret. Correlating the information from different types of wireline logs can give the geologist valuable information about porosity, permeability, and the amount of water and petroleum in a formation.

Wireline logging is cheaper, faster, and more accurate than ever due to advances in technology over the past ten years. They are useful not only for evaluating exploration wells but also for planning the best way to produce the oil or gas in a development well. Oil companies get some information from logs run on neighboring wells, while they run some logs routinely on every well.

Among the most common wireline logs are electric logs, radioactivity logs, and acoustic logs.

Electric Logs

Two types of electric logs are records of electrical currents. One type, a *spontaneous potential (SP) log*, records weak electrical currents that flow naturally in the rock next to the wellbore (natural electricity). This shows the boundaries and thickness of each layer of rock. Because the SP log is so simple to obtain and provides such basic information, it is the most common log.

In another technique, a sonde sends an electrical signal through the formation and relays it back to a receiver at the surface (induced electricity). The surface detector may measure either the formation's resistance to the current or how well it conducts the current. A *resistivity log* records resistance, and an *induction log* records conductivity (fig. 2.13). Because salt water conducts electricity much better than oil and gas, resistance and induction logs help determine how much water the formation holds, how freely the water can move (permeability), and how saturated the formation is with water rather than with hydrocarbons.

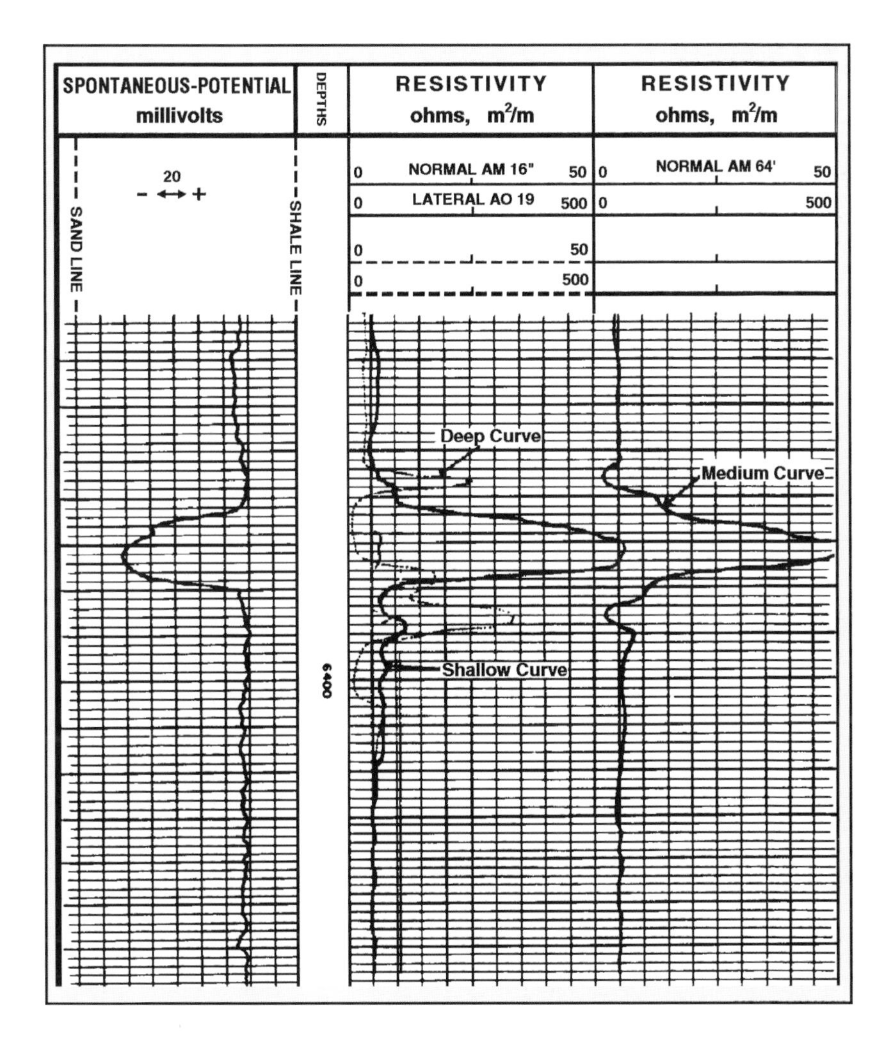

Figure 2.13 The SP log on the left records natural electrical currents flowing from sand and shale. When the curve is closer to the sandline it indicates a sandy formation; when closer to the shale line, shale is present. The resistivity log on the right records the same formation's resistivity to an induced current.

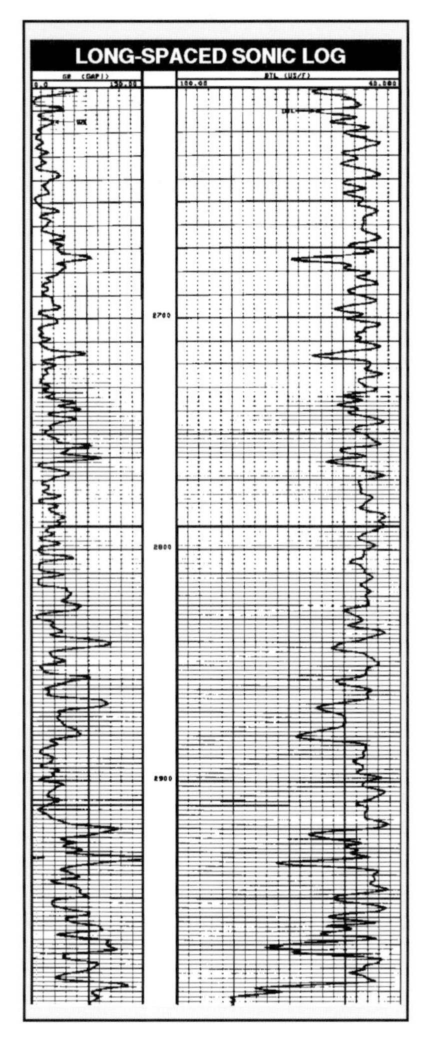

Figure 2.14 An acoustic (sonic) log is a curved line that moves horizontally to show the speed of the sound waves and vertically to show depth.

Nuclear Logs

Just as SP and resistivity logs record natural and induced electrical currents, nuclear logs (also called radioactivity logs) record natural and induced radioactivity. A *gamma ray log* records gamma particles, often called gamma rays, that the formation gives off naturally. A gamma ray log is useful in identifying impermeable formations such as shale and clay-filled sands.

To obtain a *neutron log*, a sonde sends atomic particles called neutrons through the formation. When the neutrons collide with hydrogen, the hydrogen slows them down. When the detector records slow neutrons, it means a lot of hydrogen is present; hydrogen is the main component of water and hydrocarbons, but not of rocks. So the log shows how much oil, gas, and water may be in the formation; the type of rock and its porosity; and the salt content.

Acoustic Logs

An *acoustic*, or *sonic log* is a record of sound waves sent through the rock by an acoustic sonde that transmits and receives a constant ticking sound (fig. 2.14). How fast sound travels through a rock depends on how dense it is and how much fluid it contains. As rock becomes more porous, the sound waves travel faster.

Sample Logs

Sample logs are logs of physical samples of the underground rock. The two types are core samples and cuttings samples.

Core Samples

A *core* is a slender column of rock that shows the sequence of rocks as they appear within the earth (fig. 2.15). It provides the most accurate information about the underground formations. To take a core sample, the driller substitutes a coring bit (fig. 2.16) with a hollow center for the conventional drill bit. A core ranges in length from 25 to 60 feet (7.6 to 18.3 metres).

Figure 2.15 This technician is marking the well core to maintain the order in which the pieces came out of the ground. Later they will be sent to a laboratory for analysis.

Once this core is brought to the surface, technicians carefully package and send it to a laboratory for analysis of many characteristics, including porosity, permeability, composition, fluid content, and geological age. This information helps the geologist determine the oil-bearing potential of the cored beds.

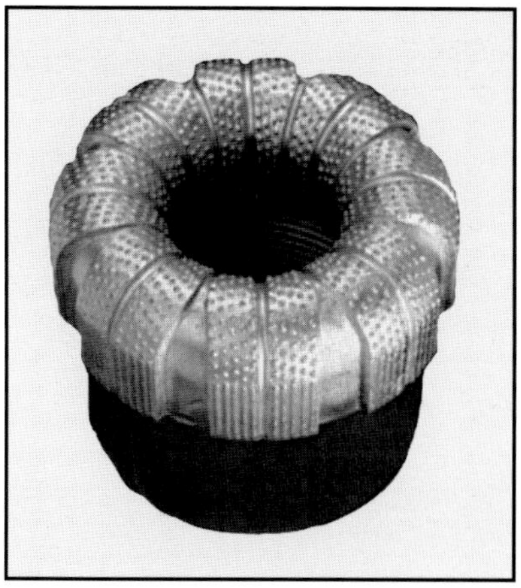

Figure 2.16 This coring bit drills out a slender column of rock.

Cuttings Samples

As a regular bit drills a hole, it breaks up the rock into pieces called *cuttings*. The cuttings flow out of the hole where geologists can use them to analyze the rock being drilled. Since cuttings are fragments of rock and do not form a continuous sample like a core, they are not as useful as cores are to the geologist. Cuttings may not all come from the bottom of the hole but may include pieces of formations that have sloughed off closer to the surface. Even with these limitations, however, cuttings can provide useful data and are regularly examined during drilling.

Drill Stem Test

Like geologic data, information from tests on the formation accumulates over the life of a field. The more that is known about a reservoir, the easier it is to test a new well more quickly and less expensively. Formation testing provides the oil company with pressure charts or logs made during the test and a report that describes the fluids in the well.

The *drill stem test* (or *DST*) is the primary way to test a formation that has just been drilled. Although generally costly, drill stem testing can bring an oil company good returns on its investment. To run the test, a DST tool is run into the hole. Formation fluids flow into a perforated pipe in the tool. A pressure recorder inside the tool and another below the perforated pipe chart the pressure. When the pressure testing is finished, valves in the DST tool close to trap a fluid sample. The DST gives accurate data about a formation's pressure and the composition of the fluids in it.

Strat Test

A useful test in reservoir development is the *stratigraphic test*, or *strat test*, as it is commonly called. *Stratigraphy* is the study of the origin, composition, distribution, and sequence of rock strata.

A strat test involves drilling a hole primarily to obtain geological information. The borehole exposes complete sections of the formations penetrated, and stratigraphers analyze the cuttings taken from the wellbore as it is drilled. The strat test differs from a cuttings sample log in that a company only runs it on an exploratory well to examine the cuttings for hydrocarbons. Stratigraphers try to recognize and follow beds of rock from one well to another, going from a well or formation whose beds and sequence of rocks they know to an area they do not know but assume to be similar.

Stratigraphic Correlation

Stratigraphic correlation is the process of comparing geologic formations. In the oil exploration industry, stratigraphers compare the geology of a known area with unknown formations in nearby locations in order to try to predict where new reservoirs are. They do this by looking at information collected in drillers' logs, sample logs, and electrical logs in order to compare fossils, the composition of formations, and electrical data from one well or area with the same information from another well or area (fig. 2.17). Rock texture and fossil characteristics in formations, particularly those under the oceans, normally change very gradually. Sudden changes in the sequence of rock types are geologic indicators, and often geologists use particular fossils as markers when they are trying to recognize the continuation of a formation in a new location. If the kind of rock is the same and the fossil markers are the same, the formation is probably a continuation of the one in the previous location.

The sequence, or order, of formations can also be used as a method of correlation. In *sequence stratigraphy*, geologists take the information from seismic surveys and analyze it to deduce the environment that existed when a rock layer first formed. For example, geologists can distinguish between sandstone laid down in a river from sandstone that formed from a beach. This method depends in part, as do the others, on drilling closely spaced wells. Stratigraphers can match the formations accurately when they correlate the data derived from several wells.

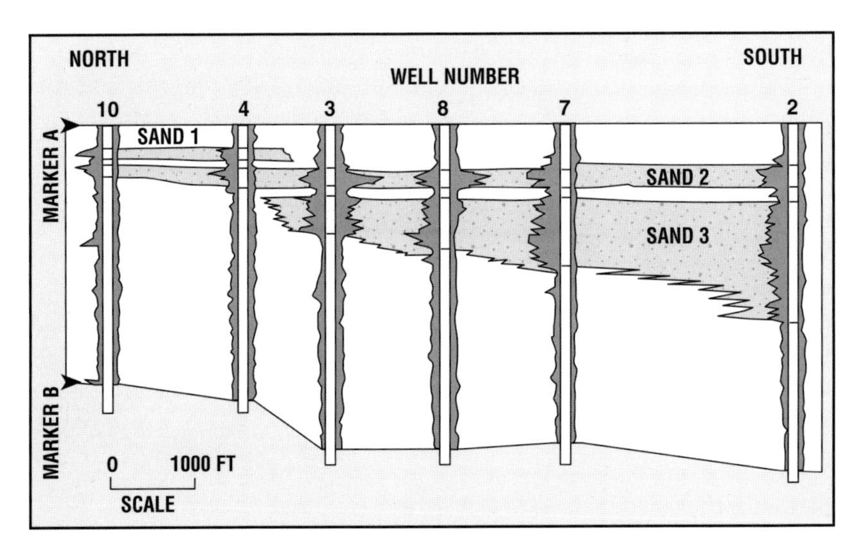

Figure 2.17 This stratigraphic cross section shows the presence of sand at various depths over an area of several thousand feet (metres). The data came from six wells near each other.

Maps

Explorationists use various kinds of maps throughout the exploration process. Geologists use *base maps* that show existing wells, property lines, roads, buildings, and other manmade surface features to recommend sites for geophysical studies, exploratory drilling, and reservoir development. *Topographic maps* show surface features such as mountains and valleys. Maps can also show information about underground formations gathered from exploration surveys. Gravity survey results are displayed on a *Bouguer gravity map*.

Contour Maps

Topographic, gravity, and other surveys can be depicted in the form of *contour maps*—those most commonly used by petroleum geologists. Contour maps show a series of lines drawn at regular intervals, often connected to enclose an area. The points on each line represent equal values, such as depth or thickness.

To understand a *topographic contour map*, imagine looking at a mountain from above it. If you could mark a dot every place on the mountain that is 1,000 feet high and then connect the dots, you would have a curve going around the mountain. Do the same for 2,000 feet, 3,000 feet, and so on, and you have a series of concentric curves. Draw these curves on a piece of flat paper, and you have a contour map.

Contour maps for exploration may depict geologic structure and thickness of formations. They can also show the angle of a fault and where it intersects with formations and other faults, as well as where formations taper off or stop abruptly.

One type of contour map is the *structural map*, which depicts the depth of a specific formation from the surface (fig. 2.18). The principle is the same as that used in a topographic map, showing instead the highs and lows of the surface of the buried layer.

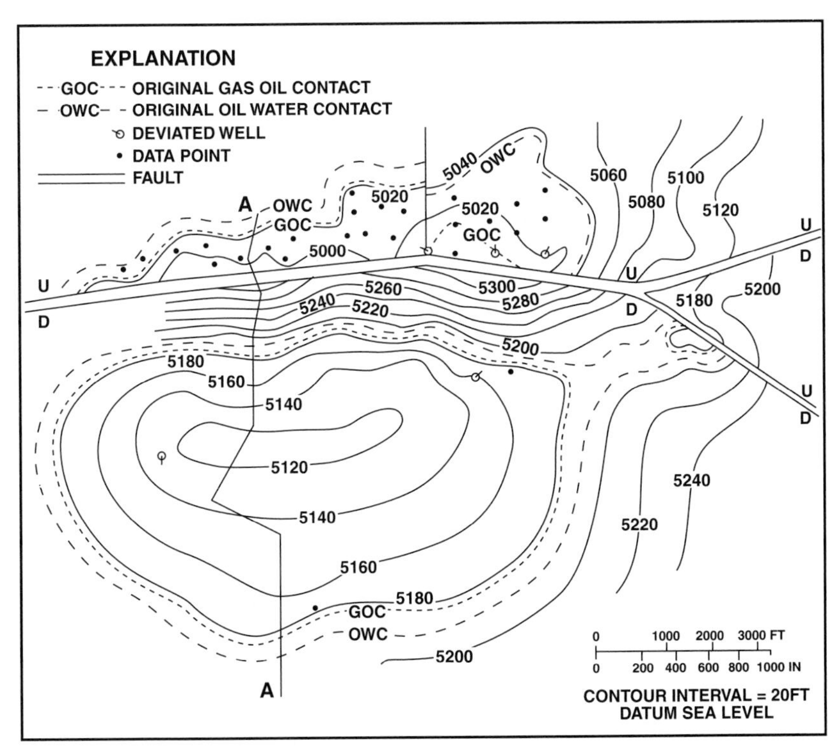

Figure 2.18 A structure contour map shows the depth of a formation from the surface. *(Courtesy of Bureau of Economic Geology, The University of Texas at Austin)*

Another type of contour map is the *isopach map* (fig. 2.19). It shows the variations in thickness of a formation. Geologists use isopach maps to calculate how much petroleum remains in a formation and to plan ways to recover it, as well as an aid in exploration work. Similar to the isopach is the *isochore map*, which shows the thickness of a layer from top to bottom along a vertical line.

A *lithofacies map* shows the character of the rock itself and how it varies horizontally within the formation. This type of map has contours representing the variations in the proportion of sandstone, shale, and other kinds of rocks in the formation. Another type of facies map, the *biofacies map*, shows variations in the occurrence of fossil types.

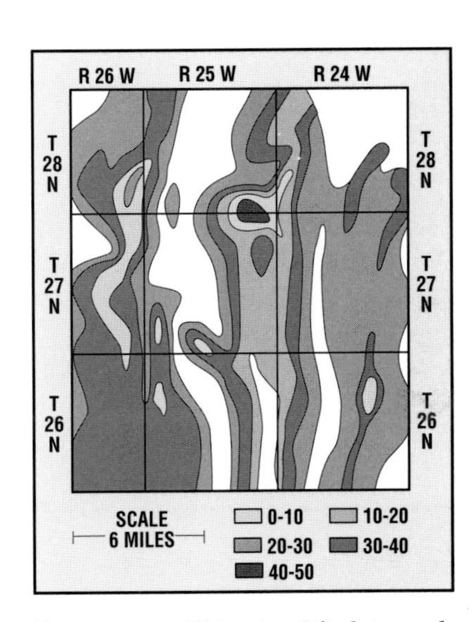

Figure 2.19 This simplified isopach map shows the thickness of a formation based on microlog surveys. The different shadings represent different depths.

Vertical Cross Sections

A *vertical cross section* is a type of map that represents a portion of the crust as though it were a slice of cake (fig. 2.20). It shows structures and fault patterns. Most cross sections show both structural and stratigraphic features (see chapter 1) together. They may show possible anticlinal and fault traps, or they may show only horizontal variations in type of rock or thickness. The fact that cross sections can show gravity anomalies in a contour form has been instrumental in spreading the use of gravity surveys in the exploration industry.

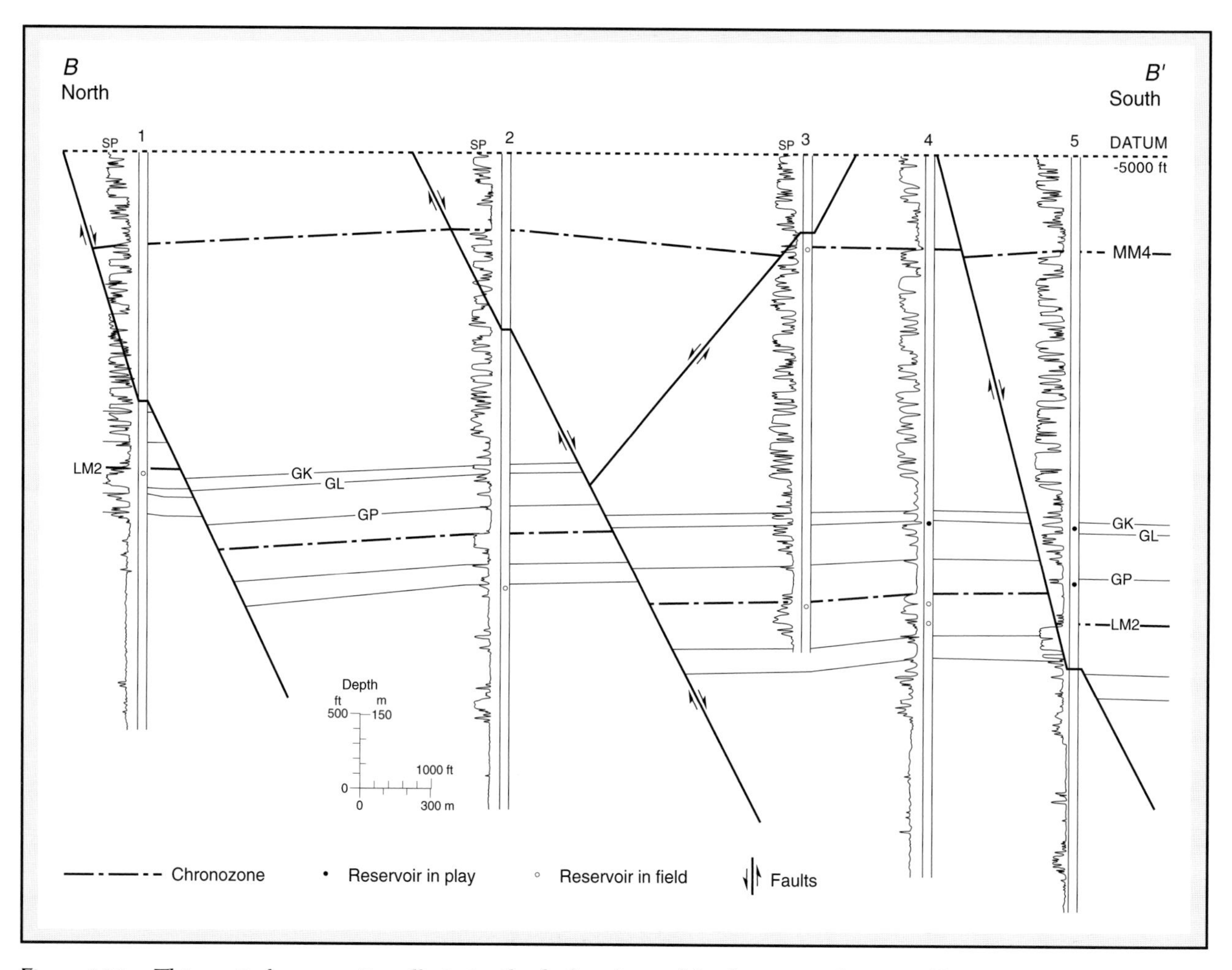

Figure 2.20 This vertical cross section illustrates the fault pattern of the formation shown in Figure 2.18. *(Courtesy of the Bureau of Economic Geology, The University of Texas at Austin)*

Computer Graphics and Models

Computer technology has become indispensable in exploration. The huge quantities of data that a computer can store and manipulate, the ease of updating data, and new software for modeling and analyzing data provide explorationists with more information, faster than ever before.

Graphics

The results of some types of surveys, such as seismic surveys, may be analyzed by so-called supercomputers, computers that are so fast they can handle thousands of times the information a PC can work with.

Many graphics that draftsmen laboriously created in the past are now handled at PC workstations running sophisticated software. Skilled computer operators can generate maps that integrate all sorts of different data. For example, they can lay a cross section of a well's path over a seismic section showing the geologic structures (fig. 2.21) or display lease boundaries on top of a porosity map (fig. 2.22). Computer operators can enhance Landsat, seismic, and other graphics to create three-dimensional images and add false color to highlight certain features with a keystroke.

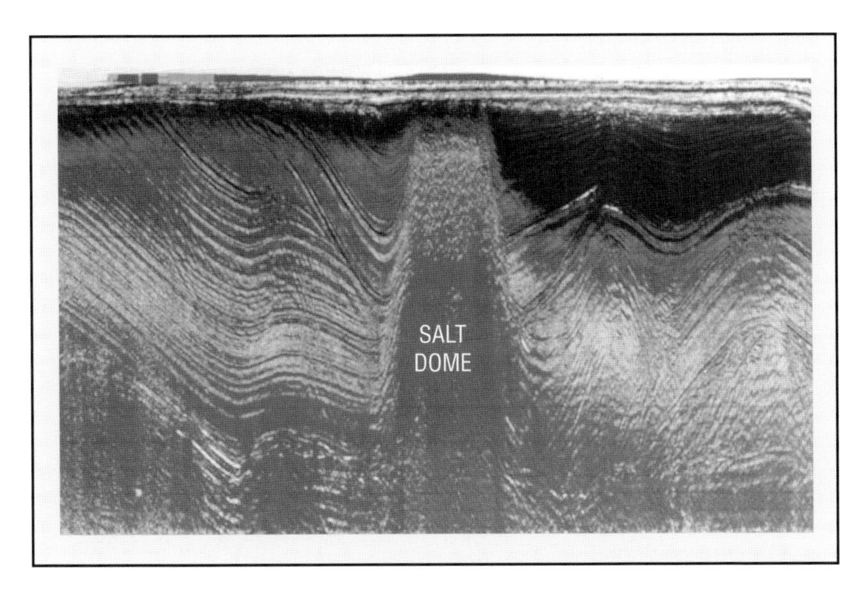

Figure 2.21 A seismic section shows a salt dome and possible petroleum traps.

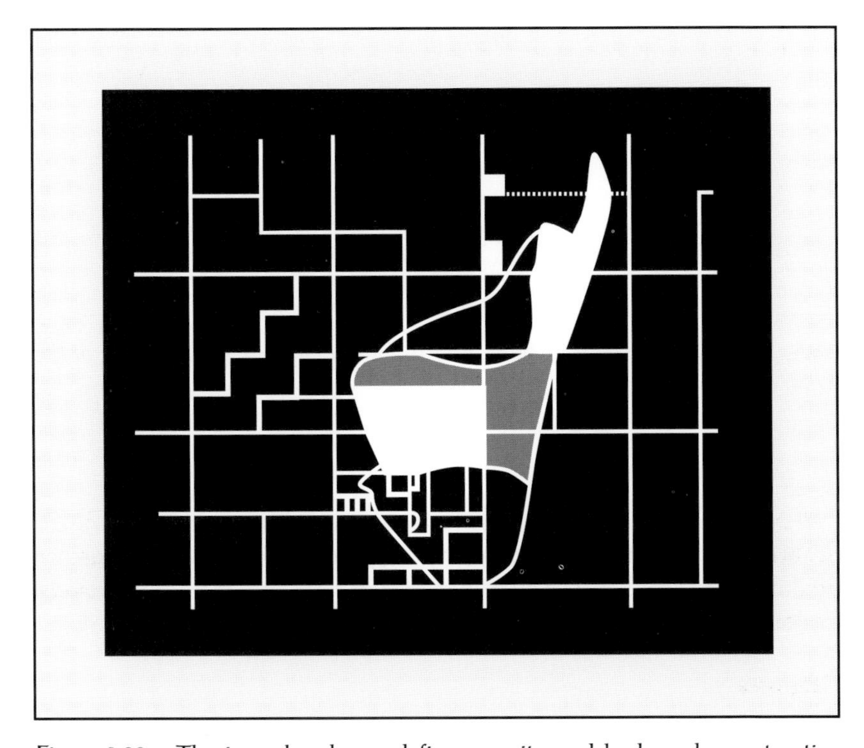

Figure 2.22 The irregular shapes define porosity and hydrocarbon saturation underground, while the straight lines represent lease boundaries on the surface.

Computers can also generate a block diagram, which is a perspective drawing of a section of the earth's crust as it would appear if cut out in a block (fig. 2.23). A block diagram shows two vertical cross sections whose faces are at right angles to each other, and the top of the block is either a subsurface view or the surface topography. Operators frequently use computers to analyze gravity and magnetic data and combine the results with seismic data to produce a complex three-dimensional image of the geological features of the area of interest.

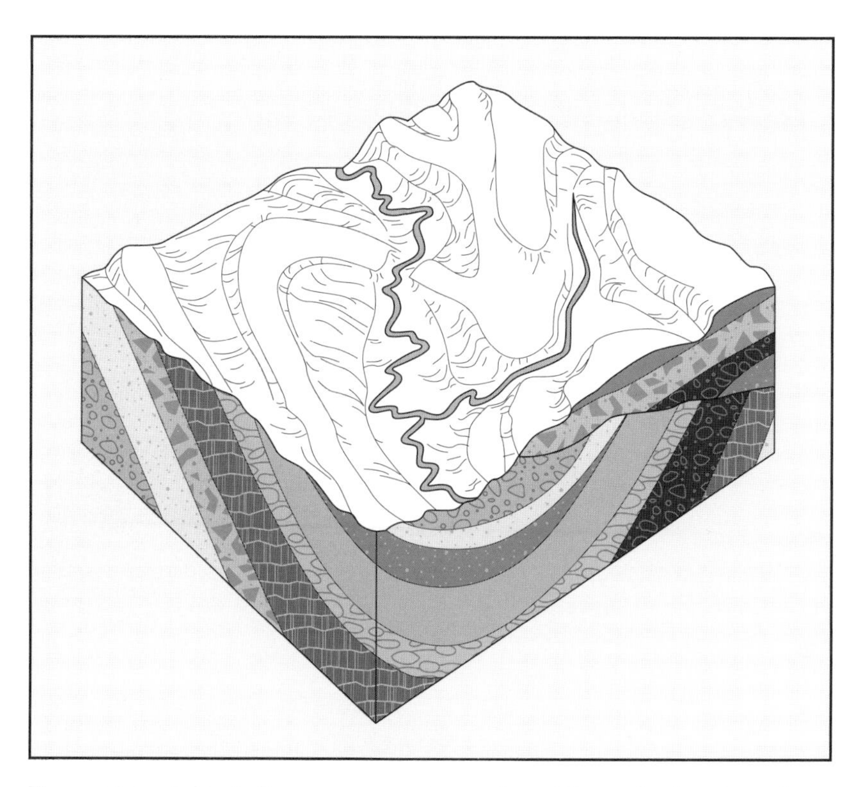

Figure 2.23 A block diagram represents a section of the earth, showing subsurface strata and surface topography.

Models

Reservoir modeling is another technique made practical by the computer. Information is manipulated using a *model*—a computer program containing a set of parameters, such as the number of wells in a section, the production rates of the wells, and so on. The operator knows the values of the parameters for the present and for the production history of the reservoir. To predict future production, the operator enters new parameters into the program in the form of mathematical "what if" questions: for example, "What if we add three wells in this area and increase the production rate by 20 barrels per day?" The computer plugs the new information into the model, performs all the calculations, and comes up with production predictions. This type of analysis is an invaluable aid to the decision maker.

3 Aspects of Leasing

Finding petroleum is not the only requisite to production. Before an oil company can develop a reservoir, it must obtain the legal rights for such exploitation. In most countries, the state or national government owns the mineral wealth, including petroleum. Companies that have the capital and expertise may negotiate a contract with representatives of the government claiming the reserves. Often, the host country retains controlling interest throughout exploration and development. Thus, extremely complex arrangements between the host country and consortiums of petroleum companies, many of which are state or nationally owned, take on an air of international policymaking.

Securing the rights to explore, drill, and produce differs from country to country (fig. 3.1). In the United States, four sources exist for the rights to petroleum: the private property owner, the state government, the federal government, and certain Native American tribes. Each may hold title, or established ownership, to a given piece of land, and only the titleholder may grant the rights to develop the resources of the land. The instrument used to grant these rights is called a *lease,* traditionally an *oil, gas, and mineral lease,* or simply an *oil and gas lease.* For a lease to be valid, ownership of the minerals must be established, and provisions of the lease must be explicit and legally executed.

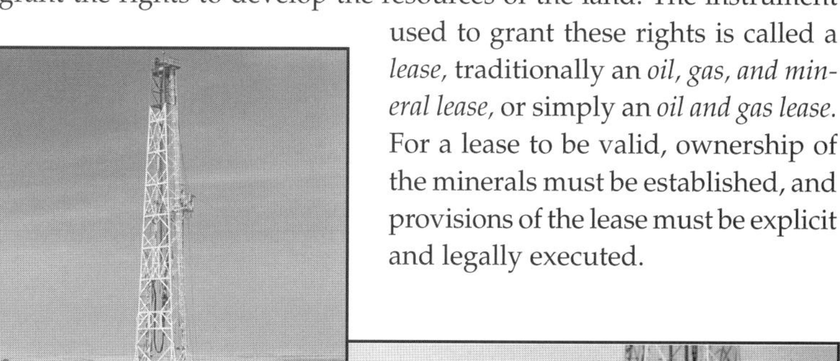

Figure 3.1 In most oil-producing nations outside the United States, the mineral rights are owned and controlled by the government.

TYPES OF PRIVATE OWNERSHIP

Establishing ownership of the oil, gas, and mineral resources—called the *mineral estate*—is not as simple as it might seem. In the United States, the mineral estate is considered as part of the real estate. Real estate law is case law, or common law, and is determined by historical precedents established in court cases. Most of the states have adopted the English common law of real property, except when contravened by statute. An exception is Louisiana, where civil law, derived from the Napoleonic code, is the rule, rather than English common law. Since each state has its own court system in the form of a separate set of cases for the precedents, each state has in effect different laws regarding the ownership of petroleum.

States that follow the doctrine that oil and gas are owned in place, underground, are sometimes termed *absolute ownership*, or ownership-in-place states. Texas is an example. However, if a reservoir lies under two properties, and one owner drills and produces but the neighbor does not, then the neighbor might end up with no oil. The title has shifted to the owner who captured the oil and reduced it to personal property. Because an owner cannot protect a shifting title, courts in some states—Louisiana and Pennsylvania, for example—find that ownership in place is not an acceptable theory.

In states that follow a *nonownership-in-place* doctrine, no one owns the petroleum until it is captured. Title can be assumed upon production, and the oil and gas become personal property. In both types of jurisdictions, the lease is a conveyance, granting the exclusive right to explore, drill, and produce.

Regardless of the doctrine of ownership followed, two thirds of onshore lands in the United States are in *private ownership*. Generally, the rights to the minerals, oil, and gas of these lands are also privately owned, although not necessarily by the same person who owns the surface. An individual or corporation may be a landowner in fee simple, a mineral estate owner, a surface estate owner, or an owner of a royalty interest in mineral production from the land. The remaining lands—both onshore and offshore—are public property, owned and administered by the individual states or by the federal government.

Fee Simple Landowner

Preserved from the Middle Ages through English legal heritage, an estate of complete ownership in real property is a *fee*. An owner of *fee simple* property (the *landowner*) owns the right to exploit what wealth the land might provide, whether above, on, or below the surface. Until the federal and state governments established certain restrictions to protect the rights of adjacent property owners and to conserve the overall hydrocarbon resources of the country, landowners could do just about anything they wanted with their land.

Today, landowners still have the exclusive right to search for and remove oil and gas from their property; however, most are not financially able to expend the enormous sums required to drill and produce. For this reason, they may grant the right to exploit the resources to someone else for a given period of time through a leasing agreement.

Landowners can sell the mineral estate or a percentage of it to someone else by use of a *mineral deed*, or they can sell the surface and retain all or part of the mineral estate. (A mineral deed is distinguished from a lease in that a lessee will lose his (her) rights to the oil and gas unless production is established within the time allowed by the lease.) In either case, fee simple ownership ends, and mineral estate and surface ownership begin.

Mineral Estate and Surface Owners

The rights of the mineral estate owner and the surface estate owner depend on (1) the state in which the property is located and (2) the minerals detailed in the sale agreement. Some states regard the mineral estate as a *possessory estate*—that is, as fee ownership of the minerals in place. Other states regard the mineral estate as a *servitude estate;* that is, it is subject to a specified use or enjoyment by one party, even though the surface is owned by another.

The owner of a possessory mineral estate cannot lose it by merely failing to explore it. On the other hand, if the state regards the mineral estate as a servitude, the mineral estate owner only has the right to explore and produce the minerals, and owns them only after reducing them to a possession. A servitude can be lost by failure to explore diligently. In Louisiana, for example, an owner has a ten-year period to exercise mineral rights to avoid their merger with the surface estate.

If the minerals in an agreement are oil and gas, the mineral estate is the dominant estate, and the surface is the servient estate. The surface owner may not prevent the mineral owner's access to and use of the mineral estate, nor may the surface owner lease the minerals. The mineral estate owner may use as much of the surface as is reasonably necessary to recover the oil and gas, with due regard to the rights of the surface owner. To avoid disputes, the lease usually contains specific agreements regarding surface use. The surface owner is normally entitled to compensation for property damage that results from exploration and development operations.

Like the fee simple owner, the mineral estate owner who cannot afford to exploit the petroleum resources himself may convey these rights to another by way of an oil and gas lease.

Royalty Interest Holder

Another type of owner is one who holds a royalty interest in the mineral estate. A royalty interest is a share or percentage of the total or gross oil and gas production. A *participating royalty owner* is one who also owns all or part of the mineral estate and has the executive rights of a mineral estate owner. A *nonparticipating royalty interest holder* is one who owns no part of the mineral estate and therefore has no right to execute leases, enter the property, nor explore or produce any minerals on the property. A nonparticipating holder only receives a share of the profits from production.

Landowners can transfer royalty interest by sale or reserve it totally or in part. They might sell both the surface and mineral estates but reserve half of the royalty interest for themselves and their heirs. On the other hand, an owner might keep his or her property and sell, by means of a *royalty deed,* some fraction of the royalty interest. He or she might even divide the royalty interest into several shares, retain a portion of the shares, and become a co-owner of the royalty interest along with several other owners. Depending on the laws of the state in which the land lies, these arrangements can be perpetual, for fixed terms, or for the lifetimes of the purchasers.

THE LEASE AND THE LAW

The most commonly used method for securing the rights to mineral production is the oil and gas lease. Because leasing is absolutely essential to the petroleum industry, a familiarity with lease laws, leasing practices, and regulatory agencies is important. The next step to understanding some of these legal aspects is to learn the language of leasing.

The Language of Leasing

A lease is a contract between a mineral estate or fee owner, called the *lessor,* and a petroleum company or other party, known as the *lessee.* For *consideration* (usually money called *bonus*) and the further consideration of receiving a share of what is produced (known as *royalty*), the lessor gives exclusive rights to the lessee. In return, the lessee agrees to explore and produce the oil and gas in a timely and prudent fashion. Under most leases, the lessee pays a *delay rental* (money) each year to prevent the lease from automatically expiring during a certain period of time (the *primary term)* in the absence of drilling. The law imposes certain *implied covenants,* or obligations, on the lessee. The lease itself contains specific or *expressed covenants* for development, and these replace the implied covenants.

The mineral interest is usually shared by the lessor and the lessee on a percentage basis. For example, if the royalty owned by the lessor is 25 percent, then the balance of 75 percent is the lessee's portion and is called the *working interest.* The party who ends up controlling the working interest and developing the resource is called the *operator.* The operator may be an individual, one of several in partnership, one or several corporations, or any combination agreed upon by all parties in a contract called the *operating agreement.*

The First Leases

Before oil was discovered and commercially produced in Titusville, Pennsylvania, leasing agreements between prospectors and landowners were extremely simple. One lease made in 1853 gave the oil company the rights to "dig or make new springs," the cost of which would be deducted "out of the proceeds of the oil, and the balance, if any, to be equally divided" between the oil company and the landowner. The concluding statement of the agreement, "if profitable," reveals the uncertainty of the whole venture.

The length of the contract is also significant: both parties agreed that five years was sufficient time to discover and develop any resources that happened to be under the land. Prospecting was luck, not science; and, if petroleum was not discovered within a relatively short time span, it was universally believed that the oil and natural gas would "migrate" somewhere else, thus making further exploration unprofitable in any case.

Court Rulings on Oil Migration

As scientists slowly discovered more about the natural laws that govern the behavior of oil and natural gas, some of the stranger theories concerning the migration of these resources were disproved. Landowners, who had thought their chances of profiting from an oil or gas lease were severely diminished if the company did not begin exploration and production immediately upon signing the lease, could not sue oil companies for forfeiture of the lease agreement on those grounds after 1875. Before that date, the courts had upheld landowners' claims that their lease rights included an "implied covenant" with the lessee that required the oil operator to develop or release the leasehold at once.

The courts, having heard endless litigation over the migration of oil and gas, finally decided that oil and gas are the property of the person who first captures the resource and reduces it to his control. The phrases "capturing" oil and "reducing it to the owner's control" comes from the old theory that oil and gas are migratory resources—they might wander onto one's property and then leave—almost like animals. The courts decided it was not their province to judge whose property the oil may have migrated from; the person who drilled a well and found oil was the legal owner. Although it seemed like a reasonable decision at the time, the consequences of the court rulings were not all fortunate.

Rule of Capture

One consequence of the decision was the landowner's freedom from liability for drainage of a common reservoir even though part of the oil may have migrated from a neighbor's land. This so-called rule of capture, especially in states that have adopted the nonownership doctrine, allows the owner of a tract to drill as many wells as he or she can, provided he or she avoids drilling diagonally onto a neighbor's property. The rule does not give the landowner the right to draw a disproportionate amount from the common reservoir because government regulations prorate the amount of oil that can be produced and because adjacent landowners can drill and produce wells from the same reservoir.

Offset Drilling Rule

The offset drilling rule, which is an outgrowth of the rule of capture, states that a landowner whose oil and gas reserves are being drained by a neighbor's wells cannot go to court and recover damages or stop the offending operator. The landowner's only recourse is to drill his or her own wells and produce as fast as possible. If the landowner leases the mineral interest, the lessee must assume the burden of the offset drilling rule. Production on an adjoining lease obligates the lessee to protect the lessor's interest by drilling a well nearby to offset the potential loss of oil.

Court Decisions on Mineral Leases

In many of the older mineral leases, oil and gas were implied as the minerals (fig. 3.2); however, iron ore, lignite, and other minerals were mined under these leases to the detriment of the surface owner. As a result of several cases,

Figure 3.2 In an "Oil, Gas, and Mineral" lease, oil and gas are *implied* as the minerals resulting from past court decisions.

Producers 88 (12/79) Revised
With 320 Acres Pooling Provision

POUND PRINTING & STATIONERY COMPANY
2325 Fannin, Houston, Texas 77002 (713) 659-3159

OIL, GAS AND MINERAL LEASE

THIS AGREEMENT made this________ day of ________, 19______, between

Lessor (whether one or more), whose address is: ________ Zip Code ______,
and

Lessee, (whether one or more), whose address is: ________ Zip Code ______.

WITNESSETH:

1. Lessor in consideration of________ Dollars ($______), in hand paid, of the royalties herein provided, and of the agreements of Lessee herein contained, hereby grants, leases and lets exclusively unto Lessee for the purpose of investigating, exploring, prospecting, drilling and mining for and producing oil, gas and all other minerals, conducting exploration, geologic and geophysical surveys by seismograph, core test, gravity and magnetic methods, injecting gas, water and other fluids, and air into subsurface strata, laying pipe lines, building roads, tanks, power stations, telephone lines and other structures thereon and on, over and across lands owned or claimed by Lessor adjacent and contiguous thereto, to produce, save, take care of, treat, transport and own said products, and housing its employees, the following described land in________ County, Texas, to-wit:

the courts reasoned that the mineral estate owner did not have the right to "appropriate the soil." Thus, minerals substantially affecting the surface belong to the surface owner.

Government Regulations

The combination of the rule of capture and the offset drilling rule led to excessive drilling and production from reservoirs. Not only was there waste from drilling unneeded wells, but the rapid production reduced reservoir pressures and altered the natural drive mechanisms (the forces that drive the oil and gas to the surface). The total amount of petroleum produced from a reservoir was often less than ideal.

When it proved impossible to efficiently extract oil and gas and equitably distribute the minerals among the owners under existing legislation, the state governments stepped in and formed regulatory commissions. Conservation or regulatory agencies, eventually established in all oil-producing states, have the authority and responsibility to prescribe and enforce sound conservation rules such as well-spacing requirements and prorated production rates. State regulatory agencies also have the responsibility for protecting landowners' correlative rights. *Correlative rights* are those rights given to the owner of each property in a pool to produce, without waste, his or her just and equitable share of the oil and gas in such a pool. *Pooling* is the practice of combining small tracts of land into a pool large enough to satisfy state spacing regulations for drilling.

Like their counterparts to the south, the Canadian provinces enact new oil and gas legislation from time to time. The laws, administered by various government agencies, control mineral development and leasing practices. In addition to the Department of Energy, Mines and Resources in Ottawa, Ontario, each province has its own regulatory agency.

The U.S. government also involves itself in regulating the production of oil and gas through its Environmental Protection Agency (EPA), Bureau of Land Management (BLM), and Minerals Management Service of the Department of the Interior (DOI). Oil and gas production is now among the most heavily regulated of American industries.

PREPARATIONS FOR LEASING PRIVATELY OWNED LANDS

Once an operating company decides to lease privately owned land, the *landman,* or *leaseman,* comes on the scene. A landman (sometimes called an oil scout) is a person in the petroleum industry who negotiates with landowners for land options, oil drilling leases, and royalties. He or she also works with producers for the pooling, or combining, of production in a field. Sometimes these preliminaries are handled by the owner and others on the staff, especially if the operating company is small.

If the operator is a large petroleum company, the company's exploration group will notify its own land department that a region is of interest, and a landman will investigate. The landman will find out who owns mineral rights in the area of the *play,* a term used in the industry to describe "geologically similar reservoirs or oilfields exhibiting the same source, reservoir, and trap characteristics."[1] If some part of the area is already leased, the landman will find out the prevailing terms of leasing, such as the amount of bonus and royalty. If the decision is made to pursue the play, the landman will contact the mineral estate owners and negotiate as the lessee, accumulating as much of the region as possible. Of course, other companies are aware of the action and are likewise investigating and leasing.

A landman can also operate independently of any one company. An independent landman is essentially a lease broker who may work with several operators or companies in the area in which he or she lives. If the area is a productive one, much bargaining and trading of the leases may take place.

Whether working for a company or as an independent, the landman's job is to acquire signed leases. Before leasing can take place, the landman must make sure that the legal ownership of the property has been established, that the owner has the capacity to legally sign a contract, and that both the lessor and lessee agree on the terms of the lease.

Determining Ownership

The landman securing an oil and gas lease takes care to determine the ownership of the land and the minerals of that land in a preliminary check of the records. This research lets the landman know the names and addresses of the current mineral interest owners so that he or she can talk with them and eventually get their signatures on a lease. The landman also finds out who owns the surface of the land and whether the property is already leased. The landman tries to look at all the documents in the land's history of ownership and briefly describes these documents in a *run sheet* (fig. 3.3). This sheet will later be used to *clear the title*—that is, establish the full legal status of the land in question—before executing the lease.

Figure 3.3 The landman checks deed records in the county or parish courthouse to establish ownership of a piece of property.

[1] D. A. White, "Assessing Oil and Gas Plays in Facies-Cycle Wedges," AAPG Bulletin, Vol. 64, No. 8, 1980.

Clearing the Title

Ideally, the recorded conveyances in the county records will show an unbroken chain of title extending from the patent of the state to the present owner (fig. 3.4). To minimize the risk of adverse claimants, the lease purchaser relies on legal advisers to establish facts that confirm the chain of title. Otherwise, some heir or long-lost relative of someone who once owned an interest in the land could show up demanding the whole well on the basis that it was leased from the wrong person.

Figure 3.4 County records should show an unbroken chain of title from the patent of the state to the present owner.

An old story illustrating the thoroughness expected in a title search concerns the Post Office Department in Washington, D.C., which was about to purchase a lot for a new post office in Louisiana. The department, dissatisfied because the title had been traced back only to 1803, requested further work on the case. The title attorney responded as follows:

> Please be advised that the government of the United States acquired the title of Louisiana, including the tract to which your inquiry applies, by purchase from the government of France in the year 1803. The government of France acquired title by conquest from the government of Spain. The government of Spain acquired title by discovery of Christopher Columbus, explorer and resident of Genoa, by agreement traveling under the sponsorship and patronage of her majesty, the Queen of Spain. The Queen of Spain received sanction of her title by consent of the Pope, a resident of Rome, and *ex officio* representative and vice-regent of Jesus Christ. Jesus Christ was the son and heir apparent of God. God made Louisiana.

Examining Titles

One way to clear a title is by obtaining *title opinions* from a qualified title examiner—an attorney who examines titles to oil and gas properties. The title examiner, often with a landman's help, must investigate every aspect of a title before legal opinions can be given. Most companies require title opinions before paying bonuses and delay rentals. A common practice is to issue a postdated check to the lessor when the lease is signed. This gives the lessor time (usually around 15 to 20 days) to clear the title before payment is made. Opinions may depend on abstracts of the property (fig. 3.5) or a takeoff. An *abstract* is a collection of all the recorded instruments affecting title to a tract of land, usually presented in a shortened form. A *takeoff* is a brief description of the documents relevant to the title of a particular piece of property. A takeoff may be purchased from an abstract company. Also, local title companies usually carry complete records of real estate transactions and may be willing to share information for a fee. Many companies are now using data banks of title information and can quickly retrieve it with in-house computers. In addition to *abstract-based opinions*, an examiner can render *stand-up opinions* based on the landman's run sheet and a careful examination of all documents in the appropriate county courthouse.

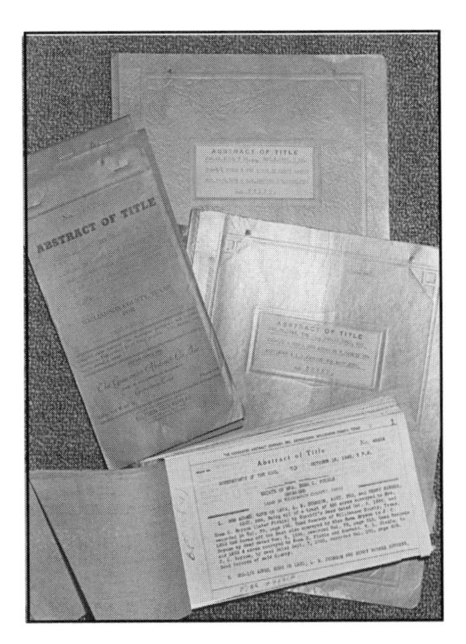

Figure 3.5 A title examiner may use abstracts of the property to help in rendering a title opinion.

Curing Titles

When the chain of title has gaps or other defects, the title examiner may obtain sworn statements or other documents that support the present occupant's title to the land, especially by possession and use. Some of these statements, sworn before a notary public and known as *affidavits*, may already be recorded because of past title transfers. In all likelihood, the title examiner will need to investigate further. The older affidavits may leave some questions unanswered. In addition, the examiner should, if possible, get affidavits from noninterested parties familiar with the facts, because a "self-serving" affidavit made by the present occupant is vulnerable to attack. Commonly used affidavits are those of *death and heirship; identity; nonproduction; use, occupancy, and possession;* and *adverse possession.*

In addition to affidavits, other documents and legal instruments that may assist the title examiner in curing a title are *tax receipts, rental receipts,* and *releases of old leases* and *mortgages.* The landman or examiner may uncover a *deed of trust* to the property and, if still in effect, the holder of such an instrument—the creditor—has rights superior to any subsequent interest in the property. A *quitclaim deed* or *disclaimer* waives any rights to property; it is often used by the surface owner or tenant to affirm that he or she has no interest in the mineral estate. It should be remembered, however, that the title is not cleared by the affidavits and documents themselves; it is the existence of the appropriate facts that makes the title satisfactory.

Validating the Owner's Capacity to Contract

The landman must also try to make sure the lessor has the capacity to contract. Simply stated, not all owners can grant valid leases. Any person or company interested in obtaining lease rights to a piece of property has to investigate the legal status and powers of the owner very carefully before investing in a drilling venture. Special arrangements are necessary to lease from life tenants and remaindermen, cotenants, married persons, illiterates, owners as agents of the state, and fiduciaries.

Life Tenants and Remaindermen

A *life tenant* is a person who is entitled to exclusive possession of a property but cannot pass it on through inheritance. A *remainderman* is the person who has title to the property and will have full possession of the property after the death of the life tenant. Because they share interests in the land, both the life tenant and the remainderman should, under ordinary circumstances, execute an oil and gas lease together.

Cotenants

Cotenants are people who own land together (co-tenants). They are entitled to equal use of the land and *may* inherit one another's interest in the land. In the majority of states, one cotenant may execute a lease on jointly owned premises without the consent of the other cotenants, but the lease is binding only on the cotenant making the agreement, although all may profit from any production.

Married Persons

The ability of married persons to make legally binding leases depends on the status of their land and the law of the state in which they reside. The estate may be *community property* (acquired after the marriage), *separate property* (owned separately by one spouse before marriage), or *homestead property* (occupied and used by the owner and his family). Some community property states (like California) and most noncommunity property states require both spouses to sign a lease.

Illiterates

The inability to read and write does not disqualify a person from making a legally binding contract. An illiterate can grant valid leases as long as the parties involved comply with proper procedures concerning witnesses, executing the legal agreement, and the illiterate making his or her mark.

Owners as Agents of the State

Some property in the state of Texas was given to its citizens who could enjoy all the rights of fee ownership except the right to lease and profit from its minerals. Later, the surface owner was given the right to grant an oil and gas lease as an *agent of the state,* with the owner and state splitting any benefits from that lease.

Fiduciaries

A *fiduciary* is a person who acts for another in a legal, financial, or other undertaking. Whether a fiduciary can execute a valid oil and gas lease depends on the laws of the state in which the property is located and the specific provisions of the empowering instrument. Among the more common fiduciaries are executors, administrators, guardians, trustees, persons given powers of attorney, and representatives of unknown or missing heirs.

Negotiating the Lease

The landman and the landowner do not always negotiate a lease over a cup of coffee. Many leases are acquired by telephone or letter, and many of those negotiated face-to-face are often sold or traded by the acquiring company. However a lease is acquired, it is always a bargaining process, with each party trying to make the best deal possible. The landman must consider the lessee company's resources and policies, be aware of the local situation, and know the prevailing prices in the area.

The landowner is concerned with making an informed evaluation of the landman's offer. He or she may contact neighbors, bankers, attorneys, accountants, or local lease brokers in order to evaluate the offer. The landman will probably not make an initial offer at the top of the scale of permissible bonuses or royalties. The landowner will probably bargain for a better price. The two usually compromise on primary term, bonus per acre, and royalty percentages. An astute landowner preparing to lease his or her land might also obtain a copy of the well-spacing and density regulations that apply to the area from the state regulatory agency.

Since real estate law rather than contract law controls leasing, an oil and gas lease must be written. An oral agreement is not sufficient. Hence, in the course of negotiations, the landowner finds herself or himself pondering a printed form that is several pages long and full of closely printed clauses filled with "hereins," "herebys," and "hereinafters." The instrument may be a lease form used by the landman or by the company he or she represents. Whether the lease is a preprinted form, such as one of the well-known "Producers 88" forms, the landowner's own form (perhaps from an association), or one drafted by an oil and gas attorney, all of its provisions are subject to mutual agreement by both the lessor and the lessee.

Because of the language of the lease and past interpretations of the courts, the prudent landowner (lessor) will have a lease examined and explained (preferably by an attorney) before signing it. It is equally important for the oil and gas operator (lessee) to thoroughly understand any special clauses and amendments resulting from final negotiations between the landowner and the landman.

PROVISIONS OF THE LEASE

The provisions essential to a lease—conveyance, term, and royalty—are contained in standard lease clauses (fig. 3.6). Most leases also contain clauses that address the relative rights of both lessor and lessee when peculiar conditions exist. In addition to these clauses, a lease contains dates, names and signatures of parties involved, and the seal and signature of a notary public.

Producers 88 (12/79) Revised
With 320 Acres Pooling Provision

POUND PRINTING & STATIONERY COMPANY
2325 Fannin, Houston, Texas 77002 (713) 659-3159

OIL, GAS AND MINERAL LEASE

THIS AGREEMENT made this__________________ day of_______________________________________, 19__________, between

Lessor (whether one or more), whose address is: ____

____ Zip Code __________ ,

and_ ____

Lessee, (whether one or more), whose address is: ____

____ Zip Code __________ .

WITNESSETH:

CONVEYANCE

1. Lessor in consideration of ____ Dollars ($____________________), in hand paid, of the royalties herein provided, and of the agreements of Lessee herein contained, hereby grants, leases and lets exclusively unto Lessee for the purpose of investigating, exploring, prospecting, drilling and mining for and producing oil, gas and all other minerals, conducting exploration, geologic and geophysical surveys by seismograph, core test, gravity and magnetic methods, injecting gas, water and other fluids, and air into subsurface strata, laying pipe lines, building roads, tanks, power stations, telephone lines and other structures thereon and on, over and across lands owned or claimed by Lessor adjacent and contiguous thereto, to produce, save, take care of, treat, transport and own said products, and housing its employees, the following described land in________________________________ County, Texas, to-wit:

This lease also covers and includes all land owned or claimed by Lessor adjacent or contiguous to the land particularly described above, whether the same be in said survey or surveys or in adjacent surveys, although not included within the boundaries of the land particularly described above. For all purposes of this lease, said land is estimated to comprise ________________________ acres, whether it actually comprises more of less.

TERM

2. Subject to the other provisions herein contained, this lease shall be for a term of_________________ ( ) years from this date (called "primary term") and as long thereafter as oil, gas or other mineral is produced from said land or land with which said land is pooled hereunder.

ROYALTY

3. The royalties to be paid by Lessee are:

(a) On oil, one-eighth of that produced and saved from said land, the same to be delivered at the well. If Lessor elects not to take delivery of the royalty oil, Lessee may from time to time sell the royalty oil in its possession, paying to Lessor therefor the net proceeds derived by Lessee from the sale of such royalty oil. Lessor's royalty interest in oil shall bear its proportionate part of the cost of treating the oil to render it marketable oil and, if there is no available pipeline, its proportionate part of the cost of all trucking charges.

(b) On gas, including all gases, liquid hydrocarbons and their respective constituent elements, casinghead gas or other gaseous substance, produced from said land and sold or used off the premises or for the extraction of gasoline or other product therefrom, the market value at the well on one-eighth of the gas so sold or used, provided that on gas sold at the well the royalty shall be one-eighth of the net proceeds derived from such sale. Lessor's royalty interest in gas, including all gases, liquid hydrocarbons and their respective constituent elements, casinghead gas or other gaseous substance, shall bear its proportionate part of the cost of all compressing, treating, dehydrating and transporting incurred in marketing the gas so sold at the wells.

(c) On all other minerals mined and marketed, one-tenth either in kind or value at the well or mine, at Lessee's election, except that on sulphur mined and marketed the royalty shall be fifty cents ($.50) per long ton.

(d) While there is a gas well on said land or on lands pooled therewith and if gas is not being sold or used off the premises for a period in excess of three full consecutive calendar months, and this lease is not then being maintained in force and effect under the other provisions hereof, Lessee shall tender or pay to Lessor annually at any time during the lease anniversary month of each year immediately succeeding any lease year in which a shut-in period occurred one-twelfth (1/12) of the sum of $1.00 per acre for the acreage then covered by this lease as shut-in royalty for each full calendar month in the preceding lease year that this lease was continued in force solely and exclusively by reason of the provisions of this paragraph. If such payment of shut-in royalty is so made or tendered by Lessee to Lessor, it shall be considered that this lease is producing gas in paying quantities and this lease shall not terminate, but remain in force and effect. The term "lease anniversary month" means that calendar month in which this lease is dated. The term "Lease year" means the calendar month in which the lease is dated, plus the eleven succeeding calendar months.

(e) If the price of any oil, gas, or other minerals produced hereunder is regulated by any governmental authority, the value of same for the purpose of computing the royalties hereunder shall not be in excess of the price permitted by such regulation. Should it ever be determined by any governmental authority, or any court of final jurisdiction, or otherwise, that the Lessee is required to make any refund on oil, gas, or other minerals produced or sold by Lessee hereunder, then the Lessor shall bear his proportionate part of the cost of any such refund to the extent that royalties paid to Lessor have exceeded the permitted price, plus any interest thereon ordered by the regulatory authority or court, or agreed to by Lessee. If Lessee advances funds to satisfy Lessor's proportionate part of such refund, Lessee shall be subrogated to the refund order or refund claim, with the right to enforce same for Lessor's proportionate contribution, and with the right to apply rentals and royalties accruing hereunder toward satisfying Lessor's refund obligations.

(f) Lessee shall have free use of oil, gas, coal, water from said land, except water from Lessor's wells, for all operations hereunder, and the royalty on oil, gas and coal shall be computed after deducting any so used.

Figure 3.6 Conveyance, term, and royalty are contained in all standard leases.

Conveyance is the granting of interest in the petroleum to a person or company for the purposes of exploring, drilling, and producing. The conveyance part of the lease is addressed in the all-important granting clause and includes the consideration, legal land description, and usually a Mother Hubbard clause.

Term is the duration of the lease and is stated in the *habendum clause.* Royalty is a share of production provided for in several *royalty clauses* dealing with payments.

Dates

Any lease should be dated in order to avoid possible disputes as to which lease (if more than one exists) is the valid one. The controlling date for a lease is the one written in the instrument, not the date on which it was signed, notarized, or recorded.

Granting Clause

The *granting clause* conveys the described mineral interest from the lessor to the lessee for a consideration. By accepting the lease, the lessee accepts the covenants expressed and implied by the granting clause (fig. 3.7). The implied covenants vary with circumstances but involve the need to exercise diligence and good faith in performance of what would be expected of an ordinarily prudent operator. Expressed covenants such as protecting and developing a lease might include drilling an offset well, drilling additional wells or recompleting wells to new depths, plugging abandoned wells, and

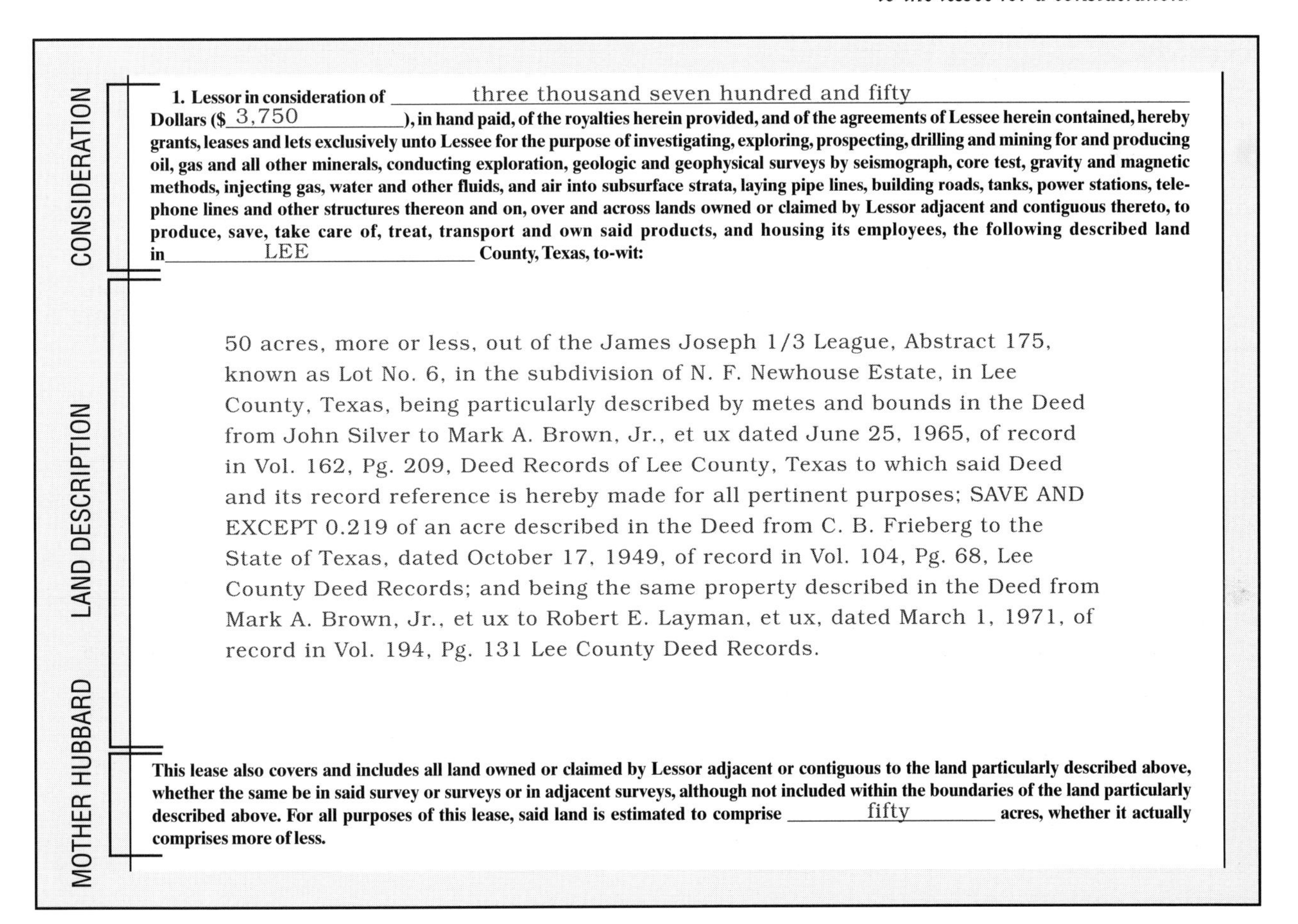
CONSIDERATION

1. Lessor in consideration of three thousand seven hundred and fifty **Dollars ($** 3,750 **), in hand paid, of the royalties herein provided, and of the agreements of Lessee herein contained, hereby grants, leases and lets exclusively unto Lessee for the purpose of investigating, exploring, prospecting, drilling and mining for and producing oil, gas and all other minerals, conducting exploration, geologic and geophysical surveys by seismograph, core test, gravity and magnetic methods, injecting gas, water and other fluids, and air into subsurface strata, laying pipe lines, building roads, tanks, power stations, telephone lines and other structures thereon and on, over and across lands owned or claimed by Lessor adjacent and contiguous thereto, to produce, save, take care of, treat, transport and own said products, and housing its employees, the following described land in** LEE **County, Texas, to-wit:**

LAND DESCRIPTION

50 acres, more or less, out of the James Joseph 1/3 League, Abstract 175, known as Lot No. 6, in the subdivision of N. F. Newhouse Estate, in Lee County, Texas, being particularly described by metes and bounds in the Deed from John Silver to Mark A. Brown, Jr., et ux dated June 25, 1965, of record in Vol. 162, Pg. 209, Deed Records of Lee County, Texas to which said Deed and its record reference is hereby made for all pertinent purposes; SAVE AND EXCEPT 0.219 of an acre described in the Deed from C. B. Frieberg to the State of Texas, dated October 17, 1949, of record in Vol. 104, Pg. 68, Lee County Deed Records; and being the same property described in the Deed from Mark A. Brown, Jr., et ux to Robert E. Layman, et ux, dated March 1, 1971, of record in Vol. 194, Pg. 131 Lee County Deed Records.

MOTHER HUBBARD

This lease also covers and includes all land owned or claimed by Lessor adjacent or contiguous to the land particularly described above, whether the same be in said survey or surveys or in adjacent surveys, although not included within the boundaries of the land particularly described above. For all purposes of this lease, said land is estimated to comprise fifty **acres, whether it actually comprises more of less.**

Figure 3.7 The granting clause conveys the mineral interest from the lessor to the lessee for a consideration.

treating and marketing the oil and gas. The granting clause is the place in which the minerals covered by the lease are clearly stated. Also, if the lessor is concerned about the restoration of his or her land after production has ceased, a stipulation may be added requiring the lessee to leave the property as it was found.

Consideration

The consideration is the benefit to the lessor and is required in some states to make the contract valid. *Consideration* is a term from contract law, often stated nominally as "$10 and other consideration." The US $10 serves as the necessary consideration of the document, and no one else needs to know what other consideration, or bonus, the lessor assumed for signing the lease. Bonus money varies from US $1 to $1,000 per acre (per .4 hectare), depending on the area and circumstances. If desired, the total amount may be written into the consideration. In Texas, no consideration is necessary for the validity of the lease, because it is a conveyance of a *present interest* in land.

Land Description

A legal description of the property involved is a necessary element in an oil and gas lease. If the lessor owns the entire undivided interest in the land being leased, the description will be straightforward. Otherwise, the lease may contain "lesser interest" or "proportionate reduction" clauses that allow lessees to reduce rents and royalties proportionately to the percentage of minerals owned. Whether the land is legally measured by *metes and bounds*, by the *rectangular survey system*, or as an *urban subdivision*, a clear and complete description of it is essential to an oil and gas lease.

Mother Hubbard Clause

Primarily in the lease to protect the working interest, the Mother Hubbard clause allows the operator to regard, as part of the lease, lands not included in the land description. A *Mother Hubbard* (a term based on the loose, rather shapeless dress worn by the nursery-rhyme character) includes odd-shaped or small wedges of land inadvertently left out or incorrectly described. A lessor will limit this coverage to 10 percent of the described land so as not to include all of the land he or she owns but only that part intended to be included by both parties.

Habendum Clause

The *habendum clause* fixes the duration of the lessee's interest (fig. 3.8). A typical habendum clause sets out the length of time for the primary term of the lease, which states the amount of time the lessee has to begin to drill a well. If not met, the lease expires. The primary term, generally used as an exploration period, varies from one to ten years; three years is common.

Figure 3.8 The habendum clause fixes the primary term of the lease.

2. Subject to the other provisions herein contained, this lease shall be for a term of _____one_____ (1) years from the date (called "primary term") and as long thereafter as oil, gas, or other mineral is produced from said land or with which said land is pooled hereunder.

The lease will continue, however, in a secondary term "as long as oil, gas, or other mineral is produced." The lessee is not bound by the lease unless the well or wells produce in "paying quantities," a term that the courts have held to mean sufficiently profitable to warrant production.

Royalty Clauses

Royalty is a share of production free of expenses, except for taxes and marketing expenses. The royalty clause (fig. 3.9) details the percentage of production given back to the lessor. Royalty varies, but one-eighth to one-fourth is common. From an oilwell, the lessor can receive royalty in *kind*—that is, in barrels of crude—or in money.

Figure 3.9 The royalty clause gives the percentage of production to be received by the lessor.

3. The royalties to be paid by Lessee are:

(a) On oil, one-eighth of that produced and saved from said land, the same to be delivered at the well. If Lessor elects not to take delivery of the royalty oil, Lessee may from time to time sell the royalty oil in its possession, paying to Lessor therefor the net proceeds derived by Lessee from the sale of such royalty oil. Lessor's royalty interest in oil shall bear its proportionate part of the cost of treating the oil to render it marketable oil and, if there is no available pipeline, its proportionate part of the cost of all trucking charges.

(b) On gas, including all gases, liquid hydrocarbons and their respective constituent elements, casinghead gas or other gaseous substance, produced from said land and sold or used off the premises or for the extraction of gasoline or other product therefrom, the market value at the well on one-eighth of the gas so sold or used, provided that on gas sold at the well the royalty shall be one-eighth of the net proceeds derived from such sale. Lessor's royalty interest in gas, including all gases, liquid hydrocarbons and their respective constituent elements, casinghead gas or other gaseous substance, shall bear its proportionate part of the cost of all compressing, treating, dehydrating and transporting incurred in marketing the gas so sold at the wells.

(c) On all other minerals mined and marketed, one-tenth either in kind or value at the well or mine, at Lessee's election, except that on sulphur mined and marketed the royalty shall be fifty cents ($.50) per long ton.

(d) While there is a gas well on said land or on lands pooled therewith and if gas is not being sold or used off the premises for a period in excess of three full consecutive calendar months, and this lease is not then being maintained in force and effect under the other provisions hereof, Lessee shall tender or pay to Lessor annually at any time during the lease anniversary month of each year immediately succeeding any lease year in which a shut-in period occurred one-twelfth (1/12) of the sum of $1.00 per acre for the acreage then covered by this lease as shut-in royalty for each full calendar month in the preceding lease year that this lease was continued in force solely and exclusively by reason of the provisions of this paragraph. If such payment of shut-in royalty is so made or tendered by Lessee to Lessor, it shall be considered that this lease is producing gas in paying quantities and this lease shall not terminate, but remain in force and effect. The term "lease anniversary month" means that calendar month in which this lease is dated. The term "Lease year" means the calendar month in which the lease is dated, plus the eleven succeeding calendar months.

(e) If the price of any oil, gas, or other minerals produced hereunder is regulated by any governmental authority, the value of same for the purpose of computing the royalties hereunder shall not be in excess of the price permitted by such regulation. Should it ever be determined by any governmental authority, or any court of final jurisdiction, or otherwise, that the Lessee is required to make any refund on oil, gas, or other minerals produced or sold by Lessee hereunder, then the Lessor shall bear his proportionate part of the cost of any such refund to the extent that royalties paid to Lessor have exceeded the permitted price, plus any interest thereon ordered by the regulatory authority or court, or agreed to by Lessee. If Lessee advances funds to satisfy Lessor's proportionate part of such refund, Lessee shall be subrogated to the refund order or refund claim, with the right to enforce same for Lessor's proportionate contribution, and with the right to apply rentals and royalties accruing hereunder toward satisfying Lessor's refund obligations.

(f) Lessee shall have free use of oil, gas, coal, water from said land, except water from Lessor's wells, for all operations hereunder, and the royalty on oil, gas and coal shall be computed after deducting any so used.

Gas Royalty

Royalty for gas is traditionally payable only in money. Many leases for gas have royalty based on the value of the actual sale by the lessee if the gas is sold at the well. And if gas is sold off the premises of the lease, royalty is usually based on market value, expressed in the lease as "market price at the well." This market price is determined by comparable sales in the field, so that in some instances the lessee may have to sell for less than the market value of the gas, yet will have to pay royalty based on a higher value. Gas royalty is complicated by federal controls over marketing and field pricing through the Natural Gas Act of 1938 and the Natural Gas Policy Act of 1978.

Nonparticipating Royalty

Royalty, or more often a fractional portion of the royalty, without other rights of the mineral interest, may be conveyed by the participating royalty owner to someone else for a definite term. Such an interest is termed *nonparticipating royalty,* and its owner cannot execute a lease or receive bonus or delay rental. He only receives a proportionate share of royalty payments. The *participating royalty owner*—one who also owns the mineral rights—can sign a lease and is said to have *executive rights.*

Shut-In Royalty

The *shut-in royalty clause* for gas wells allows the lessee to maintain the lease in force by paying money to the lessor in lieu of actual production when a well is *shut in*, or closed off and not producing. The purpose of this provision is to allow an operator to hold the lease on a well that is capable of producing in commercial amounts but is not in production for lack of a market or an available transmitting pipeline.

Pooling and Unitization Clause

A *pooling and unitization clause* is always included in the modern lease, authorizing the lessee to cross-convey interests in oil and gas (fig. 3.10). That is, two or more leases are combined to share interests in return for a proportionate sharing of royalty. For example, if John Smith had 100 acres (40 hectares) and Mary Brown had 60 acres (24 hectares) and the leases were pooled, a well drilling anywhere on the 160 acres (64 hectares) would bring John $100/160$, or $5/8$, of the royalty, and Mary would receive $60/160$, or $3/8$, of the royalty.

The terms pooling and unitization are often used interchangeably but refer to different undertakings. *Pooling* is the combining of small or irregular tracts into a unit large enough to meet state spacing regulations for drilling.

Figure 3.10 The pooling and unitization clause allows the lessee to combine leases to acquire sufficient acreage.

5. (a) Lessee, at its option, is hereby given the right and power to pool, unitize or combine the acreage covered by this lease or any portion thereof as to oil and gas, or either of them, with any other land covered by this lease, and/or with any other land, lease or leases in the immediate vicinity thereof to the extent hereinafter stipulated, when in Lessee's judgment it is necessary or advisable to do so in order properly to explore, or to develop and operate said leased premises in compliance with the spacing rules of the Railroad Commission of Texas, or other lawful authority, or when to do so would, in the judgment of Lessee, promote the conservation of oil and gas in and under and that may be produced from said premises. Units pooled for oil hereunder shall not substantially exceed 40 acres each in area, plus a tolerance of ten percent (10%) thereof, and units pooled for gas hereunder shall not substantially exceed in area 320 acres each plus a tolerance of ten percent (10%)

thereof, provided that should governmental authority having jurisdiction prescribe or permit the creation of units larger than those specified, for the drilling or operation of a well at a regular location or for obtaining maximum allowable from any well to be drilled, drilling or already drilled, units thereafter created may conform substantially in size with those prescribed or permitted by government regulations.

(b) Lessee under the provisions hereof may pool or combine acreage covered by this lease or any portion thereof as above provided as to oil in any one or more strata and as to gas in any one or more strata. The units formed by pooling as to any stratum or strata need not conform in size or area with the unit or units into which the lease is pooled or combined as to any other stratum or strata, and oil units need not conform as to area with gas units. The pooling in one or more instances shall not exhaust the rights of the Lessee hereunder to pool this lease or portions thereof into other units. Upon execution by Lessee of an instrument describing and designating the pooled acreage as a pooled unit, said unit shall be effective as to all parties hereto, their heirs, successors, and assigns, irrespective of whether or not the unit is likewise effective as to all other owners of surface, mineral, royalty, or other rights in land included in such unit. Within a reasonable time following the execution of said instrument so designating the pooled unit, Lessee shall file said instrument for record in the appropriate records of the county in which the leased premises are situated. Any unit so formed may be re-formed, increased, decreased, or changed in configuration, at the election of Lessee, at any time and from time to time after the original forming thereof, and Lessee may vacate any unit formed by it hereunder by instrument in writing filed for record in said county at any time when there is no unitized substance being produced from such unit.

(c) Lessee may at its election exercise its pooling option before or after commencing operations for or completing an oil or gas well on the leased premises, and the pooled unit may include, but it is not required to include, land or leases upon which a well capable of producing oil or gas in paying quantities has theretofore been completed or upon which operations for the drilling of a well for oil or gas have theretofore been commenced. In the event of operations for drilling on or production of oil or gas from any part of a pooled unit which includes all or a portion of the land covered by this lease, regardless of whether such operations for drilling were commenced or such production was secured before or after the execution of this instrument or the instrument designating the pooled unit such operations shall be considered as operations for drilling on or production of oil and gas from land covered by this lease whether or not the well or wells be located on the premises covered by this lease and in such event operations for drilling shall be deemed to have been commenced on said land within the meaning of paragraph 6 of this lease; and the entire acreage constituting such unit or units, as to oil and gas, or either of them, as herein provided, shall be treated for all purposes, except the payment of royalties on production from the pooled unit, as if the same were included in this lease.

(d) For the purpose of computing the royalties to which owners of royalties and payments out of production and each of them shall be entitled on production of oil and gas, or either of them, from the pooled unit, there shall be allocated to the land covered by this lease and included in said unit (or to each separate tract within the unit if this lease covers separate tracts within the unit) a pro rata portion of the oil and gas, or either of them, produced from the pooled unit after deducting that used for operations on the pooled unit. Such allocation shall be on an acreage basis—that is to say, there shall be allocated to the acreage covered by this lease and included in the pooled unit (or to each separate tract within the unit if this lease covers separate tracts within the unit) that pro rata portion of the oil and gas, or either of them, produced from the pooled unit which the number of surface acres covered by this lease (or in each such separate tract) and included in the pooled unit bears to the total number of surface acres included in the pooled unit. Royalties hereunder shall be computed on the portion of such production, whether it be oil and gas, or either of them, so allocated to the land covered by this lease and included in the unit just as though such production were from such land. The production from an oil well will be considered as production from the lease or oil pooled unit which it is producing and not as production from a gas pooled unit; and production from a gas well will be considered as production from the lease or gas pooled unit from which it is producing and not from an oil pooled unit.

(e) The formation of any unit hereunder shall not have the effect of changing the ownership of any delay rental or shut-in production royalty which may become payable under this lease. If this lease now or hereafter covers separate tracts, no pooling or unitization of royalty interest as between any such separate tracts is intended or shall be implied or result merely from the inclusion of such separate tracts within this lease but Lessee shall nevertheless have the right to pool as provided above with consequent allocation of production as above provided. As used in this paragraph 5, the words "separate tract" mean any tract with royalty ownership differing, now or hereafter, either as to parties or amounts, from that as to any other part of the leased premises.

Figure 3.10 Continued

Unitization is the combining of leased tracts on a fieldwide or reservoir-wide scale so that many tracts may be treated as one for operations such as enhanced recovery projects.

Pooling is a voluntary arrangement, although in Texas the Texas Mineral Interest Pooling Act allows the Railroad Commission of Texas to force pooling of interests in certain limited situations.

Drilling, Delay Rental, and Related Clauses

The *delay rental clause* provides the lessee with three choices (fig. 3.11). The lessee may opt to (1) drill a well, (2) pay on an annual basis to delay drilling until later but within the primary term, or (3) terminate the lease by neither drilling nor paying delay rental. Most leases written today expire one year

Figure 3.11 The delay rental clause allows extension or termination of the lease if a well is not drilled during the primary term.

6.

(a)

If operations for drilling are not commenced on said land or on acreage pooled therewith as above provided on or before one year from this date, the lease shall then terminate as to both parties, unless or before such anniversary date Lessee shall pay or tender (or shall make a bona fide attempt to pay or tender, as hereinafter stated) to Lessor or to the credit of Lessor in Monroe State **Bank at** Lexington**, Texas, (which bank and its successors are Lessor's agent and shall continue as the depository for all rentals payable hereunder regardless of change in ownership of said land or the rentals) the sum of** five hundred **Dollars ($**500.00**), (herein called rentals), which shall cover the privilege of deferring commencement of drilling operations for a period of twelve (12) months. In like manner and upon like payments or tenders annually, the commencement of drilling operations may be further deferred for successive periods of twelve (12) months each during the primary term. The payment or tender of rental under this paragraph and of royalty under paragraph 3 on any gas well from which gas is not being sold or used may be made by the check or draft of Lessee mailed or delivered to the parties entitled thereto or to said bank on or before the date of payment. If such bank (or any successor bank) should fail, liquidate or be succeeded by another bank, or for any reason fail or refuse to accept rental, Lessee shall not be held in default for failure to make such payment or tender of rental until thirty (30) days after Lessor shall deliver to Lessee a proper recordable instrument naming another bank as agent to receive such payments or tenders. If Lessee shall, on or before any anniversary date, make a bona fide attempt to pay or deposit rental to a Lessor entitled thereto according to Lessee's records or to a Lessor, who, prior to such attempted payment or deposit, has given Lessee notice, in accordance with subsequent provisions of this lease, of his right to receive rental, and if such payment or deposit shall be ineffective or erroneous in any regard, Lessee shall be unconditionally obligated to pay to such Lessor the rental properly payable for the rental period involved, and this lease shall not terminate but shall be maintained in the same manner as if such erroneous or ineffective rental payment of deposit had been properly made, provided that the erroneous or ineffective rental payment or deposit be corrected within 30 days after receipt by Lessee of written notice from such Lessor of such error accompanied by such instruments as are necessary to enable Lessee to make proper payment. The down cash payment is consideration for this lease according to its terms and shall not be allocated as a mere rental for a period. Lessee may at any time or times execute and deliver to Lessor or to the depository above named or place of record a release or releases of this lease as to all or any part of the above-described premises, or of any mineral or horizon under all or any part thereof, and thereby be relieved of all obligations as to the released land or interest. If this lease is released as to all minerals and horizon under a portion of the land covered by this lease, the rentals and other payments computed in accordance therewith shall thereupon be reduced in the proportion that the number of surface acres within such released portion bears to the total number of surface acres which was covered by this lease immediately prior to such release.**

(b)

Lessor hereby designates Monroe State **Bank at** Lexington**, Texas, and its successors as Lessor's agent to serve as the depository for any payment due with respect to any shut-in gas well. Payment of shut-in gas royalty may be made in the manner provided in paragraph 6(a) hereof for the payment or tender of rentals, including all terms with respect to the deposit of same in the designated depository bank, notwithstanding paragraph 6(a) being otherwise stricken or inoperative due to this lease having a primary term not exceeding one year, if such be the case.**

from the date on the instrument unless the lessee begins operations for drilling or makes timely payment of delay rental. Deferring drilling past the primary term of the lease generally voids the agreement. A *paid-up lease* provides for rental payment along with the cash bonus; these leases do not require the lessee to take any further action during the primary term.

A *dry hole clause* allows a lessee to keep the lease if the first hole is dry. The lessee then has a specified period of time to begin drilling a second well or resume payment of delay rentals.

If production stops, a *cessation of production statement* in the lease allows the operator a certain length of time to restore production (as in the case of a workover on a sluggish well), to drill a new well, or to return to paying delay rentals.

A *continuous development clause* is designed to keep drilling operations going steadily past the primary term. It requires the operator to develop leased land up to the maximum number of wells the government allows.

Assignment Clause

Interest in a lease can usually be transferred by either the lessor or the lessee to another party. In fact, leases frequently change hands a number of times before and sometimes after production begins. Although an assignment by one party to the lease does not materially affect the interests of the other party, both parties may be required to give notice of such transfers. The nature of this notice is provided for in the assignment clause (fig. 3.12).

Figure 3.12 The assignment clause allows the lessee to assign the lease rights to another party.

9. The rights of either party hereunder may be assigned in whole or in part, and the provisions hereof shall extend to their heirs, successors and assigns; but no change or division in ownership of the land, rentals or royalties, however accomplished, shall operate to enlarge the obligations or diminish the rights of Lessee; and no change or division in such ownership shall be binding on Lessee until thirty (30) days after Lessee shall have been furnished by registered U.S. mail at Lessee's principal place of business with a certified copy of recorded instrument or instruments evidencing same. In the event of assignment hereof in whole or in part, liability for breach of any obligation hereunder shall rest exclusively upon the owner of this lease or of a portion thereof who commits such breach. In the event of the death of any person entitled to rentals, shut-in royalty or royalty hereunder, Lessee may pay or tender such rentals, shut-in royalty or royalty to the credit of the deceased or the estate of the deceased until such time as Lessee is furnished with proper evidence of the appointment and qualification of an executor or administrator of the estate, or if there be none, then until Lessee is furnished with evidence satisfactory to it as to the heirs or devisees of the deceased and that all debts of the estate have been paid. If at any time two or more persons be entitled to participate in the rental payable hereunder, Lessee may pay or tender said rental jointly to such persons or to their joint credit in the depository named herein, or, at Lessee's election, the proportionate part of said rentals to which each participant is entitled may be paid or tendered to him separately or to his separate credit in said depository; and payment or tender to any participant of his portion of the rentals hereunder shall maintain this lease as to such participant. In event of assignment of this lease as to a segregated portion of said land, the rentals payable hereunder shall be apportionable as between the several leasehold owners ratably according to the surface area of each, and default in rental payment by one shall not affect the rights of other leasehold owners hereunder. If six or more parties become entitled to royalty hereunder, Lessee may withhold payment thereof unless and until furnished with a recordable instrument executed by all such parties designating an agent to receive payments for all.

Damage Clause

Many operators voluntarily pay for damage to the surface, but most leases include a clause that makes the lessee liable for damages or losses suffered because of drilling or production. The damage clause is usually tailored to the requirements of the landowner.

Force Majeure Clause

The *force majeure clause* (fig. 3.13) allows the lease to continue in force while the lessee is prevented from meeting the conditions of the lease during delays caused by "acts of God" or other events beyond the control of the operator. These events do not include delays due to equipment failures or labor problems. This clause also contains a reminder that the lease is subject to state and federal laws and usually excuses delays caused by government interference that cannot be blamed on the lessee.

12. When drilling, production or other operations on said land or land pooled with such land, or any part thereof are prevented, delayed or interrupted by lack of water, labor or materials, or by fire, storm, flood, war, rebellion, insurrection, sabotage, riot, strike, difference with workers, or failure of carriers to transport or furnish facilities for transportation, or as a result of some law, order, rule, regulation or necessity of governmental authority, either State or Federal, or as a result of the filing of a suit in which Lessee's title may be affected, or as a result of any cause whatsoever beyond the reasonable control of Lessee, the lease shall nevertheless continue in full force and effect. If any such prevention, delay or interruption should commence during the primary term hereof, the time of such prevention, delay or interruption shall not be counted against Lessee and the running of the primary term shall be suspended during such time; if any such prevention, delay or interruption should commence after the primary term hereof Lessee shall have a period of ninety (90) days after the termination of such period of prevention, delay or interruption within which to commence or resume drilling, production or other operations hereunder, and this lease shall remain in force during such ninety (90) day period and thereafter in accordance with the other provisions of this lease. Lessee shall not be liable for breach of any express or implied covenants of this lease when drilling, production or other operations are so prevented, delayed or interrupted.

Figure 3.13 The force majeure clause keeps the lease in force even though delays specified in the lease prevent the lessee from meeting his or her obligations.

Warranty and Proportionate Reduction Clauses

While the warranty clause seems to guarantee clear title, the *proportionate reduction clause* provides for the possibility that an owner owns less than his or her land description claims. If an owner's interest turns out to be less than he or she thought, the lessee can proportionately reduce the rentals and royalties paid. The clauses also outline what rights the lessee has in case the lessor defaults on tax or mortgage payments.

Special Provisions and Amendments

Provisions that outline specific rights of the lessee or lessor concerning the physical operation of the leasehold are included in the final lease in the form of additions, deletions, or amendments. For example, a landowner may want the right to approve the placement of equipment, storage facilities, pipelines, and roads, or may require that cattle guards be installed. The landowner may insist on a provision that wells be plugged according to state regulations and that the well site be restored as nearly as possible to its original condition. In some parts of the country, owners may bargain for gas from a gas-producing well for their own use—either free or purchased at the wellhead price.

To prevent a lessee from keeping a lease by means of producing a small amount from a shallow well, the landowner may insist on a *Pugh clause*, also called a *Freestone rider*. This provision releases nonproductive or untested zones from the lease, as well as acreage outside a producing pooled unit, if drilling or exploration does not take place by the end of a specified time.

For the operating company, or lessee, the provisions might include rights to use gas, oil, or water produced on the leasehold for its drilling operations (except, of course, the landowner's water wells or stock tanks); the right to remove its equipment when production has ceased; and the right to know in advance about pending changes in ownership or forfeiture of mortgage for nonpayment.

Lease-Terminating Provisions

The lessee's obligations concerning the termination of the lease agreement are usually covered in the delay rental and habendum clauses. Failure to comply with conditions set in these clauses can result in *forfeiture* of the property rights by the lessee. For example, the landowner can force, through the courts, the operator to relinquish his or her rights because of failure to pay delay rental or to drill within a given period of time.

On the other hand, the lessee can voluntarily give up his or her rights by neither drilling nor paying delay rental. Or, if the operator ceases production (generally because the well is no longer profitable), it may terminate the lease by *abandonment*—that is, it may physically abandon, or vacate, the premises and allow the lease to expire.

If the lease contains a *surrender clause*, the lessee must notify the lessor of its intent to surrender the lease. In some states, the lessee is required to give written notice when surrendering the lease so that the lessor has a clear title should he or she want to lease the property again. Failure to comply with surrender requirements could result in a fine or suit for damages.

EXECUTION OF THE LEASE

A properly executed oil and gas lease is a written instrument signed by the granting party, acknowledged by a notary public or other witnesses, and officially recorded in the appropriate county records. This step usually requires the help of an attorney who is licensed to practice law in the relevant state and preferably one well-versed in oil and gas law.

Signing the Lease

The lessor, or granting party or parties, must sign the lease to make it valid. The signature should conform to that used on any earlier document, such as a deed, by which the land was acquired. For example, a woman whose name has changed because of marriage since she acquired her property should show both names. People signing as a legal representative for someone else must identify their capacity and normally the party for whom they sign.

Acknowledging the Lease

Acknowledgment proves proper execution, or signing, of a lease. A notary public or other accepted officer witnesses the signing and pledges with a written certificate and an official signature that the grantor signed the lease freely. The pledge also assures the identity of the signing party.

Recording the Executed Lease

Generally, a lease is recorded as soon as possible after it is executed and consideration has been paid. Recording the lease involves officially entering it into the appropriate county records and is ordinarily handled by the lessee or its representative. The lessee keeps either the original or a copy, depending on the state, and the lessor gets a copy. This record shows other interested parties that a valid lease exists for that property and prevents possible title challenges on the grounds that the transaction has not been recorded.

TRANSACTIONS AFTER LEASING

The lessor has no further responsibility for the oil and gas after signing the lease and receiving the bonus. The lessee may do further exploration, obtain the required drilling permits, arrange for drilling a well, or negotiate other agreements that may be necessary to expedite the exploration and development of the leased property. After production begins, either the purchaser or the producer draws up a *division order*, a contract of sale for the oil or gas.

Division Orders

When production begins, a *division order* is drafted, based on the terms of the lease, the title opinion, and any other agreements affecting ownership of the oil or gas. The division order gives the names of all parties who have interests in the well (mineral owners, royalty owners, and working interest owners) and their proportionate shares of the payments. In addition to warranting title and guaranteeing correct percentages, the division order gives the purchaser certain rights in handling the oil or gas, accounting procedures, and establishing market values.

For oilwells, the purchaser of the oil usually handles the division orders and payments to the operator, royalty owners, and all parties holding a guaranteed interest in the minerals. When gas is sold, the lessee handles the division orders and payments. Whether prepared by the lessee or the purchaser, all division orders must be executed, or signed, by the operator, the royalty owners, and anyone else having an interest in the production. The division order, then, serves as a contract of sale between the mineral owners (usually represented by the operator) and the purchaser.

Support Agreements

Support can be offered in the form of money or an assigned interest in the leased property in exchange for drilling a well. A company seeking support money is probably planning to drill on one of its own leases. If it wants to drill a well on someone else's lease, it may suggest trading an assigned interest in the leased property in exchange for drilling. A lessee may farm out some of its leased acreage to a third party who wants to drill a well on it. An agreement that trades drilling obligations for an interest in the property is known as a *farmout* to the granting party (the *farmor*) and as a *farm-in* to the receiving party (the *farmee*) (fig. 3.14).

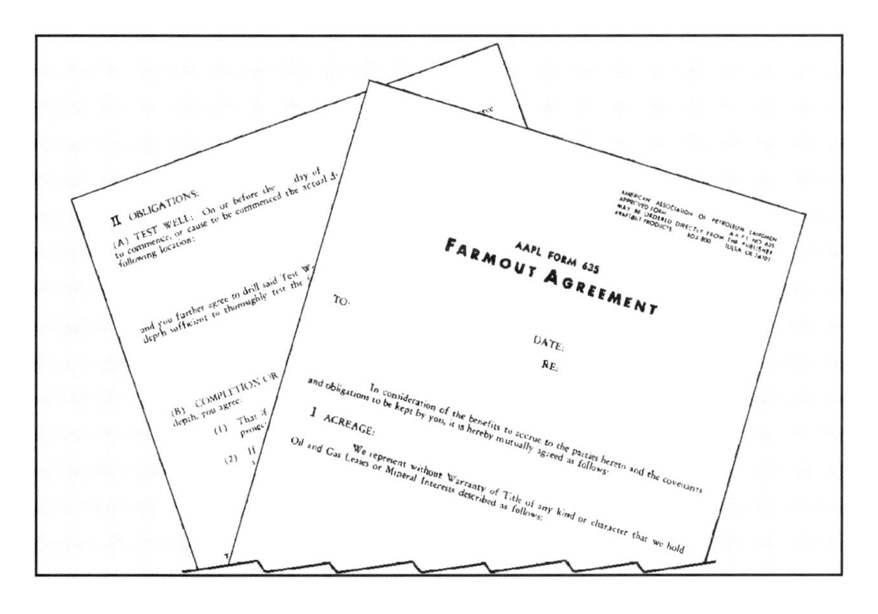

Figure 3.14 A farmout agreement trades drilling operations for an assigned interest in the leased property.

Acreage Acquisition Agreements

The simplest way to acquire acreage is to purchase the lease. In a lease purchase agreement, one company buys a block of leases from another company that has already set up files and plats (maps) of the area. Another transaction common in the industry is the agreement for acquiring acreage with the option to explore and then lease. In order to carry out seismic exploration, a company will secure blocks of acreage from one or more landowners by paying a given price per acre. After gathering exploration data, the company then usually has the option to lease selected acreage that is deemed promising for an additional fee.

Joint Operating Agreements

Two or more co-owners of the operating rights in a tract of land probably share the costs of exploration and possible development by means of a *joint operating agreement*. Usually, one of the owners serves as the operator and manages the drilling and related costs. An agreement may take many forms and is usually a complex contract. Joint operating agreements often follow farmouts when the farmor and farmee become co-owners of the rights to drill and produce. Also, joint operating agreements make possible expensive explorations that few individuals could attempt alone. In addition to exploratory operations, pooling and fieldwide unitization require operating agreements of one kind or another.

Joint Ventures

Unlike a joint operating agreement, where management is delegated and co-owners are not generally liable for certain operator actions, participants in a *joint venture* share liability for third-party claims. Of course, liability claims are settled in courts, and court decisions and legal interpretations vary from state to state and from one agreement to another. The company supporting a drilling venture may be an investment company that represents many individual investors. Most nonoperators are not liable for the operator's actions because clauses disclaiming joint liability are usually written into the agreements. Nonoperators who form a statutory *limited partnership* avoid joint liability but also relinquish any voice in operational matters.

Overriding Royalty Agreements

Overriding royalty is an expense-free share of the production and thus similar to the royalty received by the lessor, but it is paid out of the working interest rather than the royalty share. For example, a lessee might sell portions of the working interest in the lease to other operators and reserve an override. Because an override does not affect the lessor's interest or royalty, the lease does not provide for it.

LEASING PUBLIC LANDS

Extensive private ownership of mineral resources is the exception, not the rule, in most nations. Since much of the U.S. oil and gas is found on state, federal, and Native American (Indian) lands, and Canada's oil and gas largely belongs to the provinces, the lessee of these lands deals with a government leasing agency instead of individuals representing their own interests. These agencies can also provide information on how to lease from cities, counties, school districts, and other political units within the state or province.

State Ownership

Each state has a board or agency that governs the leasing of its lands. One source for a list of these agencies is the American Association of Professional Landmen. Laws vary from state to state.

Most coastal states control submerged lands and *inland waters* within three miles (4.8 kilometres) of their coast. The exceptions are Texas and Florida, which own the area 3 leagues, or 10.5 miles (17 kilometres), from their shores.

Leasing State Lands

The procedures for leasing land owned by a state depends upon the state. For example, Texas holds title to lands in various categories such as riverbeds, estuaries, Gulf Coast areas, public school lands, university lands, park lands, and so on. An operator or landman files a request with the state for an area of interest, and the Texas General Land Office determines whether to lease it and what the terms will be. The Land Office periodically distributes a *notice for bids* offering certain tracts and describing procedures and limitations to development. An applicant mails in the bid with payment, and the Land Office awards the lease to the highest bidder (fig. 3.15).

Figure 3.15 In some states, state-owned land is leased to the highest bidder on a mail-in basis.

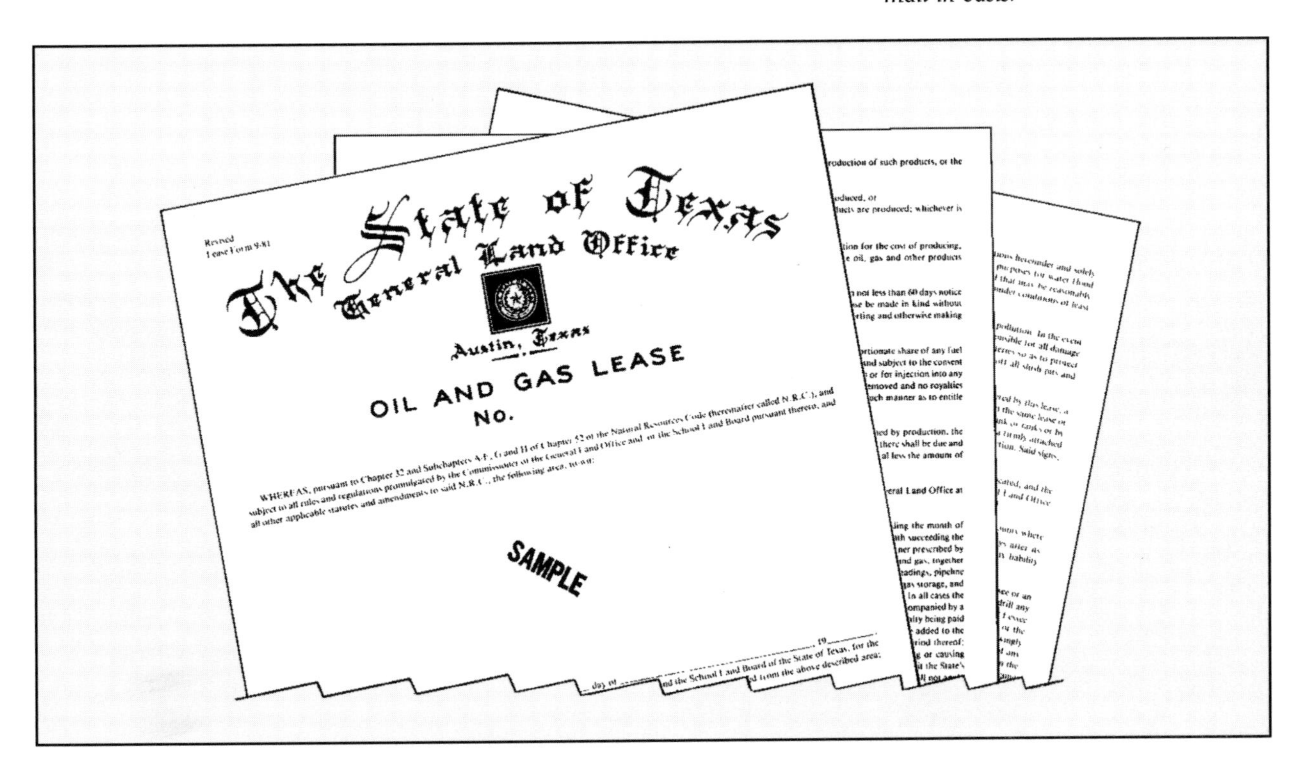

The State of Texas
General Land Office
Austin, Texas

OIL AND GAS LEASE
No.

SAMPLE

In Alaska, where only 2 percent of the land is in private ownership, the Division of Mineral and Energy Management, Department of Natural Resources (DNR), administers the leasing process, which takes into account the social, economic, and environmental impacts of the potential development and adds stipulations to modify them. Offshore, for example, certain activities must stop during bowhead whale migration. In some areas, the lessee must train employees in local customs and concerns. DNR approves specific tracts for leasing and informs potential lessees of the stipulations and bidding procedures. The highest bid gets the lease, unless the Commissioner of the Department of Natural Resources rejects the bid as too low.

Federal Ownership

The federal government is a landholder of astonishing size; however, much of its land is unavailable for oil and gas production. Land set aside for military uses, national parks, wildlife refuges, and so on is not generally leased to the petroleum industry. The federal government does lease tracts of public domain land, acquired land, certain Native American (Indian) land, and offshore land.

Both the mineral and the surface estates of land in the public domain belong to the nation. The federal government has also reserved the mineral rights to some property patented to individual citizens. *Public domain* land is land that the federal government originally owned and did not subsequently sell, while *acquired federal lands* were acquired by deed from earlier owners.

Leasing Federal Onshore Lands

The Secretary of the Interior makes all decisions to open or close federal lands to leasing, and may refuse to lease certain lands based on conservation principles or wildlife protection. The Bureau of Land Management (BLM) owns the mineral rights of federal lands and administers leasing and drilling. All BLM public lands and national forests are open to oil and gas leasing unless prohibited by law or administrative decision. For example, national parks and wildlife refuges are generally excluded. Various federal land management agencies, in particular the Forest Service, regulate the surface environmental consequences of drilling, so in reality these agencies can veto decisions made by BLM.

Like leases between landowners and operators, a lease from the federal government does not convey title but grants the rights to explore, drill, and produce.

Bidding for a Lease

Procedures for leasing federal onshore lands for oil and gas in the United States were first set up in the Mineral Lands Leasing Act of 1920 and the Acquired Lands Act of 1947. Under the Mineral Leasing Act, lands that had geologic structures known to contain oil or gas were leased by competitive bidding. Lands which had never been leased before and had no such geological structure (the vast majority) were leased on a first-come, first-served basis. A prospective lessee of such land could simply file for a tract, pay the filing fee, and submit the necessary forms—over the counter or by mail.

The first abuse connected with the Mineral Leasing Act was the Teapot Dome scandal. In 1922, the Secretary of the Interior, Albert Fall, secretly leased the Teapot Dome reserve in Wyoming, which was known to contain oil, to Harry Sinclair's Mammoth Oil Company. Fall later granted similar secret leases to E. L. Doheny's Pan American Petroleum Company. When word leaked out of the leases, a Senate investigation revealed that Fall, never very prosperous, had suddenly paid eight years of back taxes on his New Mexico ranch, built an irrigation system and a hydroelectric plant, and bought an adjacent ranch. The leases were eventually canceled, and Fall was convicted of accepting bribes.

More recently, two drawbacks of the system that the Mineral Leasing Act set forth became controversial. First, determining whether a geological structure that contained oil or gas was under a certain tract of land was more an art than a science, and lessees frequently challenged the Department of Interior's decisions. Second, many lessees held leases purely on speculation. When they expired because the lessee failed to drill, the scramble to be the first qualified applicant to refile became crazed. The DOI responded to this situation by considering all applications to be filed simultaneously and awarding the lease by lottery. Oil companies were not satisfied with this solution either.

As a result of pressures to reform the Mineral Leasing Act, Congress enacted the Federal Onshore Oil and Gas Leasing Reform Act in 1987. The Reform Act, as it is known, abolished the two types of leases and created a competitive bidding system, with a minimum bid of US $2 per acre. If nobody puts forth a minimum bid, the lands become available for noncompetitive leasing for two years, after which they return to the competitive bidding system. Competitive bidding takes place in an oral auction in each BLM state office at least four times per year. The Reform Act also requires that a lessee reclaim the drill sites and imposes severe penalties for those who do not.

Leasing Terms

The primary term for federal leases is five years for competitive leases and ten for noncompetitive. The lease expires at the end of the primary term if the lessee has not produced any oil or gas, but may be extended if drilling has begun. Once production ends or if the lease is otherwise surrendered, all rights revert to the government.

In addition to the fee per acre, a lessee that produces oil or gas pays royalties to the federal government. The minimum royalty collected by the BLM is 12.5 percent.

Leasing Indian Lands

Native American tribes control the leasing of their own lands. BLM supervises these leases and collects any royalties. Addresses for the agencies that lease Indian lands may be obtained from the American Association of Professional Landmen.

Leasing Federal Offshore Tracts

The federal government controls the area from the states' inland waters out to 200 miles (322 kilometres) or 8,200 feet (2,500 metres) of water depth (fig. 3.16). This region is known as the *Outer Continental Shelf*, or OCS. OCS leasing is guided by the OCS Lands Act of 1953 as amended in 1978. The Minerals Management Service (MMS) of the Department of the Interior is responsible for OCS leasing and production programs and royalty management.

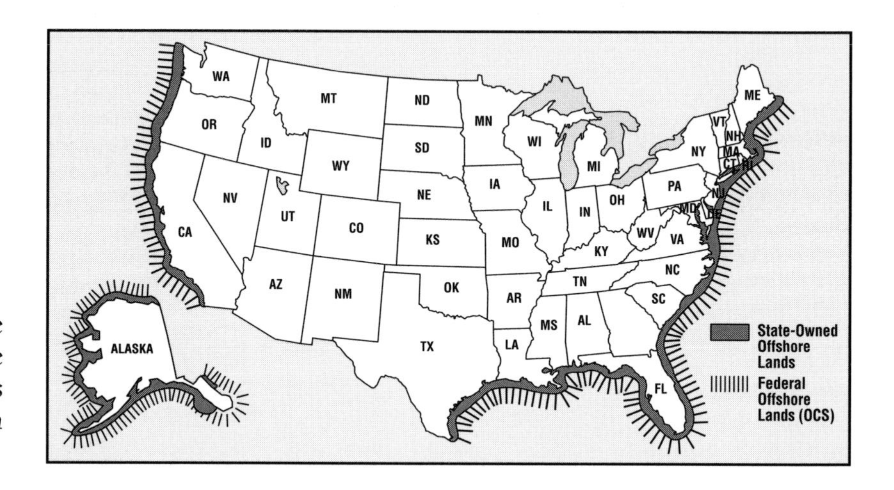

Figure 3.16 In the United States, the coastal states own offshore lands from the shoreline to 3 miles seaward, except Texas and Florida, which own 10½ miles from their Gulf Coast shores.

Unlike the highly regulated onshore leasing system, the DOI has great latitude in choosing how to lease offshore lands. Competitive bidding sets the prices. DOI sets out a five-year schedule of leases it expects to sell. These schedules may change with political or economic changes. Each sale follows a series of procedural steps. The first step is the call for information. DOI takes suggestions for which areas should or should not be leased. An *environmental impact statement (EIS)* may be prepared. DOI then gives public notice of proposed leases, and the affected states comment on the proposals. Finally, DOI selects the tracts and bidding procedures and issues the final notice of sale. A typical bid includes a cash bonus and a royalty agreement. It is not unusual for a consortium of major oil companies to submit one bid on their collective behalf.

Foreign companies may participate in federal lease sales where reciprocal agreements allow U.S. nationals to lease in that country. Securing the legal rights for resource development is becoming more sophisticated and international, reflecting global concerns and affecting everyone.

Ownership in Canada

With some exceptions, each of the ten Canadian provinces—Newfoundland, Nova Scotia, New Brunswick, Prince Edward Island, Quebec, Ontario, Manitoba, Saskatchewan, Alberta, and British Columbia—owns and manages its own mineral resources. The Canadian federal government owns and manages the oil and gas of the Northern Territories and offshore lands.

Some subsurface minerals are privately owned. In the four western provinces, a number of companies were granted land with attached mineral rights in exchange for their work in building railroads and thus opening up the provinces for settlement. Some early settlers of these provinces also acquired mineral rights with their homesteads, which are still in private ownership. Canada's Native Americans own and regulate other mineral properties.

4 Drilling Operations

Once an oil company's exploration geologists and geophysicists have obtained and analyzed the data from a prospective petroleum site, the landman has secured a lease, and drilling permits and other preliminary papers are in order, the company turns its attention to drilling. It negotiates a contract with a drilling contractor, who drills the well (known as *making hole* in the industry) according to the company's specifications. A drilling contractor is a company that owns drilling rigs and hires crews whose primary job is the drilling of wells.

HISTORY OF DRILLING FOR OIL

A brief history of drilling reveals some of the obstacles that early U.S. oil entrepreneurs faced in getting the oil out of the ground. As they drilled new areas and new problems arose, they developed new drilling methods. Engineers are still devising special tools and techniques to solve the problems of drilling for today's oil and gas.

Drake's Well

In 1857, James M. Townsend, a New Haven banker and the new president of the Pennsylvania Rock Oil Company of Connecticut, decided to send someone to Titusville, Pennsylvania—someone who could turn a headache into a money-making proposition. Townsend had been on the company's board of directors for several years. During this time, the company had leased the oil rights to an island that lay in Oil Creek, about a mile south of Titusville. An oil seep, or spring, bubbled to the surface on the property, but it released oil only in very small amounts.

At this time, the value of petroleum (also called *rock oil*) as a lubricant and illuminant was beginning to be recognized. Whale oil, the premium lubricant and main source of oil for illumination during this time, was scarce and expensive. So it seemed to Townsend and others in his company that if they could recover ample quantities of rock oil from the Oil Creek site, they could make a profit by selling it as a substitute for whale oil. The problem was that all efforts to accumulate the oil once it bubbled to the surface had failed. Entrepreneurs had dug trenches, constructed dams, and excavated holes by hand, but in every case either rainwater or groundwater had washed out the trenches, holes, and dams and had allowed the oil to run off into the creek.

What Townsend wanted was someone in Titusville to supervise a scheme that he believed would solve the problem. Townsend proposed that the company drill for oil just as others in the area for many years had been drilling for salt water, or brine. These wells allowed large volumes of brine to be pumped to the surface. Once on the surface, the water was evaporated to leave salt—a precious commodity in those days. By December 1857, Townsend had found his man, an unemployed railroad conductor named Edwin L. Drake (fig. 4.1).

Figure 4.1 The United States' first commercial oilwell was drilled near Titusville, Pennsylvania. Edwin L. Drake, in top hat and frock coat, oversaw the operation of this primitive rig in 1859.

Drake did not have an easy time of it. Almost two years passed before he was able to carry out Townsend's idea. For one thing, Drake had a difficult time locating a driller who was willing to give up a profitable saltwater drilling business to drill for oil—a substance that as far as most drillers were concerned served only to contaminate salt. Then, when Drake finally hired a driller, technical problems began to crop up. The worst had to do with the way in which brine wells were drilled at that time. It was customary for workers to use picks and shovels to dig out the topsoil at the site until they reached bedrock. Then the driller erected a rig over the hand-dug hole, or *cellar*, and drilled the well. But at the Oil Creek site, groundwater persisted in filling up and caving in the cellar long before the driller reached bedrock, thus making it impossible to fully excavate the cellar.

To solve the problem, Drake got his drilling team—a blacksmith named Billy Smith and his son—to hammer steel pipe into the ground. They drove pipe, called *casing*, down to bedrock, which prevented the topsoil from caving in. Then they built the rig and started drilling. By now it was April 1859. To drill the well, Drake used a steam-powered cable-tool rig. The rig's steam engine turned large pulleys and belts, which in turn caused a large wooden beam—the *walking beam*—to move up and down, much as the head of a child's rocking horse nods up and down when being ridden. A drill bit hung from the end of a rope or cable attached to the front of the walking beam. The team lowered the bit into the hole with the cable and set the walking beam in motion. As the walking beam rocked up and down, the bit rose and dropped, rose and dropped, over and over. Each time the bit dropped to bottom, it pierced the rock and made hole (fig. 4.2).

In spite of Drake's being able to overcome the problems and get the well started, the citizens of Titusville remained thoroughly convinced that he was wasting his time and openly referred to the project as "Drake's Folly." Even Townsend had begun to lose faith. The company's stockholders had long since given up hope and refused to sink any more money into the project. Townsend had been the only source of capital for months, and in August 1859, he too decided to call it quits. He sent a letter to Drake instructing him to abandon the well.

Figure 4.2 Using large pulleys and belts, the oak walking beam (shown near the top of this photo of the Drake well reconstruction) alternately raised and dropped the bit to drill a hole.

Meanwhile, in Titusville, William "Uncle Billy" Smith and Drake continued to drill, unaware that Townsend's fateful letter was wending its way by stagecoach from New Haven. One Sunday afternoon before the letter arrived, Uncle Billy decided to visit the well to check on its progress. The hole was about 69 feet deep and had been drilled to that depth only with a great deal of difficulty. It is easy to imagine the anticipation he must have felt when he saw something glimmering in the pipe lining the well. Could it be oil? Sure enough, it was, and the boom began. They had drilled the first commercial oilwell in the United States. Within a few years, the area in and around Oil Creek was covered with derricks (fig. 4.3).

Figure 4.3 This oilfield on the Benninghoff farm in Pennsylvania was one of many that sprang up during the 1865 oil boom. *(Courtesy of the American Petroleum Institute)*

Figure 4.4 This portable cable-tool unit can drill and service holes up to 5,044 feet (1,537 metres) deep.

Since the days of Drake's discovery, drilling methods have changed drastically; indeed, newer methods have almost totally replaced the techniques used in Drake's time. But the fact remains that Drake's well marked the beginning of a boom in oil that has not ceased. You could even say that Drake opened a new era: the petroleum era.

Cable-Tool Drilling

One of the earliest methods of drilling, and the one that Drake used, is *cable-tool drilling*. Cable-tool drilling can be very effective, especially in hard-rock formations. As described above, the drill bit hangs in the hole on the end of a rope or cable. A powered walking beam raises and drops the cable and attached bit. This up-and-down motion is repeated over and over, and each time the bit drops, it hits the bottom of the hole with great force to pierce the rock. As it strikes the rock at the bottom of the hole, the chisel point of the bit usually penetrates quite deeply. Since the rate of penetration is high, the drilling rate is fast.

However, cable-tool drilling has two disadvantages. One is that the driller must stop drilling frequently and pull the bit out of the hole to remove pieces of rock, or *cuttings*, that the bit has chipped away. If not removed, the cuttings will impede the bit's ability to drill ahead. The other disadvantage of cable-tool drilling is that the method cannot drill soft-rock formations. The splintered rock fragments tend to close back around the bit and wedge it in the hole.

Figure 4.5 A wooden derrick was typical of cable-tool drilling rigs from the late 1800s to the 1920s. *(Courtesy of the Humanities Research Center, The University of Texas, at Austin)*

In spite of its disadvantages, modern versions of cable-tool rigs are still used occasionally, particularly to drill shallow wells or for various service operations, such as knocking loose stuck tools and straightening collapsed casing (fig. 4.4). The rig still makes hole by the impact action of a bit suspended from steel drilling cable, and drilling still stops when it becomes necessary to bail, or remove, cuttings from the hole. Its advantages are that it is cheaper to operate and it can perform certain operations faster.

Although cable-tool rigs are not very common anymore, in their heyday they drilled a large number of wells. Stationary cable-tool drilling rigs, with their pyramid-shaped wooden derricks (fig. 4.5), were a common sight in the oil patch from the 1860s to the 1920s. Then, the portable cable-tool rig, which was smaller and contained many steel components, became standard. But by the late 1950s, even portable cable-tool rigs had all but disappeared from the scene. In spite of the early popularity of the cable-tool method, rotary drilling has largely replaced it.

Rotary Drilling

The first rotary drilling rig was developed in France in the 1860s, but it did not catch on at first. Because drilling companies erroneously believed that most petroleum lay in hard-rock formations, which they could very effectively drill with cable tools, cable-tool rigs dominated the scene. Then, in the 1880s, two brothers named Baker gained a reputation for drilling successful water wells in the soft formations of the Great Plains of the United States, an area where cable-tool rigs were not having much success. The rig the Bakers used was a rotary unit with a fluid-circulating system. The rotary technique proved equally successful in the unconsolidated soft rocks of Texas in the Corsicana oilfield, which drillers discovered while they were searching for water.

Finally, around 1900, several unsuccessful attempts were made with cable tools to drill the great Lucas well at Spindletop, near Beaumont, Texas. Anthony Lucas, an Austrian-born mining engineer, was convinced that oil did indeed exist under the dome of Spindletop; the problem was how to drill for it. Rotary drilling provided the answer (fig. 4.6). After Spindletop, where historians estimate that 80,000 to 100,000 barrels of oil per day (1 barrel equals 42 gallons or 159 litres) gushed from the single well in the first nine days, rotary drilling took off in a big way (fig. 4.7).

In *rotary drilling*, the drilling action comes from pressing the teeth of a bit firmly against the ground and turning, or rotating, it. At the same time the bit is rotating, a fluid, usually a liquid concoction of clay and water called *drilling fluid* or *drilling mud*, shoots out of special openings, or nozzles, in the bit with great velocity (fig. 4.8). These jets of mud move cuttings made by the bit teeth away from the teeth, and so continuously expose fresh, uncut rock to the teeth. Once the mud lifts the cuttings off bottom, it carries them

Figure 4.6 Rotary drilling at Spindletop, near Beaumont, Texas, revolutionized the drilling industry.

Figure 4.7 This 1920 oilfield is typical of early oil drilling and producing operations. *(Courtesy of American Petroleum Institute)*

Figure 4.8 Drilling fluid, or mud, circulates down through pipe, out through the bit, and back up the hole.

up the hole and to the surface for disposal. Since the drilling fluid continuously removes the cuttings from the hole, drilling does not have to stop to remove cuttings. Further, because the cuttings, whether from soft or hard rock, are constantly removed, they do not impede the bit's ability to drill ahead. For these reasons, rotary drilling has virtually replaced cable-tool drilling.

DRILLING TODAY

Rotary rigs are by far the most common rigs in the oil patch today (fig. 4.9). Most land rotary rigs are portable—that is, a drilling company can easily move the rig in and erect it to drill the hole and then take it down and move it onto another drilling site. The discovery of oil offshore has led to the development of several different types of rotary rigs for use in the sea. Some are mobile; a boat tows the rig onto the drill site, and the drilling company drills the well and then tows the rig to the next site. Some offshore rigs are immobile; the drilling company erects the rig on the site and leaves it there throughout the life of the field. Drilling in arctic regions of the world has led to the development of specially designed rotary rigs that are able to withstand the rigors of extreme cold and the effects of moving packed ice.

Figure 4.9 A modern rotary rig

DRILLING CONTRACTS

Regardless of where the well is to be located—on land, offshore, or in polar regions—the oil company, called the *operator,* hires a company known as a *drilling contractor* to drill the well. Because the operating company concentrates on finding and producing oil and gas, it usually does not own drilling rigs. Therefore, it hires a drilling company that has the personnel, rigs, and expertise to do the job. However, since the operator holds the rights to any oil and gas on a lease and is responsible for producing it, the operator owns the drilled well and thus sets the specifications for it.

Usually, the operating company has a person on the site at all times. This *company representative* works closely with the contractor's top manager to assure that the contractor drills the well to the operator's specifications.

While it is true that a few operating companies own their own rigs and drill their own wells, drilling contractors drill most wells.

The Drilling Contractor

Typically, a drilling contractor's crew consists of a toolpusher, a driller, a derrickhand, and two or three rotary helpers, or roughnecks. Offshore, the contractor also hires several roustabouts.

The *toolpusher* is the contractor's top manager on the drill site. Toolpushers are responsible for the rig's overall operation and performance and must see to it that the crew drills the well to the operator's specifications. The *driller* is subordinate only to the toolpusher and is the one person who actually operates the rig. The driller also manages the day-to-day activities of the derrickhand and rotary helpers.

The *derrickhand* has two jobs. First, he or she monitors and records the condition of the drilling mud. Second, when drill pipe is being removed from or put into the hole, the derrickhand handles the top of the pipe from a small platform high in the derrick or mast of the rig. *Rotary helpers,* or *floormen,* handle the bottom of the pipe on the rig floor when pipe is being removed from or put into the hole. At other times, they maintain and repair the tools and equipment on the rig. Offshore, *roustabouts* assist in the loading and unloading of equipment and supplies that a boat brings to the rig. They are also responsible for keeping the entire rig painted, cleaned, and repaired.

Bid Proposals and Specifications

A drilling bid proposal and contract is the usual document that begins the process of getting a well drilled. The operator commonly sends the bid proposal and contract to several drilling companies who work in the area where the well is to be drilled. Generally, the operator selects the contractor who responds with the lowest bid. However, the operator also takes a contractor's past performance and proven capability to drill into consideration.

In some cases, the operator and the contractor form an alliance. An alliance is an agreement between the two companies that allows the contractor to drill several wells for the operator in a particular area. With an alliance, the operator does not ask for a bid on each well.

Once the operator accepts a contractor's bid, both parties sign the bid, and it becomes a contract. The *contract* is an agreement between the operator and the contractor that spells out what each is expected to do and provide in order to get the well drilled to specifications. Clauses in the contract cover such items as the location of the well, the date on which the drilling is to begin, the well's depth, the basis of determining the amounts payable to the contractor, the time the amounts are payable, and so on.

One of the more important parts of the contract concerns the specifications of the well as determined by the operator. The contract lays out such specifications as the diameter and depth of each part of the hole, the drilling muds to be used, and the equipment and services that each party will furnish. In short, the operator states precisely what the contractor will do, and the contractor can then respond with a price that provides not only a well to meet specifications but also, if all goes well, a profit. As you can imagine, the business of bidding on wells requires skill, patience, experience, and no small measure of luck. Contract drilling is a highly competitive field.

Operators and contractors use one of four basic types of contracts: the footage contract, the daywork contract, the turnkey contract, and the combination agreement. Regardless of the type of contract, both the operator and the contractor are always concerned about the time needed to complete the job; the safety of the personnel, equipment, and property throughout the operation; and the ability of both to do acceptable work.

Footage Contract

Under a *footage contract*, the operator agrees to pay the contractor a certain amount for each foot (metre) of hole drilled. A footage contract is riskier for the contractor than for the operator because the operator pays the contractor the same amount regardless of how long it takes to drill the well. Should something that is not the operator's fault happen that makes it impossible to drill the well, the contractor loses money.

Daywork Contract

The most commonly used contract is the *daywork contract*. In this case, the operator pays the contractor so much per day for the use of the rig, regardless of what work the rig is performing. Another way to look at a daywork contract is to think of the contractor as being paid by the hour instead of by the foot or metre. Usually, daywork contracts stipulate rates of pay based on the type of operation the rig is performing. For instance, one rate may apply while the rig is actually drilling, and another may apply while the rig is capable of drilling but is shut down awaiting further orders from the operator.

In general, contractors use daywork contracts when the well to be drilled is in a high-risk area—one in which the formations are, for various reasons, more difficult than usual to drill. Since difficult-to-drill formations may present unusual delays and risks to the contractor and thus cost more money, the daywork contract compensates for the extra costs. Another use for a daywork contract is when the contractor is drilling a wildcat well. A *wildcat well* is a well in an area that has not been previously drilled. In such cases, the risks are pretty much unknown.

Turnkey Contract

A *turnkey contract* requires the operator to pay an agreed-upon amount to the drilling contractor when the well is finished. In this type of contract, the contractor furnishes all the equipment, material, and people needed to drill the well and controls the entire drilling operation without any on-site supervision by the operator. The contractor assumes all the risks and adjusts the price to reflect these risks. The operator benefits by not having to assume any risks and by receiving only one bill for the entire operation, which eliminates accounting expenses. Turnkey contracts are sometimes awarded to contractors with whom the operator has worked closely in a particular area.

Combination Agreements

Often, the final contract combines payment methods. For example, the operator may pay footage rates to a certain depth and then pay daywork rates for any drilling below that depth. Another way to structure a combination agreement is to include clauses that provide for a daywork rate for particular operations. For instance, a standby time rate compensates the contractor for days when the rig and crew are on the site and able to drill, but for reasons beyond their control cannot drill. This situation can occur when the contractor is waiting for permission from the operator to start testing operations, for the arrival of equipment or material from the operator, for muddy roads to become passable, and so forth. Most footage contracts contain clauses concerning daywork rates, making them combination agreements.

ROTARY DRILLING SYSTEMS

To drill a hole that might easily be 3 miles (4.8 kilometres) or deeper into the earth is a monumental job. The collection of equipment and machinery that does it is a rotary drilling rig. Basically what happens when a rig drills a hole is that a bit is pressed hard against the ground and turned. The rotating bit scrapes and gouges the rock out to make a hole. Above the bit, long sections of pipe, the *drill pipe,* screw together in a *drill string* to connect the equipment in the hole to the equipment on the surface. The drilling crew keeps adding lengths of drill pipe as the hole gets deeper.

A drilling rig is a sort of portable factory whose sole purpose is to make holes in the ground, or make hole. The many pieces of equipment and machinery necessary to do this job can be divided into four main systems: hoisting, rotating, circulating, and power (fig. 4.10).

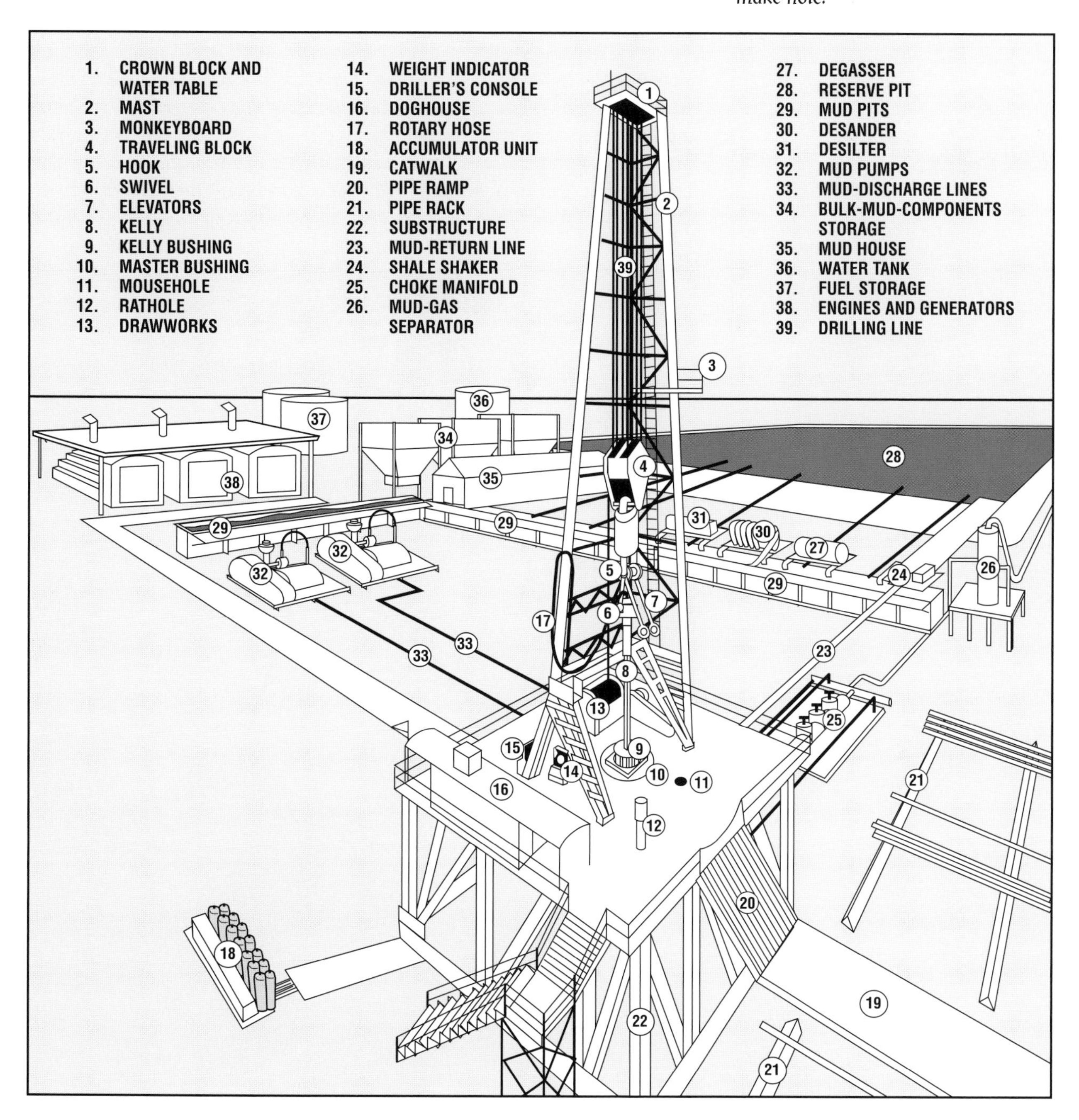

Figure 4.10 The major components of a rotary drilling rig work together to make hole.

Hoisting System

The *hoisting system* has two main purposes: to hoist the drill pipe in and out of the hole and to pull up on the drill pipe while drilling in order to keep it in tension.

To understand the second purpose, imagine a wire with a heavy weight tied to one end and a box of sand. If you hold the wire by the free end above the box, the wire will be stretched taut. If you set the weight down onto the box of sand and push down, the wire will bend. To allow the weight to make an indentation in the sand without bending the wire, you must pull up on the wire slightly, against gravity. In the same way, heavy pipes called *drill collars* attached to the bottom of the drill string apply weight by pressing down on the bit. However, because drill pipe has relatively thin walls, it would buckle, like the wire in our example, if it were used to add weight to the bit. Therefore, the driller must maintain an upward pull on the drill pipe during drilling, to keep it in tension.

The hoisting system of a drilling rig works like an old-fashioned *windlass* or *winch*. In a windlass, a *drum*, or *spool*, sits horizontally between two posts with one end of a rope attached to it (fig. 4.11). The other end of the rope is attached to something to be lifted, such as a bucket. Turning the drum with a crank winds the rope around the drum and lifts the bucket. The hoisting system of a rotary drilling rig consists of the mast or derrick, traveling and crown blocks, the drilling line, and the drawworks (fig. 4.12).

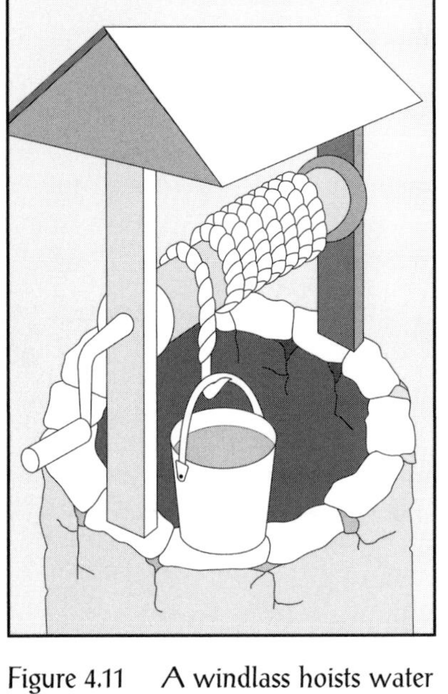

Figure 4.11 A windlass hoists water from a well.

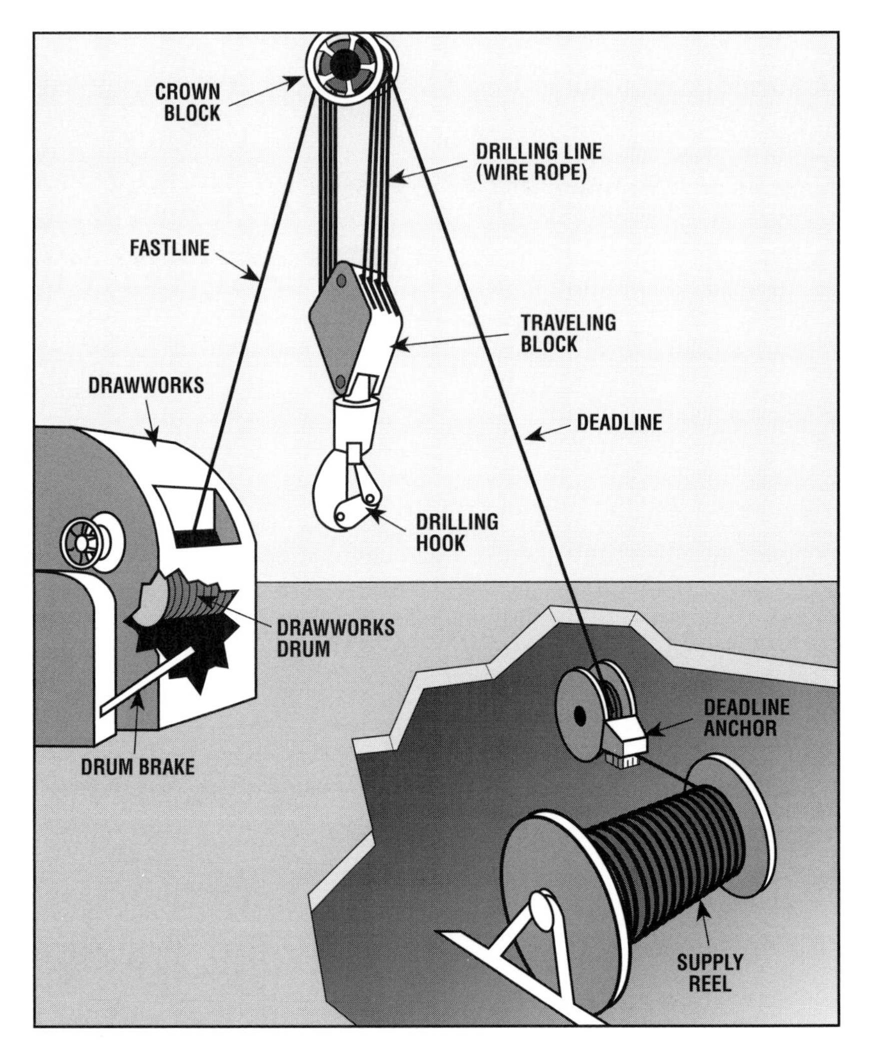

Figure 4.12 The hoisting system of a rotary rig is shown without the derrick.

Derricks and Masts

Derricks and masts look much the same and do the same job, but they are used at different types of drilling sites (fig. 4.13). The derrick or mast is the universally recognized symbol of oilwell drilling, a steel tower that may rise 120 feet (36 metres) above the rig floor (fig. 4.14). A *derrick* is a more or less permanent structure. Its legs sit on the corners of the rig floor, and the crew must disassemble it to move it. A *mast*, on the other hand, is portable. It fits into an *A-frame* that may sit on the rig floor or on the ground. The crew can fold or telescope it down to move it. The purpose of the derrick or mast is to support the traveling and crown blocks and the enormous weight of the drill stem. A derrick or mast also supports the drill pipe and drill collars when they are pulled out of the hole and set back.

Most land rigs use masts because masts are so much easier to move than derricks. Offshore rigs, however, often use derricks because the entire rig, along with the derrick, moves, and assembling and disassembling the derrick is unnecessary.

Figure 4.14 This mast supports the hoisting system of the rig.

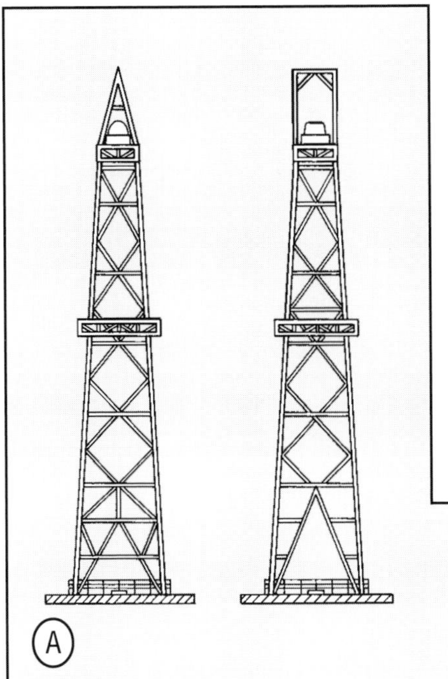

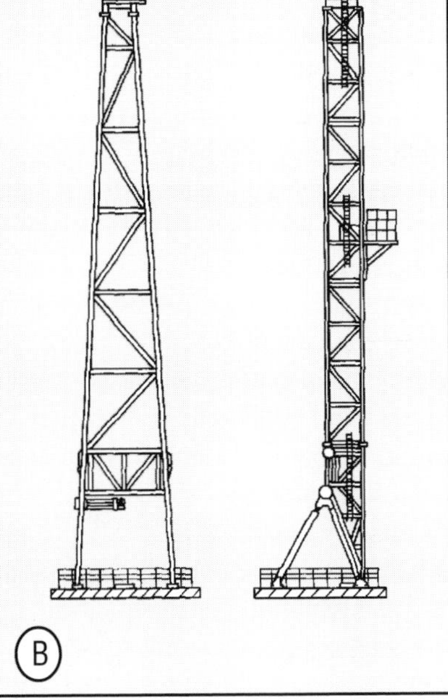

Figure 4.13 The legs of a derrick rest on the rig floor (A). A mast fits onto an A-frame, which may be on the rig floor or at ground level (B).

Blocks and Drilling Line

The *crown block* and the *traveling block* are each a set of pulleys, or *sheaves* (pronounced "shivs"). The crown block sits at the top of the derrick or mast (the crown) and never moves. The traveling block, as its name implies, travels up and down, in the center of the derrick or mast. A *drilling hook* extends from the bottom of the traveling block. The swivel or top drive, which holds the drill stem, hangs from the drilling hook.

Drilling line looks like an ordinary fiber rope, except that it is made of woven steel wire. It ranges in diameter from ⅞ to 2 inches, or about 22 to 51 millimetres. It comes on a spool called a *supply reel* (fig. 4.15). This reel, depending on the length of line wrapped on it, can be quite large—6 feet (almost 2 metres) in diameter.

Figure 4.15 This skid-mounted supply reel supplies wire rope for the rig.

The drilling line runs from the supply reel to the crown block and passes through one sheave (fig. 4.16). Then it goes down to the traveling block and wraps around it through one of its sheaves and heads back up to the crown block. To multiply the strength of the hoisting system, and therefore the amount of weight it can hoist, the crew threads, or *reeves,* the line back and forth several times between the two blocks. Even though the line is one continuous piece, the effect is that of several lines—8, 10, or 12. Finally, the end of the line coming from the traveling block goes to the *drawworks drum,* where it is anchored (see fig. 4.18A). The drum spools the line in or out, thus lifting or lowering the traveling block.

The part of the drilling line running from the drum through the traveling block and the crown block is called the *fastline* because it moves rapidly on and off the drum. The part of the drilling line from the crown block to the supply reel is the *deadline*. It does not move at all during hoisting. In fact, a special anchor that is usually fastened to the rig's substructure secures the deadline (fig. 4.17).

With perhaps 2,200 feet (671 metres) of wire rope in use as drilling line, the driller uses the remainder on the supply reel in scheduled *slip-and-cut* maintenance programs. A slip-and-cut program changes the locations of wear and stress by slipping the line through the system at a rate so that it

Figure 4.16 Drilling line passes several times through the traveling block (A) and the crown block at the top of the mast (B) through grooved sheaves in each.

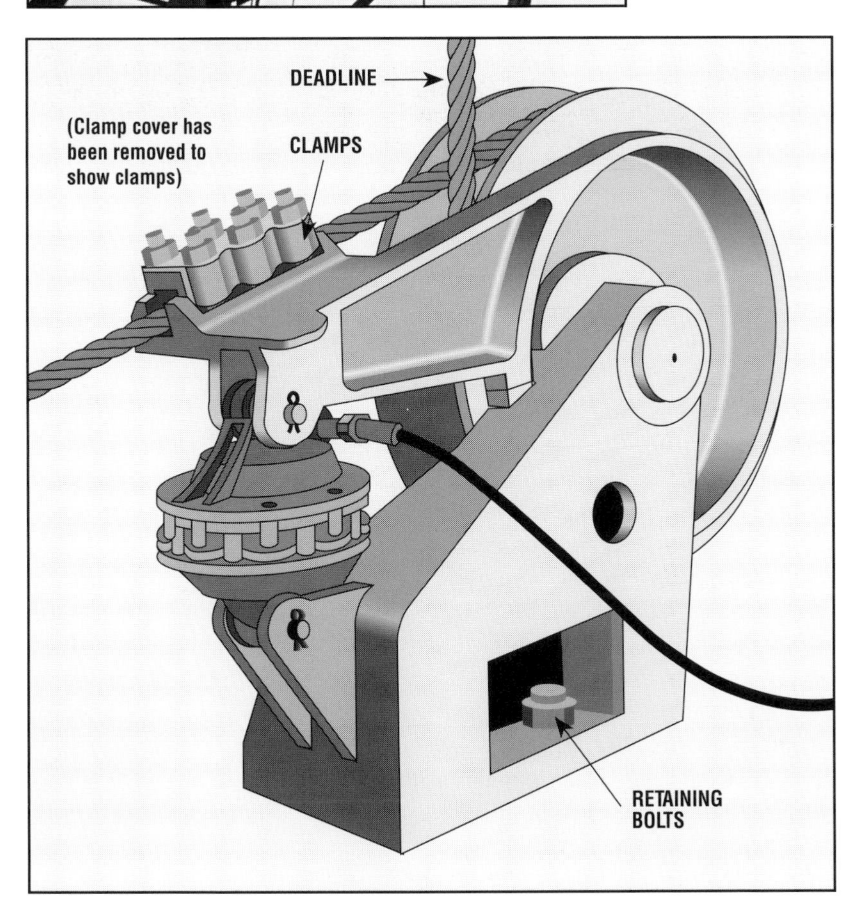

Figure 4.17 A deadline anchor on the rig's substructure holds the deadline firmly in place.

wears evenly and then cutting it off at the drum when it has reached the end of its usefulness. When the line has moved a ton of load over the distance of one mile, it is said to have given a *ton-mile* of service. (The metric unit for this measurement is the *megajoule*; one ton-mile equals 14.317 megajoules.) The driller carefully keeps ton-mile records in order to plan the slip-and-cut program for good service life.

Drawworks

The drawworks houses the large drum around which the drilling line is wrapped (fig. 4.18). As the drum rotates one way or the other, the drilling line spools on or off the drum. This raises or lowers the traveling block and the drill stem hanging from it.

A. Front

B. Rear

Figure 4.18 The drawworks, on the floor of the rig, contains the drum (A) and other equipment inside a steel housing (B).

Figure 4.19 The driller's console, on the left side of the drawworks, has controls for the power, transmission, and brakes of the hoisting system.

The *drawworks* is one of the largest and heaviest pieces of equipment on a drilling rig. The driller operates the drawworks at the *driller's console,* on the front left side of the drawworks, with controls for brakes, clutches, and a transmission (fig. 4.19). One set of brakes on each end of the drum holds it stationary and sustains the great weight of the traveling block, rotating equipment, and drill string. By releasing the brake or applying power to the drum, the driller can raise and lower the drill string.

Extending out of each end of the drawworks above the drum is a powered shaft called a *catshaft.* On each end of the catshaft is a spool-shaped *cathead.* On one side of the drawworks is the *makeup cathead* (fig. 4.20), and on the other side is the *breakout cathead.* The crew uses the makeup cathead to apply tightening force to large wrenches called *tongs* to *make up*, or screw together, and tighten joints of drill pipe and drill collars. A chain runs from the end of the tongs to the makeup cathead, and when the driller rotates the cathead, the chain is pulled tight. Pulling in the chain pulls on the tongs and causes them to tighten the pipe on which the tongs are latched.

The breakout cathead is used to apply loosening force to *break out*, or unscrew, joints of pipe. A wire rope runs from the end of the tongs to the breakout cathead, which loosens the pipe as the cathead rotates. The makeup and breakout catheads are often called *automatic*, or mechanical *catheads.*

Figure 4.20 The makeup cathead is attached to a catshaft coming out of the top of the drawworks.

Figure 4.21 The crew uses an air hoist to move heavy equipment around the rig floor.

Next to each automatic cathead is a small spool also attached to the catshaft. The spool forms a *friction cathead* around which large cloth-fiber rope, or soft line, can be wrapped. With a friction cathead and soft line, a crew member can move fairly heavy items of equipment on the rig floor. However, because using friction catheads is dangerous, many contractors install small, pneumatically operated hoists called *air hoists,* or *air tuggers,* to perform many of the light hoisting tasks on the rig floor that formerly were done with friction catheads (fig. 4.21).

Rotating System

A conventional rotating system includes all the equipment that makes the bit turn. The main part of the rotating system is a powerful machine called the *rotary table.* Located on the floor of the rig, the rotary table is capable of creating a strong rotating force, or *torque.* Since the bit is at the bottom of a hole that could be thousands of feet (metres) deep and the rotary table is on the surface, additional equipment transmits the torque from the table to the bit.

Rotating equipment from the surface to the bottom of the hole consists of the swivel, the kelly, the rotary table, the drill pipe, the drill collars, and the bit (see fig. 4.34). The kelly, the drill string, and the drill collars are collectively called the *drill stem.* Frequently, however, drilling personnel refer to the entire drill stem as the drill string.

Swivel

The *swivel* hangs from the drilling hook on the bottom of the traveling block by means of a large *bail,* or handle (fig. 4.22). Even though the swivel does not rotate, it allows everything below it to rotate.

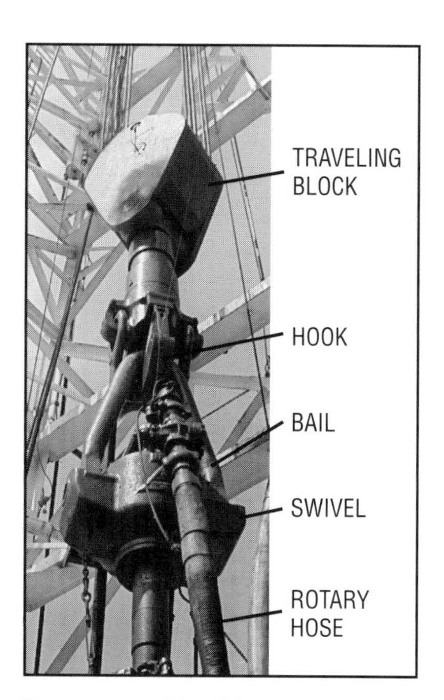

Figure 4.22 The drill stem is attached to the bottom of the swivel, which hangs from the hook on the traveling block.

Kelly

Attached to a threaded connection on the bottom of the swivel is the kelly. The *kelly* is a special section of pipe that is available in lengths of 40 and 54 feet (about 12 and 16 metres). Usually, it is 40 feet (12 metres) long. Unlike most pipe, the kelly is not round; rather, it has flattened sides so that in cross section it has a square or hexagonal shape.

The four- or six-sided kelly fits inside a corresponding square or hexagonal opening in a device called the *kelly bushing* (fig. 4.23). The kelly bushing in turn fits into a part of the rotary table called the *master bushing.* Powered gears in the rotary table rotate the master bushing. As the master bushing rotates, the kelly bushing also rotates. The square or hexagonal opening in the kelly bushing fits against the square or hexagonal kelly and causes the kelly to turn. The flat sides of the kelly and kelly bushing act like the head of a bolt and a wrench that can grip and turn it. The turning kelly rotates the drill stem and thus the bit. Since the kelly slides through the opening in the kelly bushing, the kelly can move down as the bit drills the hole deeper.

The Rotary Table

The *rotary table,* with its master bushing and kelly bushing, supplies the necessary torque to turn the drill stem. In addition, when the crew removes the kelly and kelly bushing, the hole left in the master bushing accommodates a special set of gripping devices called *slips.* Slips have teethlike gripping elements called *dies* that fit around the drill pipe to keep it from falling into the hole when the kelly is disconnected (fig. 4.24).

Figure 4.23 The kelly passes through the kelly bushing, which fits into the master bushing of the rotary table. In this photo, the rathole, a hole for storing the kelly, is out of the picture to the right; and the mousehole, a hole for storing a length of pipe, is in the foreground.

Figure 4.24 Crewmen grasp the slips by the handles as they set them in the master bushing.

Figure 4.25 The rotary table locks in place to allow a new bit to be installed.

The driller can lock the rotary table to keep it from turning when necessary, such as when installing a new bit (fig. 4.25).

Top Drive

On large rigs, the owners replace the conventional swivel with a powered swivel called a *top drive*. The top drive hangs from the traveling block and has its own heavy-duty motor (fig. 4.26). The motor turns a threaded drive shaft that connects directly to the top of the drill stem to turn it.

Rigs with top drives do not need a kelly, kelly bushing, or master bushing. They do retain the rotary table, but only as a place for the crew to hold the drill stem in place with the slips.

The main advantages of a top drive over the conventional kelly and rotary table system are that it is safer and more efficient. Its disadvantage is its cost—it is too expensive to be practical on smaller rigs. However, manufacturers have recently developed a portable top drive that an operator of a smaller rig can rent to drill a difficult section of the hole or to drill horizontally (see "Directional Drilling" later in this chapter).

Figure 4.26 A top drive hangs from the traveling block in place of the conventional swivel.

Drill Pipe and Drill Collars

Although aluminum drill pipe is available, most drill pipe and drill collars are steel tubes. Drilling mud flows to the bottom of the hole through them. *Drill pipe* and *drill collars* come in sections, or *joints*, about 30 feet (about 9 metres) long (fig. 4.27). Other dimensions of interest are the outside diameter and weight. Drill pipe is available in many different diameters, but the most commonly used diameters are 4, 4½, and 5 inches (101, 114, and 127 millimetres). Drill pipe weight varies with its diameter, but a commonly used 4½-inch (114-millimetre) drill pipe weighs 16.6 pounds per foot (6.9 kilograms per metre). Thus, a 30-foot (9-metre) joint of such drill pipe weighs 498 pounds (226 kilograms).

The purpose of using drill collars is to put extra weight on the bit, so they are usually larger in diameter than drill pipe and have thicker walls.

Figure 4.27 Available in 30-foot (9-metre) joints, this drill pipe is being set back for storage on the rig floor.

Figure 4.28 Tool joints are wider than the body of the drill pipe.

For example, one available type of drill collar joint is 7 inches (about 18 centimetres) in diameter and weighs 125 pounds per foot (186 kilograms per metre). Thus, a 30-foot (9-metre) joint of a drill collar this size weighs 3,750 pounds (1,701 kilograms), or over 1I tons (nearly 1I tonnes).

Since several joints of drill pipe and drill collars must be joined together, or *made up*, as the hole gets deeper, they have threaded connections on each end. On drill pipe the threaded connections are called *tool joints* (fig. 4.28). Tool joints are steel rings that are welded to each end of a joint of drill pipe. They are also called collars; do not confuse connection collars with drill collars. One tool joint is a pin (male) connection, and the other is a box (female) connection (fig. 4.29). The pin of one joint fits into the box of another joint, thus allowing the crew to make up long strings of pipe.

Drill collars do not need tool joints. Their walls are so thick that the manufacturer can machine the pin and box connection threads directly into the steel of the wall.

Figure 4.29 On one end of each pipe is a pin and on the other end is a box.

The Bit

At the bottom of the drill stem is the *bit*, which drills the formation rock and dislodges it so that drilling fluid can circulate the fragmented material back up to the surface. In general, the driller chooses the bit according to the hardness of the formation to be drilled. The most common types of bits are roller cone bits and diamond bits (fig. 4.30). They range in size from 3¾ inches (9.5 centimetres) in diameter to 26 inches (66 centimetres) in diameter, but some of the most commonly used sizes are 17½, 12¼, 7⅞, and 6¼ inches (44, 31, 20, and 16 centimetres). Bits are available in many different sizes to drill different sizes of holes.

Roller Cone Bits

Roller cone bits usually have three (but sometimes two or four) cone-shaped steel devices that are free to turn as the bit rotates (fig. 4.31). Several rows of teeth, or *cutters*, on each cone scrape, gouge, or crush the formation as the teeth roll over it. The cutters may be teeth machined out of the steel alloy of the cone or they may be very hard pellets of tungsten carbide, called *inserts*, placed into holes drilled into the cones (see fig. 4.30B).

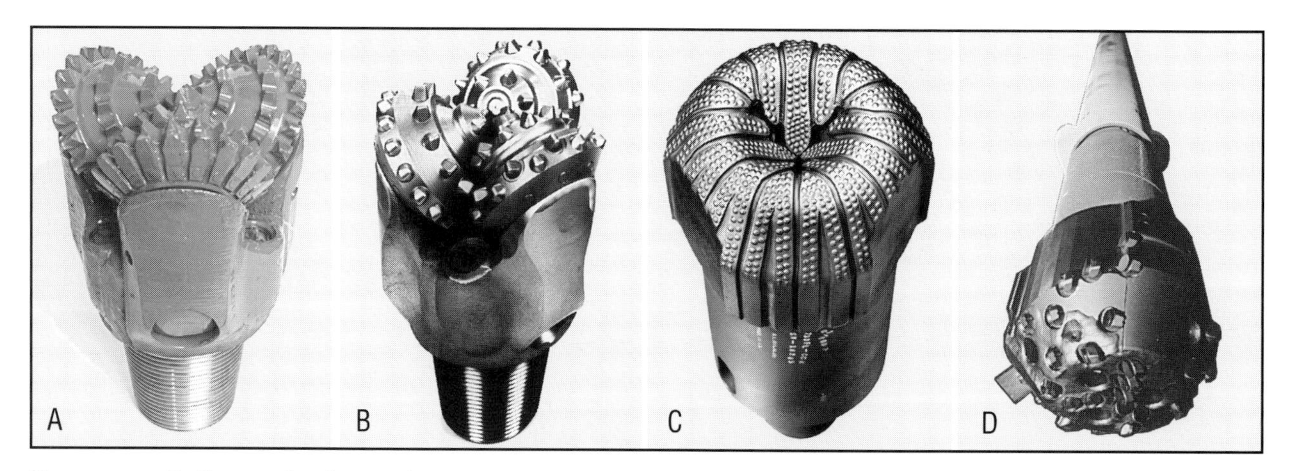

Figure 4.30 Roller cone bits have either milled teeth (A) or tungsten carbide inserts (B). A natural diamond bit has dozens of tiny diamonds on its face (C). A PDC bit has synthetic diamond studs in the face (D).

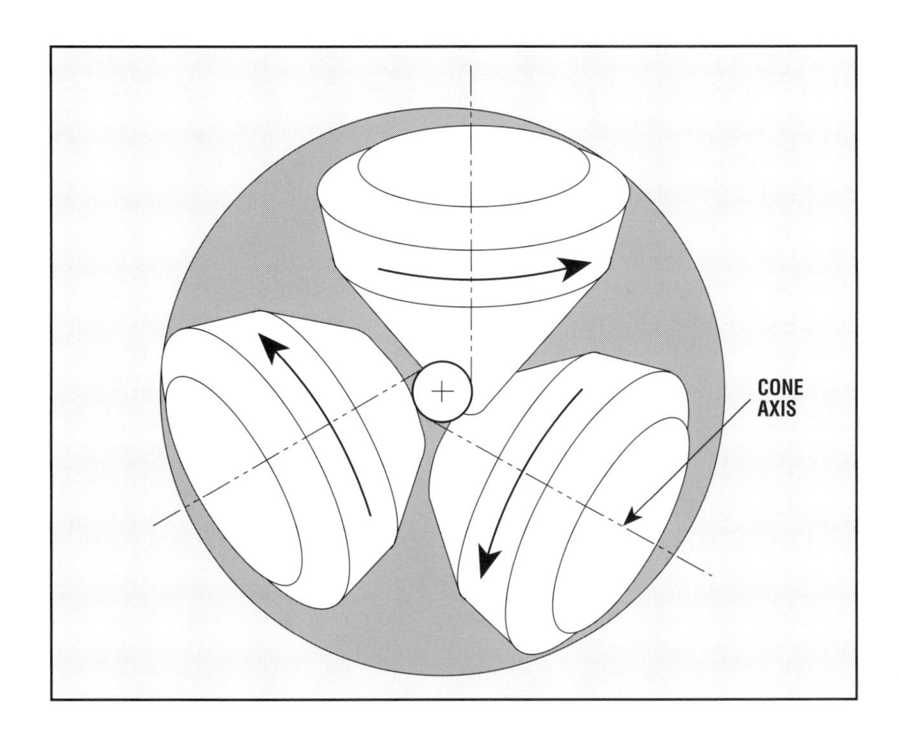

Figure 4.31 Each cone of a roller cone bit rotates on its own axis.

The cones rotate on bearings. Some bits use ball and roller bearings (fig. 4.32A); others have journal, or plain, bearings; and some use a combination of both (fig. 4.32B). Journal bearing bits generally do not wear out as fast as ball and roller bearing bits but are more expensive. The bearings may be either sealed or unsealed. In unsealed bits, drilling mud provides the only lubrication to the bearings. Some ball and roller bearing bits and all journal bearing bits are sealed. The seals keep drilling mud away from the bearings, which protects them from abrasion. But since drilling mud cannot lubricate sealed bearings, these bits have a small reservoir of grease built in.

Figure 4.32 Each cone rotates on ball and roller bearings (A) or journal bearings or both (B).

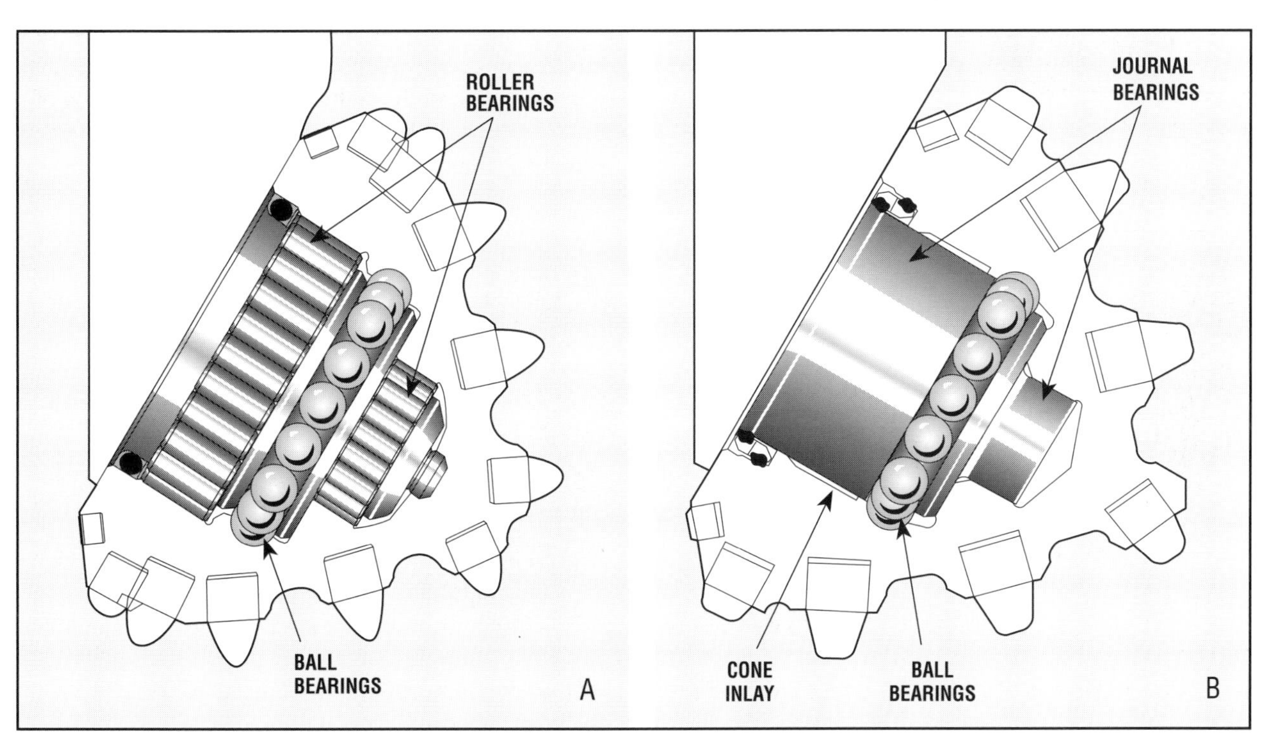

Most roller cone bits are *jet bits*; that is, drilling fluid exits from the bit through nozzles between the cones. Just as a nozzle on a garden hose creates a high-velocity stream of water, the nozzles on the bit create high-velocity jets of mud. The jets of mud strike the bottom of the hole and help lift cuttings away from the bit so that they will not impede drilling.

Diamond Bits

Diamond bits do not have roller cones. Instead, small industrial diamond cutters are embedded into the sides and bottom of a single *fixed head* that rotates as one piece with the drill string (see fig. 4.30C). Diamond bits work by shearing, or slicing, the formation, rather than gouging and crushing it like roller cone bits. They are available for drilling soft, medium, and hard formations. Their initial cost is higher than roller cone bits, but sometimes the higher cost is worth it because a diamond bit may not become dull as quickly as a roller cone bit.

The diamonds in a diamond bit may be either natural or manmade. The rarity and the expense of mining natural diamonds led scientists to try to create, or synthesize, diamonds from carbon. They succeeded, but synthetic diamonds are even more expensive to manufacture than natural diamonds are to mine. However, manufacturers can control the size and shape of synthetic diamonds: they may be larger than natural diamonds and are cylindrical or triangular. The synthetic diamonds now in use are one of two types: the *polycrystalline diamond compact* (PDC) and the *thermally stable polycrystalline* (TSP).

The main disadvantage of diamonds is that temperatures high enough to damage them are easy to reach in drilling. For this reason, mud circulation to cool the bit is very important, and designers and operators are always trying to improve the bit hydraulics. PDC bits have jet nozzles similar to those on roller cone bits. Natural diamond bits and TSP bits do not use jet nozzles. Instead, they have a single outlet for the drilling mud in the center of the bit. The outlet leads the mud into channels on the face of the bit between the rows of diamonds (fig. 4.33) to cool them.

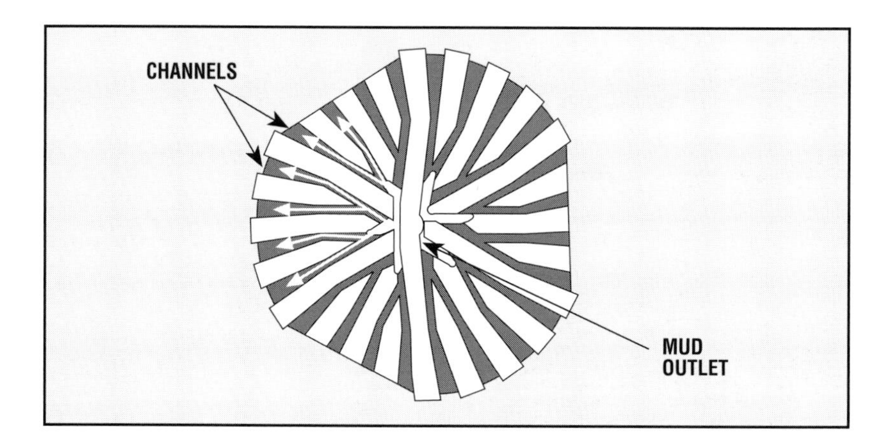

Figure 4.33 Channels on the face of a diamond bit direct the drilling mud next to each row of diamonds to cool them.

Hybrid Bits

Hybrid bits combine natural and synthetic diamonds and sometimes even tungsten carbide inserts on a fixed-head bit. The diamonds may be placed in the form of individual cutters, or they may be used in a pad made by heating grit-sized diamonds with tungsten carbide powder. Bit designers place the different types of cutters on different parts of the hybrid bit in order to use each where it will work best and last longest.

Circulating System

The circulating system pumps a special liquid called *drilling fluid* or *drilling mud* down the hole through a series of pipes, out of the bit on the end of the pipe at the bottom of the hole, and back to the surface (fig. 4.34).

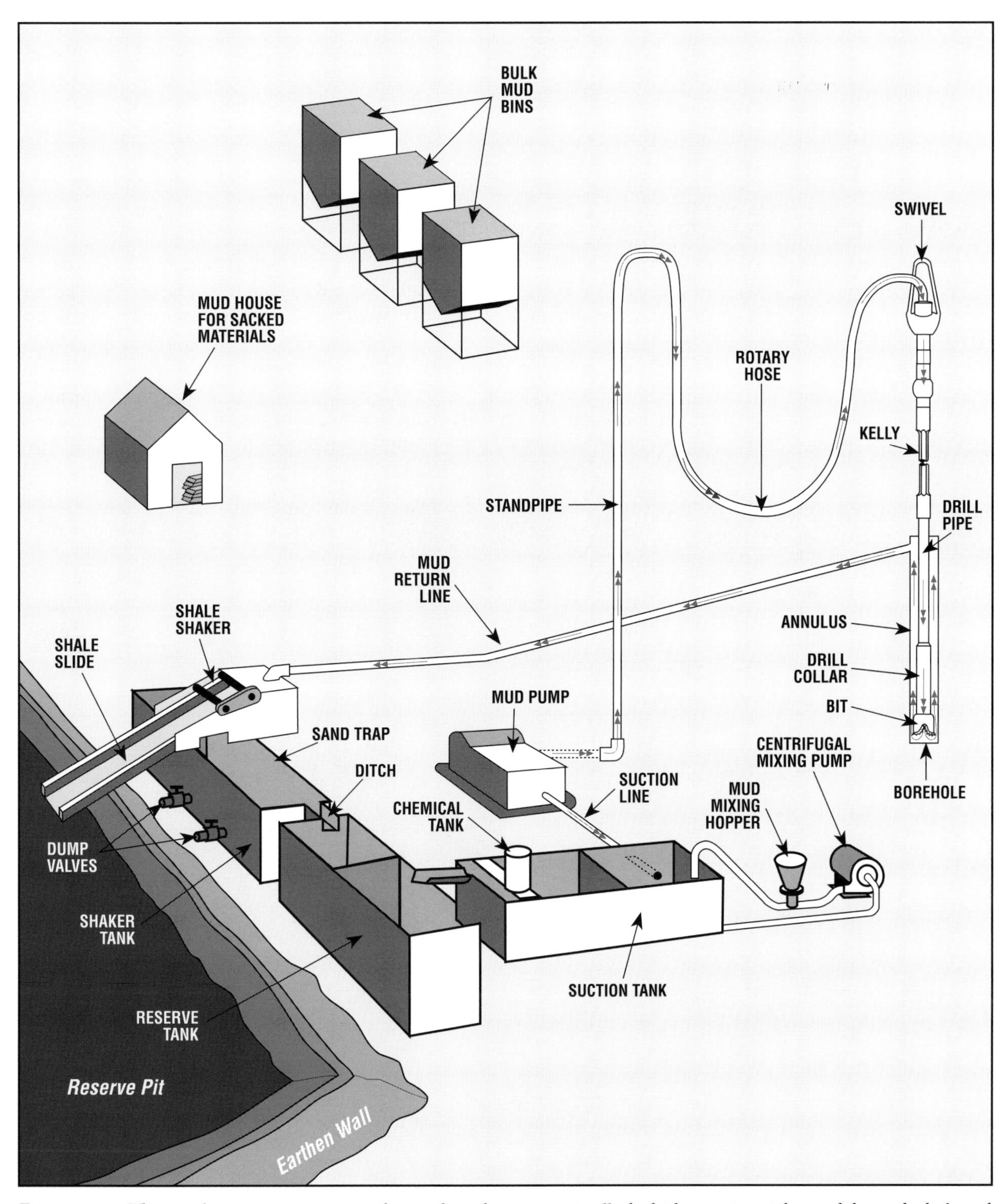

Figure 4.34 The circulating system consists of a number of components, all of which serve to get the mud down the hole and back to the surface.

Figure 4.35 The mud pump pumps drilling fluid into the hole.

Circulating Equipment

Large, heavy-duty pumps called *mud pumps* are the heart of the circulating system (fig. 4.35). Most rigs have two pumps, even though only one at a time is normally in use during drilling. The second serves as a backup if the first requires repair. However, if drilling requires a large volume of mud (for example, when the hole is very deep), the driller can use the two pumps together, compounding them, to increase pump capacity.

Usually, however, one pump or the other picks up mud from steel tanks, or *mud pits*, in which the mud is stored and sends it through a *standpipe* and *rotary hose* (fig. 4.36). The standpipe is a rigid pipe that conducts mud from the pump, up one leg of the derrick, and to the rotary hose. The flexible rotary hose, or kelly hose, is connected to the swivel or top drive. The rotary hose is flexible because it must move downward as the hole is drilled deeper and the swivel moves closer to the rig floor. It must also be able to move upward when the crew adds sections of pipe to the drill string.

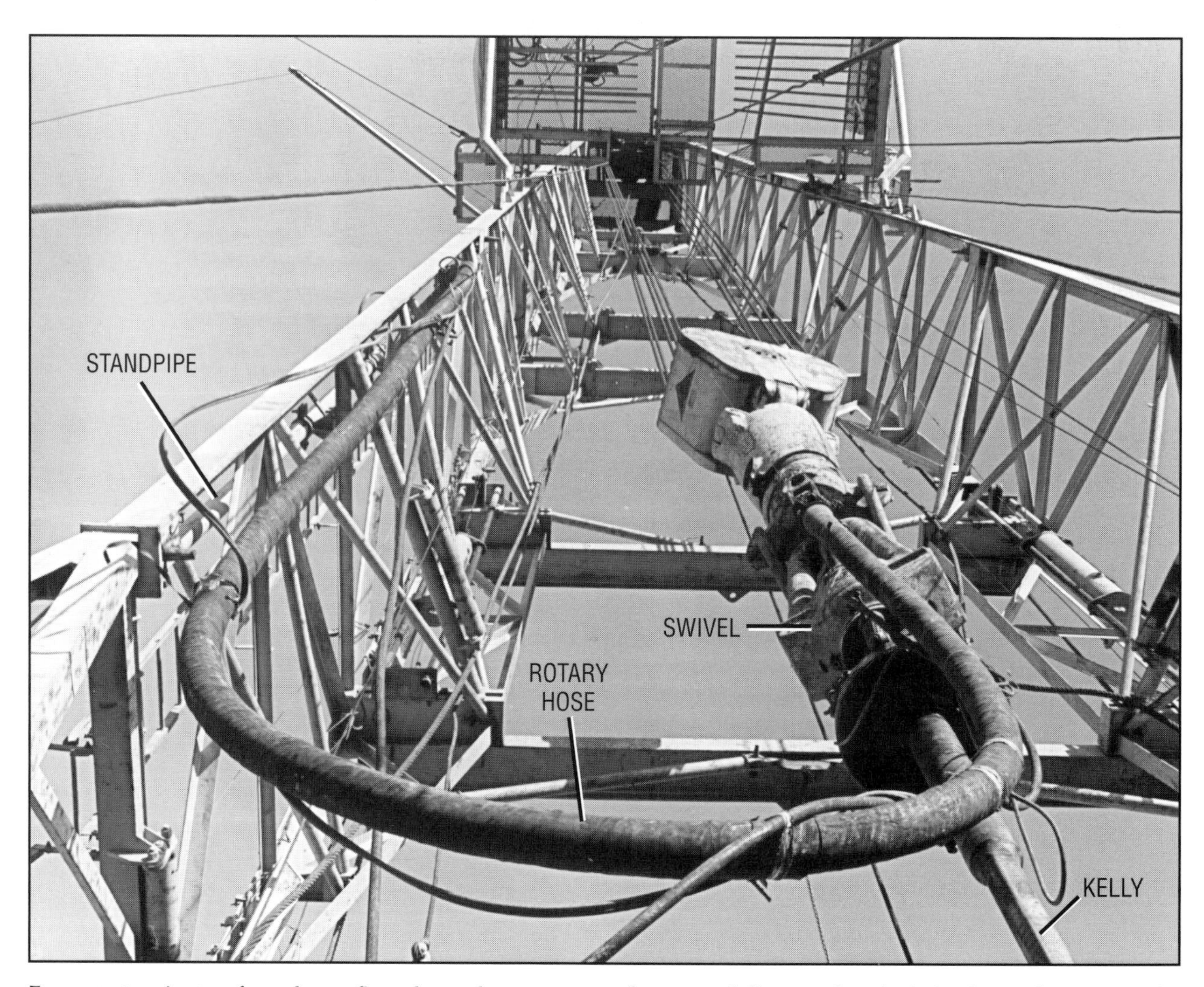

Figure 4.36 A view from the rig floor shows the components that carry drilling mud to the hole: the mud travels up the standpipe, through the rotary hose, into the swivel (or top drive), and down through the kelly.

Exiting the rotary hose, mud goes into the swivel, then down the kelly, and enters the drill pipe that is connected to the kelly. The mud goes down the drill string, through the drill collars, and out of the bit nozzles. When the mud shoots out of the rotating bit, it lifts the cuttings off the bottom, and then returns to the surface, carrying the cuttings with it. It moves back up the hole to the surface through the *annulus,* or annular space, between the outside of the drill collars and drill string and the inside of the hole.

At the surface, the mud and cuttings leave the well through a pipe called the *mud return line.* The mud and cuttings flow out of the return line and onto a vibrating screen called the *shale shaker* (fig. 4.37). The cuttings fall onto the screen of the shaker, but mud falls through the screen and back into the mud pits. The pumps again pick up this clean mud and send it back down the hole. Normally, mud circulation continues as long as the bit is on bottom and drilling.

Figure 4.37 Here, two shale shakers remove cuttings carried to the surface by the mud.

The cuttings that fall onto the shaker vibrate off and move down a slanted trough called the *shale slide.* They fall off the shale slide and into a pit dug into the earth. This earthen-walled pit, which is often lined with plastic to protect the surrounding ground, is called the *reserve pit.* The reserve pit serves mainly as a disposal area.

Sometimes the mud contains very small particles of rock, called *solids,* along with the larger cuttings. One way to clean the mud of solids is to circulate it through the reserve pit instead of the mud pits because the reserve pit is much larger in area than the mud pits. The solids settle out in the reserve pit and fall to the bottom before the mud is recirculated back into the hole. If the solids cannot settle out in the pits,

Figure 4.38 Additional circulating equipment can include a degasser, desilter, and desander, which are located over the mud pits downstream from the shaker.

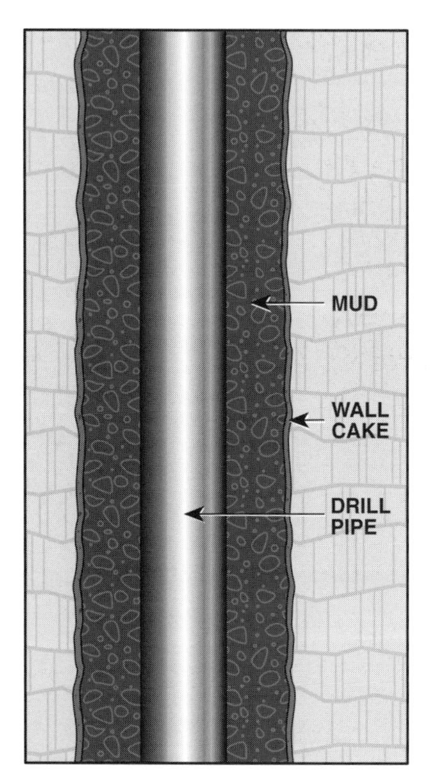

Figure 4.39 Solid particles in the drilling mud plaster the wall of the hole and form an impermeable wall cake.

the contractor may install special equipment such as a *desander* or *desilter* to clean the mud (fig. 4.38). If the mud contains gas, the circulating system may include a *degasser* to remove it.

Drilling Fluid

Drilling fluid, or drilling mud, is very important to the rotary drilling process. Basically a mixture of water, clay, and special minerals and chemicals, mud performs many important jobs. Besides removing cuttings from the hole, mud cools and lubricates the bit as it turns on bottom. Mud also exerts pressure inside the hole. This pressure keeps fluids that may be in the formation from entering the hole and perhaps blowing out to the surface. In addition, pressure in the hole forces solid particles of clay in the mud to adhere to the sides of the hole as the mud circulates upward on its way to the surface. The solids form a thin, impermeable cake on the walls of the hole. This *wall cake* plasters the hole and prevents the liquid part of the mud from seeping into the formations that the hole penetrates. If the liquid in the mud passed into the formation, the derrickhand would need to continually add water or other liquid to the mud. Wall cake also stabilizes the hole; that is, it prevents the hole from caving in (fig. 4.39).

Because it is so important, a mud engineer carefully formulates mud to do the best job on the particular well being drilled. Once formulated, the mud engineer monitors and adjusts its properties as necessary during drilling. Some of the mud properties that the engineer closely monitors are its viscosity, or resistance to flow; its weight, or density; its filtration rate, or water-loss properties; and its solids content.

The mud's viscosity is important because viscosity affects the ability of the mud to carry cuttings. Its weight affects its ability to prevent formation fluids from blowing out, and its filtration rate affects its ability to build an effective wall cake. Its solids content is important because the amount of solids in the mud affects the rate of penetration of the bit—the more solids in the mud, the slower the rate of penetration. This happens because the solids add weight to the mud, which increases the mud pressure on bottom, and this pressure holds the cuttings on the bottom, which interfere with the bit's ability to drill the formation.

Power System

A drilling rig needs power to run the circulating, rotating, and hoisting systems. Usually, this power comes from internal-combustion engines as *prime movers* (prime power sources). The rig also needs some method of transferring the power from the engines to a particular component, such as the mud pumps, the drawworks, or the rotary table.

Engines

The power requirements of a rig are large enough to require several engines—one engine is not enough. Most rigs need two or more engines to provide from 1,000 to 3,000 horsepower (hp), depending on well depth and rig design (fig. 4.40). Shallow or moderate-depth drilling rigs need from 500 to 1,000 hp for hoisting and circulation. Heavy-duty rigs for deep, 20,000-foot (6,100-metre) holes are usually in the 3,000-hp class. Auxiliary power for lighting and the like may be from 100 to 500 hp.

Figure 4.40 Diesel engines power the rig.

Figure 4.41 A heavy-duty electric motor is mounted on or near each component requiring power. The cables on the left receive electric power from the generator.

Today, most prime movers are diesel engines. Diesel engines are compression-ignition engines; that is, they do not use spark plugs to ignite the fuel-air mixture in the engine cylinders. Rather, compressing the fuel-air mixture generates heat, which ignites it.

Natural gas and liquefied petroleum gas (LPG) engines are spark-ignition engines; they use spark plugs as a car does. Diesel engines have become more popular than gas or LPG engines because diesel fuel is easier to transport than gas or LPG.

Power Transmission

In general, two methods are used to transmit power from the engines to the components that require the power. Both methods have advantages and disadvantages, and the choice of which type to use is mainly a matter of the contractor's preference.

One is the *electric drive*, in which diesel engines drive electric generators that are mounted directly on the engine (fig. 4.41). As the engine runs, the generator generates electric power. Cables transmit this power to electric motors that are mounted on or near the component requiring the power. The motors drive the equipment. The driller controls the amount of electric power from each engine-generator to the motors. If a particular component needs more power than one engine can provide, the driller can engage controls at the console to compound the engines—that is, assign the power from two or more engines to the motors on the component.

The second method of transmitting engine power to components is by means of a *mechanical drive*. In a mechanical-drive rig, a huge collection of pulleys, belts, and chains connect each diesel or gas engine to the rig components requiring power (fig. 4.42). This collection is called the *compound*. As with electric-drive rigs, the driller controls the amount of power and where it goes.

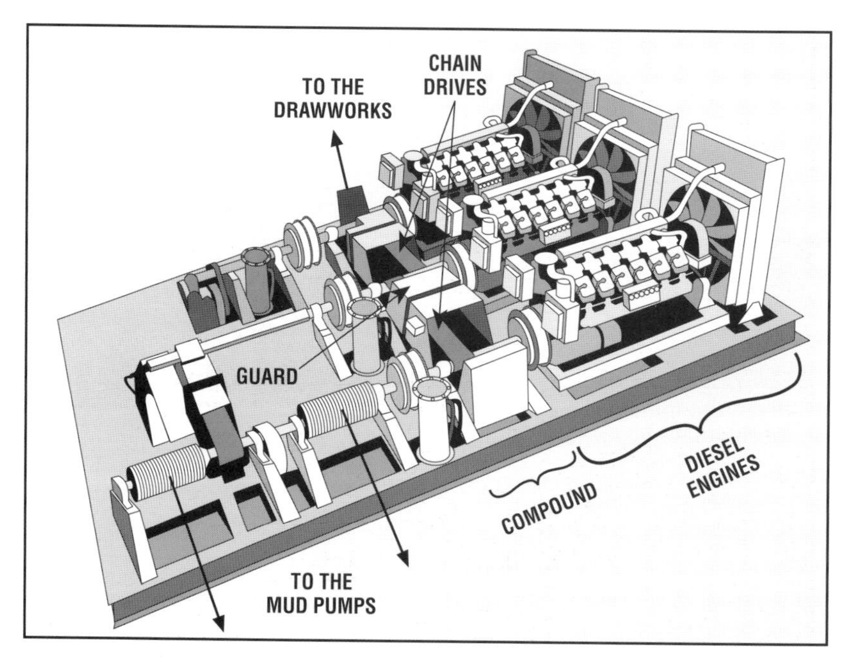

Figure 4.42 Three diesel engines and the compound send power to the drawworks and mud pumps.

ROUTINE DRILLING OPERATIONS

Once the operating company has taken care of the legal aspects of drilling a well—acquired the leases, signed the contracts, and so on—it determines an exact spot on the surface to drill the well. Now is the time to prepare the drill site, rig up, and begin making hole.

Preparing the Drill Site

The drill site must be prepared to accommodate the rig and equipment. If necessary, the operator clears and levels land, builds access roads, and digs the reserve pits. In areas where laws to protect the environment are in effect, such as Alaska and California, operators make a large effort not to alter the surface area of the drill site more than is necessary. For example, they may not dig reserve pits. Instead, they may place large steel bins on the site to receive the cuttings and other materials that are normally dumped into the reserve pits. The company arranges to truck these bins away from the site and dispose of the material inside them properly. Even in areas where the law permits reserve pits, the operator will often line the pits with thick plastic sheeting to prevent any contaminated water or other materials from seeping into the ground.

Since the contractor needs a source of fresh water for the drilling mud and other purposes, the operator sometimes drills a water well before moving the rig onto the location. If other sources are available, the water may be piped or trucked to the site.

At the exact spot on the surface where the well will be, the crew digs a rectangular pit called a *cellar* or drives a culvert-like pipe into the ground. The crew may line the cellar with boards or build forms and pour concrete to make walls for the cellar (fig. 4.43). The cellar provides room beneath the rig to install drilling equipment that will be needed later.

Figure 4.43 The cellar is like a basement under the rig.

In the middle of the cellar, the crew starts the top of the well, sometimes with a small truck-mounted rig. This hole—the *conductor hole*—is large in diameter, perhaps as large as 36 inches (9 centimetres) or more, about 20 to 100 feet (6 to 30 metres) deep, and is lined with pipe called *conductor casing* or conductor pipe. If the topsoil is soft, the crew may drive the conductor pipe into the ground with a pile driver instead of using a small rig. In some cases the main rig itself drills the conductor hole. In any case, the conductor casing keeps the ground near the surface from caving in. Also, it conducts drilling mud back to the surface from the bottom when drilling begins; thus the name *conductor pipe*.

Usually, the crew also digs another hole considerably smaller in diameter than the conductor hole next to the cellar and also lines it with pipe. Called the *rathole*, it is a place to store the kelly when it is temporarily out of the main hole during certain operations (see fig. 4.23). Sometimes on small rigs, the crew digs a third hole, called the *mousehole*. The mousehole, also lined with pipe, extends upward through the rig floor. It holds a joint of pipe ready for makeup. On large rigs, it is not necessary to dig a mousehole because the rig floor is high enough above the ground to set a joint through a hole in the floor.

Rigging Up

With the site prepared, the contractor moves in the rig and related equipment. The process, known as *rigging up*, begins by centering the base of the rig—the *substructure*—over the conductor pipe in the cellar. The substructure supports the derrick or mast, the pipe, the drawworks, and sometimes the engines. If the rig uses a mast, the crew places the mast into the substructure in a horizontal position and hoists it upright (fig. 4.44).

Figure 4.44 A mast is raised into the upright position, using the drawworks.

If using a derrick, they assemble it piece by piece on the substructure. Meanwhile, they move other drilling equipment such as the mud pumps into place and prepare them for drilling.

Other rigging-up operations include erecting stairways, handrails, and guardrails; installing auxiliary equipment to supply electricity, compressed air, and water; and setting up storage facilities and living quarters for the toolpusher and company representative. Finally, the contractor must bring drill pipe, drill collars, bits, mud supplies, and many other pieces of equipment and supplies to the site before the rig can make hole.

Drilling the Surface Hole (Spudding In)

With all the preparations complete, drilling is ready to begin; in oilfield parlance, the hole is ready for *spudding in.* To spud in, the crew attaches a large bit, say 17½ inches (444 millimetres) in diameter, to the first drill collar and lowers it into the conductor pipe by adding drill collars and drill pipe one joint at a time until the bit reaches the bottom. With the kelly attached to the top joint of pipe, the driller begins making hole by starting the pump to circulate mud, engaging the rotary table or top drive to rotate the drill stem, and setting the drill stem down with the drawworks to apply weight on the bit. Normally, when rotating the bit, the rotary table turns to the right. Thus, when the bit is on bottom and making hole, the expression "on bottom and turning to the right" indicates that drilling is proceeding.

As the bit drills ahead, the kelly moves downward through the kelly bushing. Since the ground near the surface is normally fairly soft, the kelly is soon *drilled down;* that is, its entire length reaches a point just above the bushing.

To drill the hole deeper, the crew must add more pipe to the string to make it longer. To add pipe, the driller uses the hoisting system to pick up the kelly and attached drill string off bottom. When the tool joint of the topmost joint of pipe clears the rotary—passes through the opening in the rotary table to a point just above the rig floor—the crew sets the slips around the pipe and into the opening in the master bushing. The slips grip the pipe and keep it from falling back into the hole while the crew *breaks out*, or unscrews, the kelly from the drill string.

To break out the kelly requires two sets of tongs. The rotary helpers latch one set of tongs around the bottom of the kelly where it joins the drill pipe. These tongs are the *breakout tongs*. They latch the other set, the *backup tongs*, around the tool joint of the drill pipe. A line attached to the end of the breakout tongs goes to the breakout cathead on the drawworks. When the cathead is engaged, it pulls on the line to cause the tongs to apply loosening torque on the kelly. Another line attached to the backup tongs fastens onto the rig and keeps the backup tongs (and the pipe on which they are latched) from turning as the breakout tongs apply torque.

Once the kelly is broken out, the crew removes the tongs and the driller spins the drill pipe out of the kelly by turning the rotary table. When the kelly spins out of the drill pipe and releases, they move it over to a 30-foot joint of drill pipe resting in the mousehole. They stab the pin of the kelly into the box of the new joint and screw the two together, or make them up. Often, the crew uses a *kelly spinner,* which is a motor mounted near the top of the kelly, to make up the kelly on the joint of pipe. They latch backup tongs onto the joint of pipe in the mousehole to keep it from turning as the kelly spinner turns the kelly into the joint.

Once the kelly is made up tightly to the joint, the driller picks them up and moves them from the mousehole to the rotary table. The crew stabs the bottom of the new joint of pipe into the top of the joint of pipe coming out of the borehole and again uses the kelly spinner to make up the joints. With the new joint made up, they pull the slips, and the driller lowers the pipe until the bit nears the bottom. Then he or she starts the pumps, begins rotation, applies weight to the bit, and drills another 40 feet (12 metres) or so of hole, depending on the length of the kelly. The crew repeats this process each time the kelly is drilled down.

When the rig uses a top drive, the crew follows essentially the same procedures to make up the drill string, often making up two, three, or four joints at a time instead of one. The multiple made-up joints, called a *stand*, sit in a rack on the rig floor to the side of the mast or derrick.

Eventually, at a depth that could range from hundreds of feet (metres) to a few thousand feet (metres), drilling comes to a temporary halt, and the crew pulls the drill stem from the hole. This first part of the hole is known as the *surface hole*. Even though the formation that contains the hydrocarbons may lie many thousands of feet (metres) below this point, the toolpusher stops drilling temporarily to take steps to protect and seal off the formations close to the surface. For example, drilling mud could contaminate zones containing fresh water that nearby towns use for drinking. To protect such zones, the crew runs special pipe called *casing* into the hole and cements it in place.

Tripping Out

The first step in running casing is to pull the drill stem and the bit out of the hole. Pulling the drill stem and the bit out of the hole in order to run casing, change bits, or perform some other operation in the borehole is called *tripping out*.

To trip out, the driller stops rotation and circulation. Then, using controls on the drawworks, he or she raises the drill stem off the bottom of the hole until the top joint of drill pipe clears the rotary table and holds it there. Then, the rotary helpers set the slips around the drill pipe to suspend it in the hole. Next, using the tongs, the rotary helpers break the kelly out of the drill string and put it into the rathole (fig. 4.45). Since they leave the kelly bushing, the swivel, and the rotary hose on the kelly when placing it in the rathole, the area above the rotary where the top of the drill string protrudes from the hole is clear. Only the traveling block hangs above the drill pipe suspended in the hole.

Attached to the traveling block are a set of drill pipe lifting devices called *elevators*. The elevators usually remain attached to the traveling block at all times and swing downward into position when the crew removes the swivel from the hook. Elevators are clamps that can latch onto the tool joints of the drill pipe (fig. 4.46). The crew latches the elevators around the drill pipe, and the driller raises the traveling block to pull the pipe upward. When the third joint of pipe clears the rotary table, the rotary helpers set the slips and use the tongs to break out the pipe. The pipe is usually removed from the hole in stands of three joints. The crew guides the stand to the rack on the rig floor.

Once the bottom of the stand of pipe is set down on the rig floor, the derrickhand goes into action. Standing on a small platform called the *monkeyboard* that is about 90 feet (metres) high in the mast or derrick,

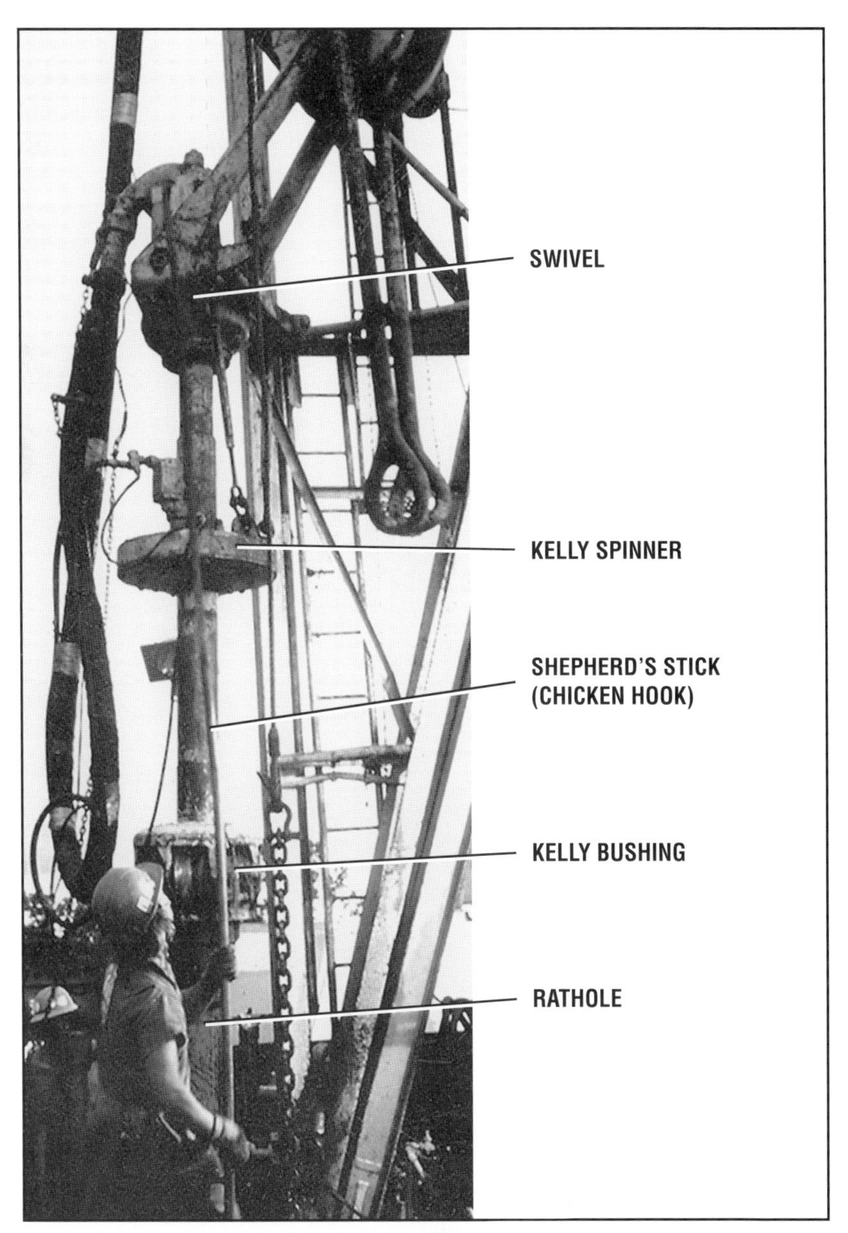

Figure 4.45 While tripping out, the kelly and related equipment rest in the rathole.

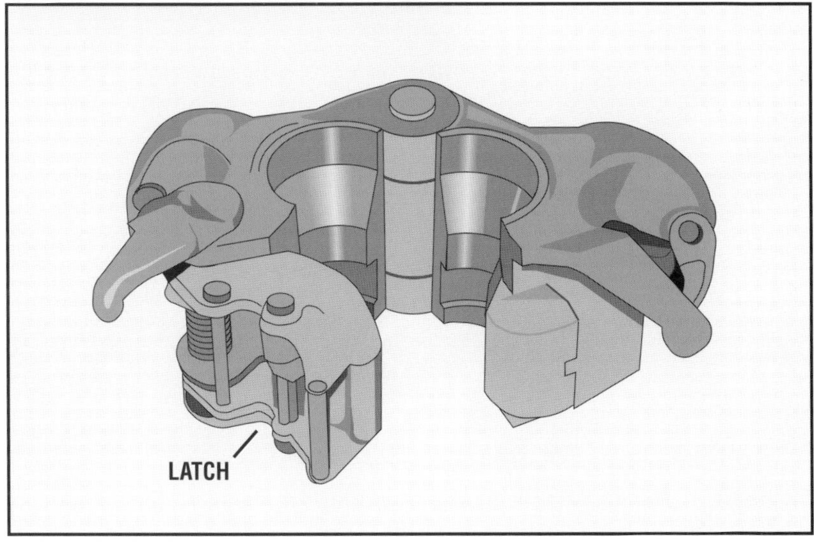

Figure 4.46 This open elevator has a latch and comes in several sizes and weight ratings.

he or she unlatches the elevators from the top of the stand, and guides the stand back into the *fingerboard*, the rack that supports the tops of the stands (fig. 4.47). Safety belts prevent the derrickhand from falling. Working as a close-knit team, the driller, rotary helpers, and derrickhand continue tripping out until all the drill pipe, drill collars, and the bit are out of the hole. At this point the only thing in the hole is drilling mud because mud was pumped into the hole as the pipe was tripped out to keep the walls from caving in.

Figure 4.47 Standing on the monkeyboard high in the mast, the derrickhand guides a stand of pipe into the fingerboard.

Running Surface Casing

Once the drill stem is out, often a special casing crew moves in to run the surface casing. Casing is large-diameter steel pipe. The crew runs it into the hole using special heavy-duty casing slips, tongs, and elevators. Casing accessories include centralizers, scratchers, a casing or guide shoe, and a float collar.

Centralizers keep the casing in the center of the hole so that the cement distributes evenly around the outside of the casing. *Scratchers* help remove wall cake from the side of the hole so that the cement can form a better bond (fig. 4.48). A *casing shoe,* or *guide shoe,* fits onto the end of the last joint. The casing shoe is a short, heavy, cylindrical section of steel filled with concrete and rounded on the bottom. It prevents the casing from snagging on irregularities in the borehole as it is lowered. It has an opening in its center out of which cement can exit the casing.

The *float collar* is a special coupling used one or two joints above the bottom of the casing string. It contains a valve that allows fluids to pass down the string but prevents them from flowing up. It keeps drilling mud from entering the casing as it is being lowered, so the casing floats, and later prevents cement from flowing up the casing.

The casing crew, with the drilling crew available to help as needed, runs the surface casing into the hole one joint at a time (fig. 4.49). Like drill pipe, casing is available in joints of about 30 feet (9 metres). Once the hole is lined from bottom to top with casing, the casing is cemented in place.

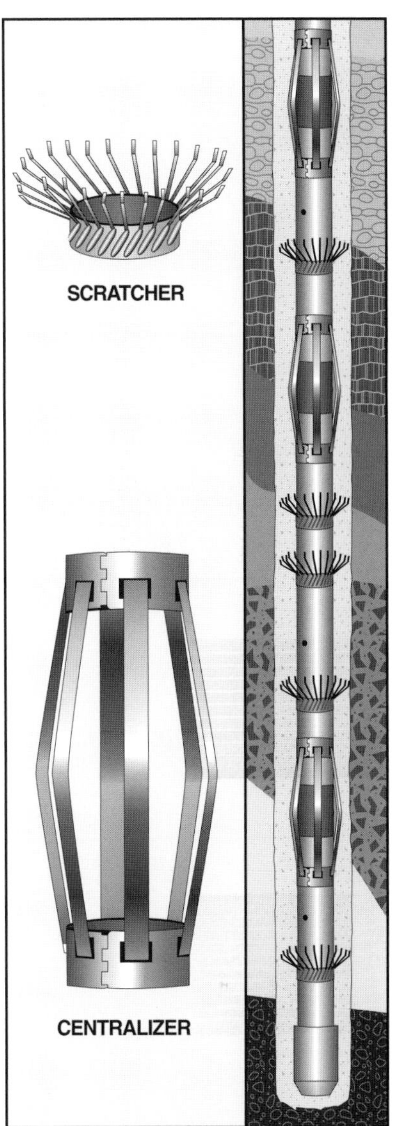

Figure 4.48 Several centralizers and scratchers are installed on the casing to aid in cementing.

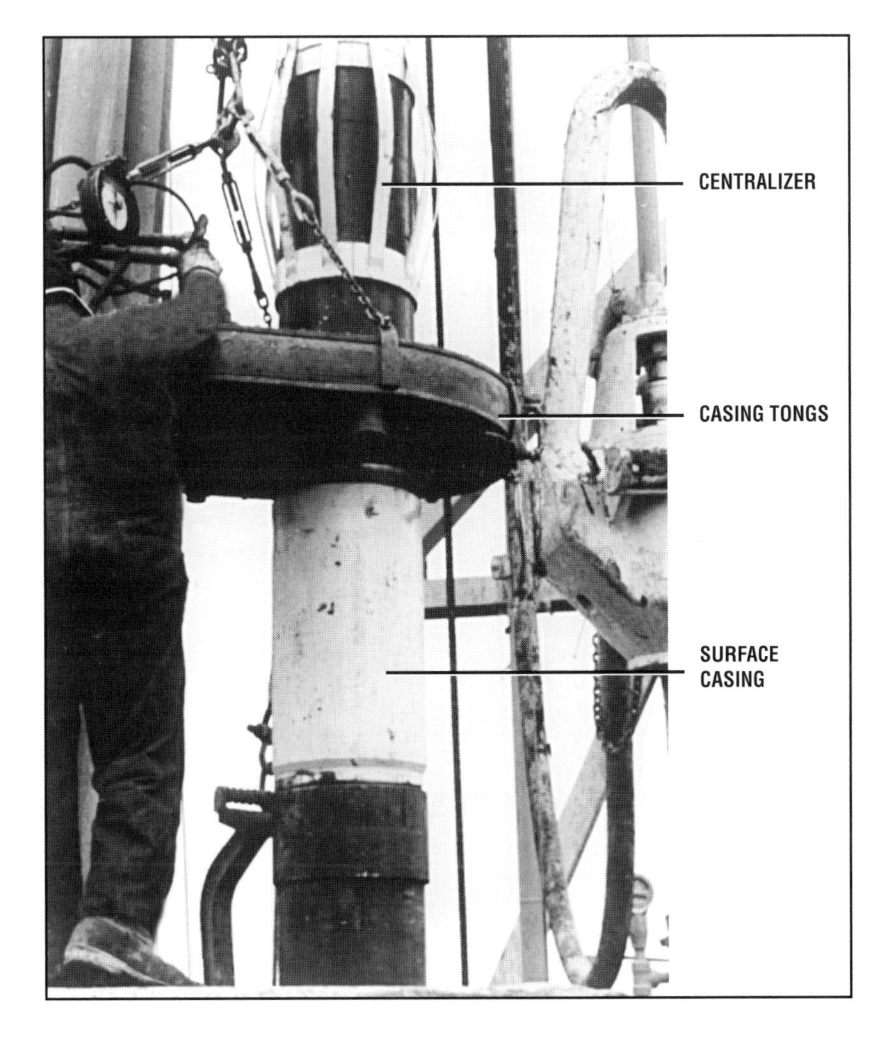

Figure 4.49 The crew runs surface casing into the hole one joint at a time. Notice the centralizer above the casing tongs.

Cementing the Casing

An oilwell cementing service company usually performs the job of cementing the casing in place. As in running casing, the personnel and equipment of the drilling contractor are usually available for the cementing operation. The cement used to cement oilwells is not too different from the cement used as a component in ordinary concrete. Basically, oilwell cement is *portland cement*. It usually contains special additives to make it suitable for cementing a particular well. For example, the temperature in the well may be high, which causes the cement to set (harden) faster than normal. Adding a retarder slows down the setting time of cement and makes it possible to successfully cement high-temperature wells.

Cementing service companies stock various types of cement and use special trucks to transport the cement in bulk to the well site (fig. 4.50). Bulk cement storage and handling at the rig location make it possible to mix the large quantities needed in a short time. The cementing crew mixes the dry cement with water, often using a *recirculating cement mixer* (RCM). This device thoroughly mixes the water and cement by recirculating previously mixed liquid cement (called a *slurry*) with dry cement and more water.

Powerful cementing pumps move the slurry through a pipe to a special valve made up on the topmost joint of casing (fig. 4.51). This valve is called a *cementing head*, or *plug container*. As the cement slurry arrives, the crew releases a plug—the bottom plug—from the cementing head, which precedes the slurry down the inside of the casing. The bottom plug keeps

Figure 4.50 This cement truck carries an RCM and powerful pumps.

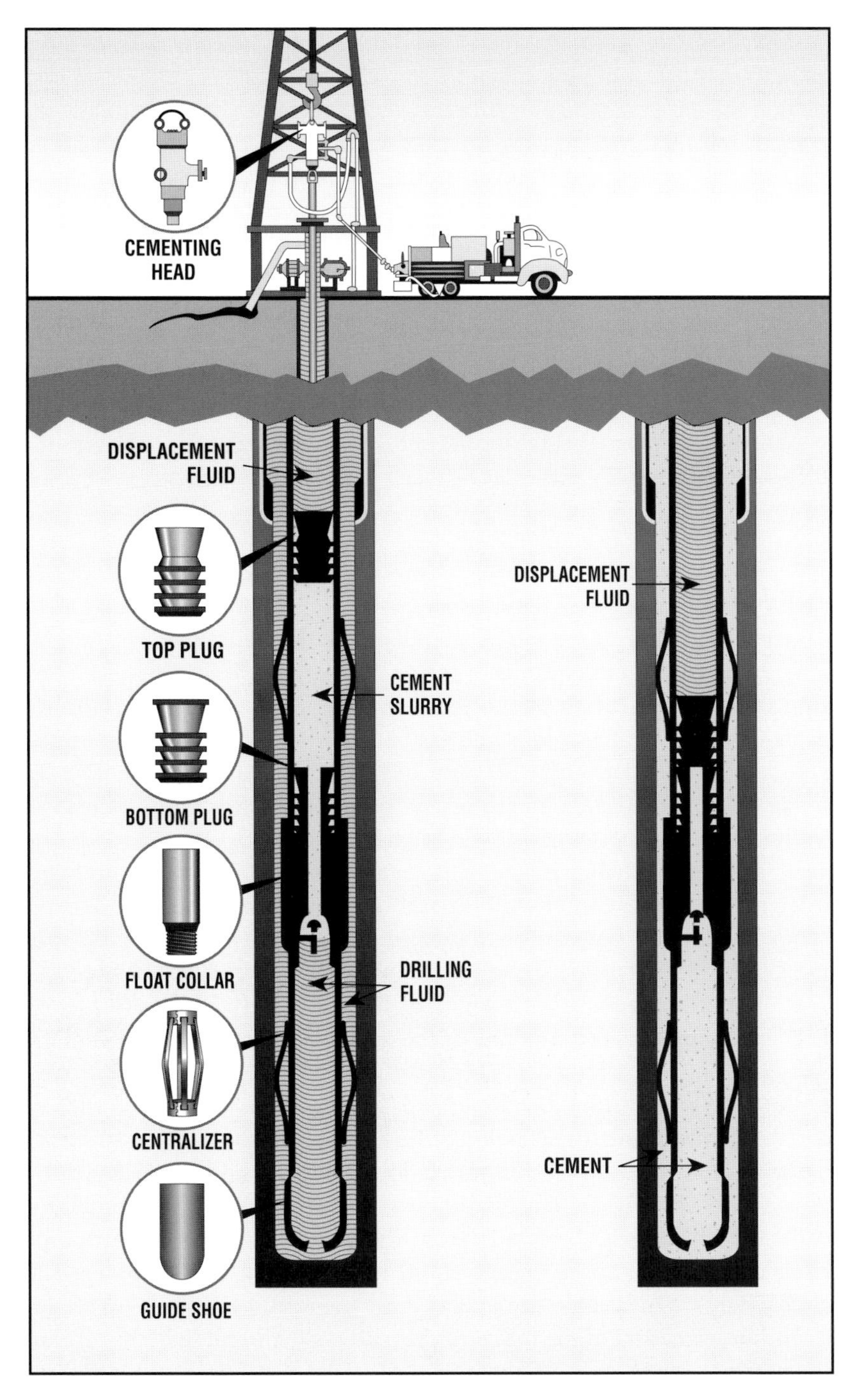

Figure 4.51 This diagram of a casing cementing job shows the route of the slurry as well as the role played by the various cementing accessories.

any mud that is inside the casing from contaminating the cement slurry where the two liquids meet. Also, the plug wipes off mud that adheres to the inside wall of the casing and prevents it from contaminating the cement.

The plug travels ahead of the cement until it reaches the float collar. At the collar the plug stops, but continued pump pressure breaks a seal in the top of the plug and allows the slurry to pass through it. The slurry flows out through the float collar and guide shoe and starts up the annulus between the outside of the casing and the wall of the hole until the annulus is filled.

When enough cement has entered the casing, the crew releases a top plug from the cementing head and pumps a liquid, called *displacement fluid*, behind the top plug. The plug and this liquid, usually salt water, follow the cement and fill the casing. The top plug keeps the displacement fluid from contaminating the cement slurry. When the top plug comes to rest on the bottom plug in the float collar, the crew shuts down the pumps and allows the slurry to harden.

Allowing time for the cement to set is known as *waiting on cement (WOC)*. In some cases, WOC may be only a matter of a few hours; in other cases, it may be 24 hours or even more, depending on well conditions. Allowing adequate WOC time is important so that the cement sets properly and bonds the casing firmly to the wall of the hole. After the cement hardens and tests indicate that the job is good—that is, that the cement has made a good bond and no voids exist between the casing and the hole—drilling resumes.

Tripping In

To resume drilling, the crew begins tripping the drill stem and a new, smaller bit that fits inside the surface casing back into the hole. They first make up the bit on the bottommost drill collar. Then, working together, the driller, the rotary helpers, and the derrickhand make up the stands of drill collars and drill pipe and trip them back into the hole. To make up a stand of pipe, the rotary helpers set the slips around the pipe in the hole. The derrickhand then latches the elevators to a stand in the rack, and as the driller picks up the stand, the rotary helpers swing it over to the pipe in the hole and stab it. Next, they latch the backup tongs around the pipe in the hole, and spin up the new stand. To spin up the stand, they use a spinning chain.

The *spinning chain* is a Y-shaped chain. The tail of the Y is attached to the makeup cathead, and one of the arms is attached to the end of the makeup tongs. The final arm is carefully wrapped around the tool joint of the pipe in the rotary just above the backup tongs. With a deft flick of the wrist, one of the rotary helpers flips the chain upward, causing it to unwrap from the joint in the rotary and wrap around the joint of the new stand (fig. 4.52). The driller then sends power to the makeup cathead, which pulls the chain from the stand. As the chain unwraps from the stand, it spins up the stand into the joint in the rotary. The rotary helpers then latch the makeup tongs around the spun-up stand. Continued rotation of the makeup cathead pulls the chain on the end of the makeup tongs, causing them to apply final *makeup torque* to the connection (tighten it). The backup tongs prevent the suspended pipe from turning as the cathead applies makeup torque. The driller then picks up the made-up stand, the rotary helpers remove the slips, and the driller lowers the pipe into the hole. This making-up process continues until the bit reaches bottom.

Figure 4.52 Using a spinning chain, rotary helpers spin a new stand of pipe into the stand suspended in the hole.

When the drill bit reaches bottom, the driller resumes circulation and rotation, and the bit drills through the small amount of cement left in the casing, the plugs, and the guide shoe, and into the new formation below the cemented casing. As drilling progresses and hole depth increases, formations tend to get harder. As a result, the crew will need to make several *round trips* (trips in and out of the hole) to replace worn bits.

Controlling Formation Pressure

While drilling the next section of hole, and indeed during all phases of drilling, the drilling crew must take steps to prevent a blowout. A *blowout* is the uncontrolled gushing of fluids—oil, gas, water, or all three—from a formation that the hole has penetrated. The crew controls the well and prevents it from blowing out using proper procedures and equipment.

Blowouts are relatively rare events, but when they happen they can be spectacular. Blowouts often threaten lives and property and pollute the environment. The great Lucas well drilled at Spindletop in 1901 is one of the more famous blowouts of all time because it revealed the presence of previously unheard of amounts of oil and gas (fig. 4.53). Since Spindletop, the oil industry has made large strides in understanding what causes blowouts and how to prevent them from happening. Today, rig crews receive extensive training in how to recognize impending blowouts and what to do should one threaten.

Figure 4.53 The Lucas gusher at Spindletop blew out over 80,000 barrels (over 120,000 cubic metres) of oil a day for nine days in 1901 before the well could be controlled.

The key to well control is understanding pressure and its effects. Pressure exists in the borehole because it contains drilling mud and in some formations because they contain fluids. All fluids—drilling mud, water, oil, gas—exert pressure. The denser the fluid (the more the fluid weighs), the more pressure the fluid exerts. A heavy mud exerts more pressure than a light mud. For effective control of the well, the pressure exerted by the mud in the hole should be higher than the pressure exerted by the fluids in the formation. That is, the mud must be so heavy that it holds back the formation fluids.

Pressure exerted by mud in the hole is called *hydrostatic pressure.* Pressure exerted by fluids in a formation is called *formation pressure.* The hydrostatic pressure and formation pressure vary depending on the depth at which these pressures are measured and the density, or weight, of each fluid. Regardless of the depth, hydrostatic pressure must be equal to or slightly greater than formation pressure. The well *kicks*—formation fluids enter the hole—whenever hydrostatic pressure falls below formation pressure. For this reason, one of the crew's main concerns during all phases of the drilling operation is to keep the hole full of mud that is heavy enough to overcome formation pressure.

Sometimes, however, in spite of efforts to prevent it, a well kicks for one of several reasons. The drilling operation can come to a formation with unexpectedly high pressure. The pistonlike action of the bit as pipe is tripped

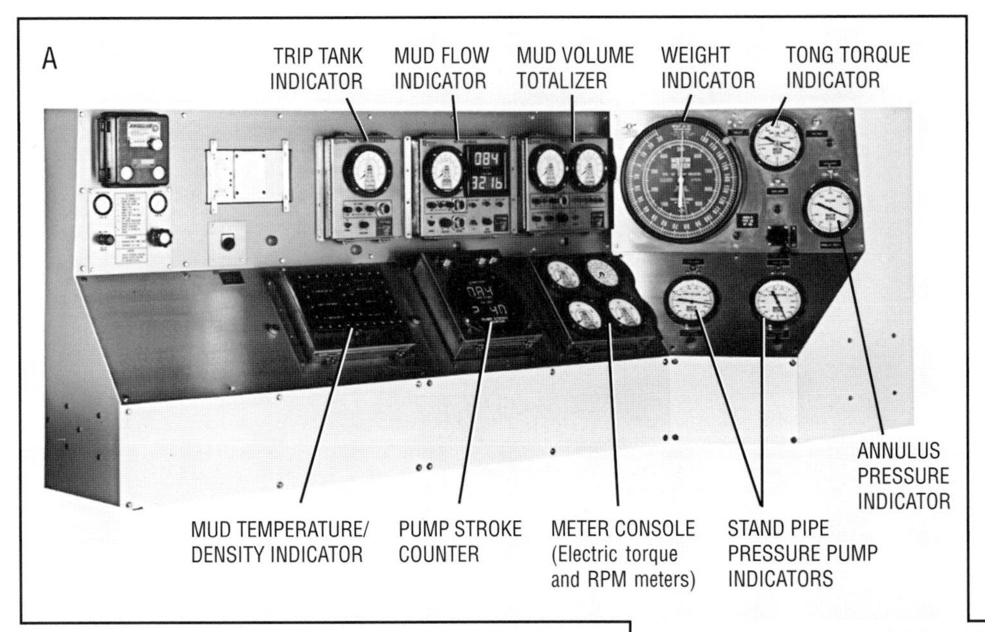

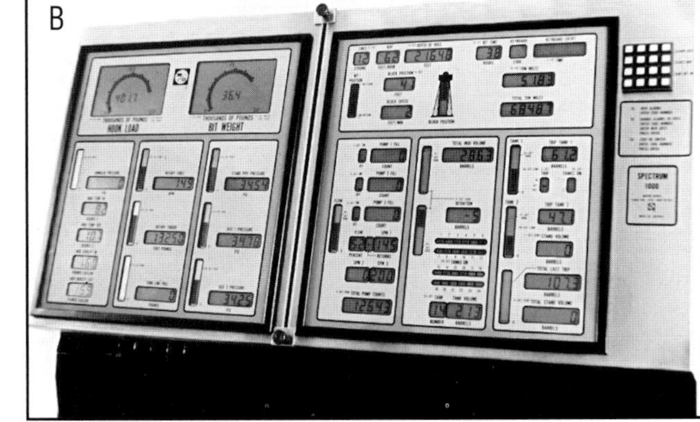

Figure 4.54 Instruments linked to various parts of the rig let the driller keep up with downhole pressure, mud volume, weight on bit, and other vital signs by reading recordings on his console. The instrument panel readouts may be analog (A) or digital (B). *(Courtesy of M-D Totco)*

out of the hole can *swab*, or pull, formation fluids into the hole. The mud level in the hole can fall so that the hole is no longer full of mud. Whatever the reason, crew members must take quick and proper action to prevent the kick from becoming a blowout.

An experienced crew can recognize a kick from the behavior of the mud circulating out of the hole. The driller's console has various control instruments located on it that help the crew keep an eye on the rig's operation (fig. 4.54). Some rigs also have data processing systems that use computer monitors, on the rig floor, in the mud logging trailer, in the toolpusher's trailer, and in the company representative's trailer. When the well exceeds the pressure limits programmed into the system, the system sets off an alarm.

Whether the kick warning signs come from electronic monitors, a computer printout, or from the behavior of the mud returning from the hole, an alert drilling crew detects the signs and takes proper action to shut the well in. To shut a well in, the crew closes large valves called *blowout preventers,* which they have installed on top of the cemented casing, to prevent further entry of formation fluids into the hole (fig. 4.55). Once the well is shut in, they begin procedures to circulate the intruded kick fluids out of the hole. Also, they add weighting material to the mud to increase its density to the proper amount to prevent further kicks, and circulate this weighted-up mud into the hole. When the mud has been weighted the proper amount, they can resume normal operations.

Figure 4.55 A blowout preventer stack is used to shut in the well if it kicks.

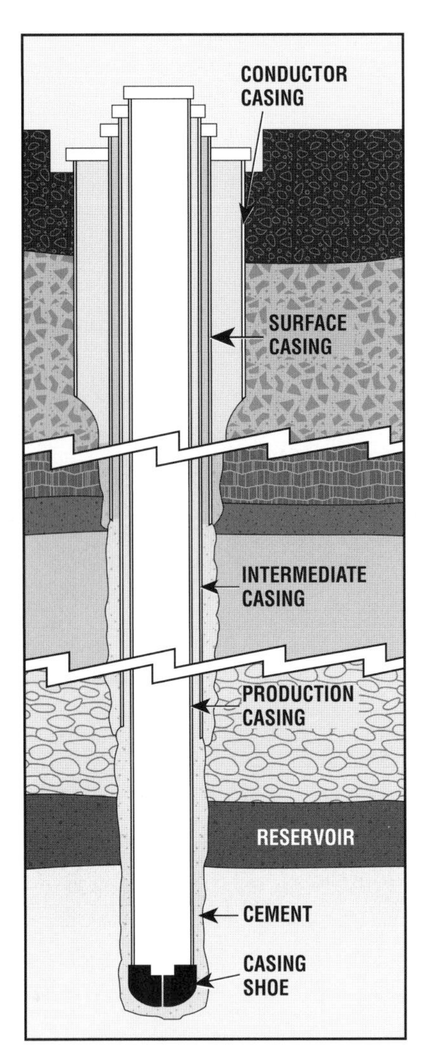

Figure 4.56 Intermediate casing fits inside the surface casing, and the production casing fits inside the intermediate casing.

Running and Cementing Intermediate Casing

At a predetermined depth, drilling stops again in order to run another string of casing. Depending on the depth of the hydrocarbon reservoir, this string of casing may be the final one, or it may be an intermediate one. In general, wells in relatively shallow reservoirs—say 10,000 feet (3,048 metres) or less—only require one more casing string. On the other hand, wells where the reservoir is deep—perhaps up to 20,000 feet (6,096 metres) or more—usually need at least one intermediate casing string. Intermediate casing is smaller than surface casing because it fits inside the surface casing to the bottom of the intermediate hole (fig. 4.56). The crew runs and cements it in much the same way as surface casing.

Other reasons for running intermediate casing in a well are to seal off so-called troublesome formations. Troublesome formations are those that could cause a blowout due to abnormal pressure or lost circulation, or that contain shale that sloughs off the walls of the hole and fills the hole. *Lost circulation* is a condition where quantities of mud are lost to a formation that contains caverns or fissures or that is coarsely permeable. Loss of mud lowers the pressure in the hole, which could allow the formation pressure to overcome it. Operators can successfully drill troublesome formations by carefully controlling the properties of the drilling mud. However, the crew must case off and cement troublesome zones so that they do not cause problems later, when they drill the well to final depth.

Drilling to Final Depth

Using a still smaller bit that fits inside the intermediate casing, the crew drills the next part of the hole. Often, the next part of the hole is the final section unless the well needs more than one intermediate string. In either case, they trip in the bit and drill stem, drill out the intermediate casing shoe, and resume drilling, usually with the *pay zone* in mind—that is, a formation capable of producing enough gas or oil to make it profitable for the operating company to complete the well.

If the company is drilling in an already existing field, they pretty well know that they are drilling a well that will be a producer. If, however, the well is a *wildcat* (a well being drilled in an area where no oil or gas is known to exist), the company must determine whether the well has indeed struck oil or gas. To make this determination, the operator orders tests to evaluate the well.

Figure 4.57 Cuttings retrieved from the shale shaker are examined by the mud logger.

Evaluating Formations

Several techniques are available to help the operator decide whether to abandon the well or to set a final string of casing. A geologist thoroughly examines the cuttings to determine whether the formation contains sufficient hydrocarbons. The geologist takes cuttings from the shale shaker and analyzes them in a portable laboratory at the well site. The geologist often works closely with a *mud logger*—a technician who monitors and records information brought to the surface by the drilling mud as the hole penetrates formations of interest (fig. 4.57).

Wireline logging is another valuable method of analyzing downhole formations. Using a mobile laboratory, well loggers lower sensitive tools to the bottom of the well on a wireline and then pull them back up the hole.

As they pass back up the hole, the tools measure and record certain properties of the formations and any fluids (oil, gas, and water) that may be there. Loggers can generate many types of logs, as detailed in the exploration chapter, for experienced geologists and engineers to study and interpret to determine the presence of oil or gas (fig. 4.58).

Another evaluation technique is the *drill stem test* (DST). The logging crew makes up a DST tool on the bottom of the drill stem and sets it down on the bottom of the hole. Formation fluids flow into it and pressure recorders chart the pressure. When the pressure testing is finished, valves in the DST tool close to trap a fluid sample, and the crew pulls the tool out of the hole. The recovered fluids and the recorded pressure graph are analyzed for indications of the presence and quantity of hydrocarbons.

Besides well logging and drill stem testing, a geologist can take core samples of the formation from the hole and examine them in a laboratory.

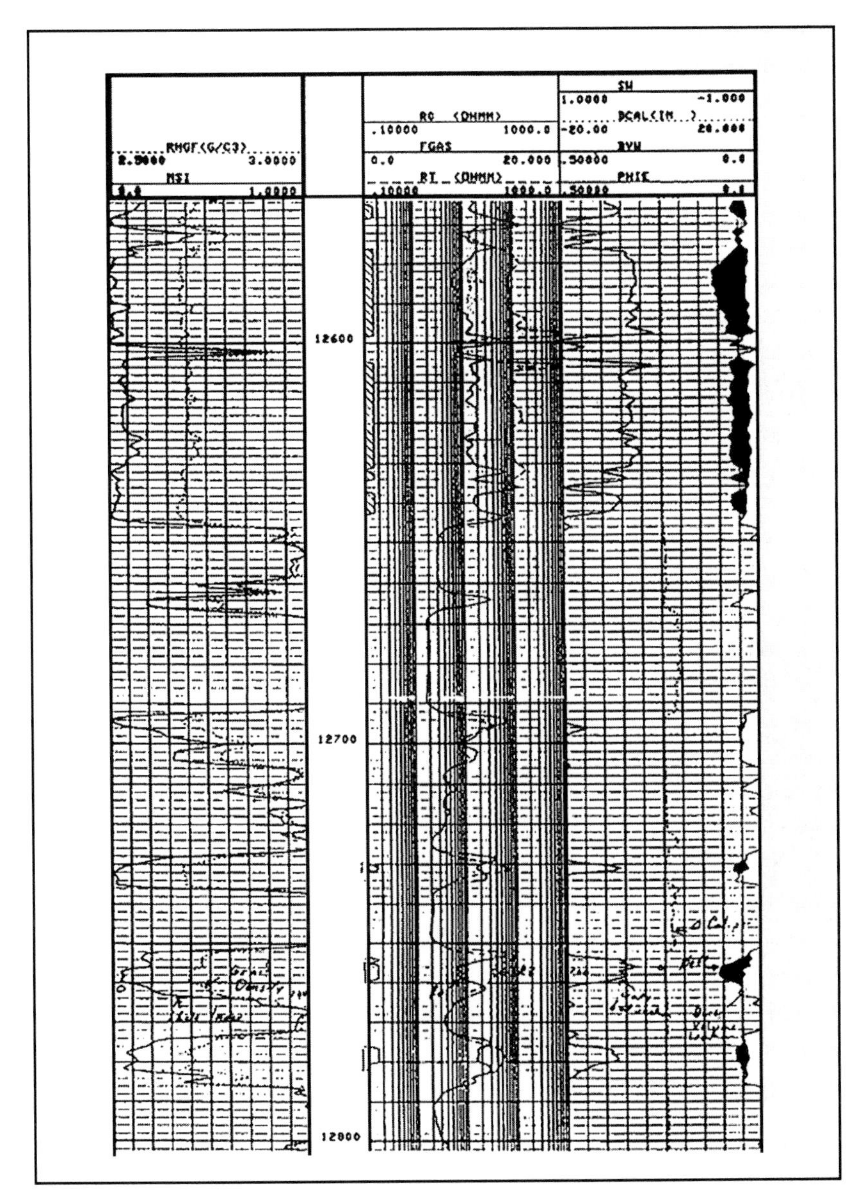

Figure 4.58 An electric log is one type of log that helps determine whether oil and gas are present in a drilled formation.

Setting Production Casing

After the drilling contractor has drilled the hole to final depth and the operating company has evaluated the formations, the company decides whether to set the final string of casing, the *production casing*, or *plug and abandon* the well (see fig. 4.56). The company may doubt that a well will produce enough oil or gas to pay for completing the well, which can be a costly proposition. If the well is judged to be a dry hole—that is, not capable of producing oil or gas in commercial quantities—the company will plug the well, by putting in several cement plugs to seal it permanently, and abandon it. Plugging and abandoning a well is considerably less expensive than completing it.

On the other hand, if evaluation reveals that commercial amounts of hydrocarbons exist, the company may decide to set casing and complete the well. Casing will be hauled in; the drilling crew will pull the drill stem from the hole and lay it down one joint at a time so that it can easily be transported to the rig's next drilling location; and a cementing company will run and cement the production casing in the well.

The drilling contractor's job is nearly finished after drilling the hole to total depth and setting and cementing production casing. Sometimes, the rig and crew remain on the location and *complete* the well. Completion involves running *tubing* (a string of small-diameter pipe inside the casing through which the hydrocarbons flow out of the well) and setting the wellhead (steel fittings that support the tubing and a series of valves and pressure gauges to control oil flow). In other cases, the drilling contractor moves the rig and equipment to the next location after cementing the production casing. In such cases, the operator hires a special completion rig and crew to finish the job.

OFFSHORE DRILLING

Equipment and techniques for drilling wells from offshore locations are very similar to those used for drilling wells on land. The biggest differences are in the appearance of the rigs used to drill offshore wells and in the specialized methods of operation that have been developed to deal with the problems that marine environments present.

History of Offshore Drilling

Offshore operations in the petroleum industry began as extensions of land operations. The first offshore well in the United States was drilled in 1897 off the coast of southern California. A wooden pier that extended about 300 feet (91 metres) into the Pacific Ocean was built from the shore. Near the end of the pier a drilling rig was erected, and a well was drilled to tap oil and gas that lay in a subsurface reservoir below the water. Because oil companies had already drilled and produced several successful wells on the beach, it was only natural for them to extend their drilling seaward.

Early Barges and Platforms

By the late 1930s geologists had made seismic surveys in the coastal marshlands, bayous, and shallow bays next to the Gulf of Mexico, and many of the surveys showed underground formations that could contain hydrocarbons. To drill exploratory, or wildcat wells, into these potential reservoirs, oil companies dredged a channel 4 to 8 feet (1.2 to 2.4 metres) deep in the marshes and bays, and towed a barge into the channel. They submerged the barge and secured it in place by wooden pilings, and then erected a rig on the deck of the barge, which remained above the waterline.

Another method of drilling wildcats involved building a wooden platform on timber piles and erecting a rig on the platform. A barge brought supplies to the platform, or trestles built from the shore to the platform created a road for trucks. The first specifically designed steel platform was installed in 1947 in the Gulf of Mexico at a depth of about 20 feet (6 metres). Operators anchored surplus barges from World War II, called *tenders*, alongside the platform. Tenders carried supplies and provided living quarters and circulation equipment. These platforms worked in waters of 60 feet or less, and after drilling the well, the operator moved the rig to another location. In the following decades, more than 2,000 fixed platforms sprang up in the Gulf, in depths ranging up to 1,200 feet (366 metres) (fig. 4.59).

Figure 4.59 A fixed platform and its tender drill in the Gulf of Mexico.

The fabrication and erection techniques perfected in the Gulf influenced platform construction around the world.

As offshore exploration techniques began to reveal the presence of possible reservoirs in deeper waters, from 50 to 300 feet (15 to 91 metres), the existing barges and platforms began to show limitations. For example, the operator could not submerge a barge in waters deeper than about 10 feet without its deck being covered by water. Mooring and anchoring problems in less sheltered ocean areas caused damage to the platform. Moreover, while operators could erect platforms in deeper waters than barges, the platforms were not mobile; once built, an operator could not move a platform without totally disassembling it. Mobility was important to the drilling of wildcat wells because usually the oil company had to drill several wells in an area before striking oil or gas. Since most wildcat wells turned out to be dry holes, building a new platform for each well was not economical.

First Mobile Drilling Rig

Because drilling in deeper waters required some type of mobile, or movable, offshore drilling rig, engineers and naval architects put their minds to the problem, and in 1948 the first mobile offshore rig was built (fig. 4.60). This rig consisted of a barge with several steel beams, or posts, attached to its deck. On top of the posts was an upper deck with drilling equipment on it.

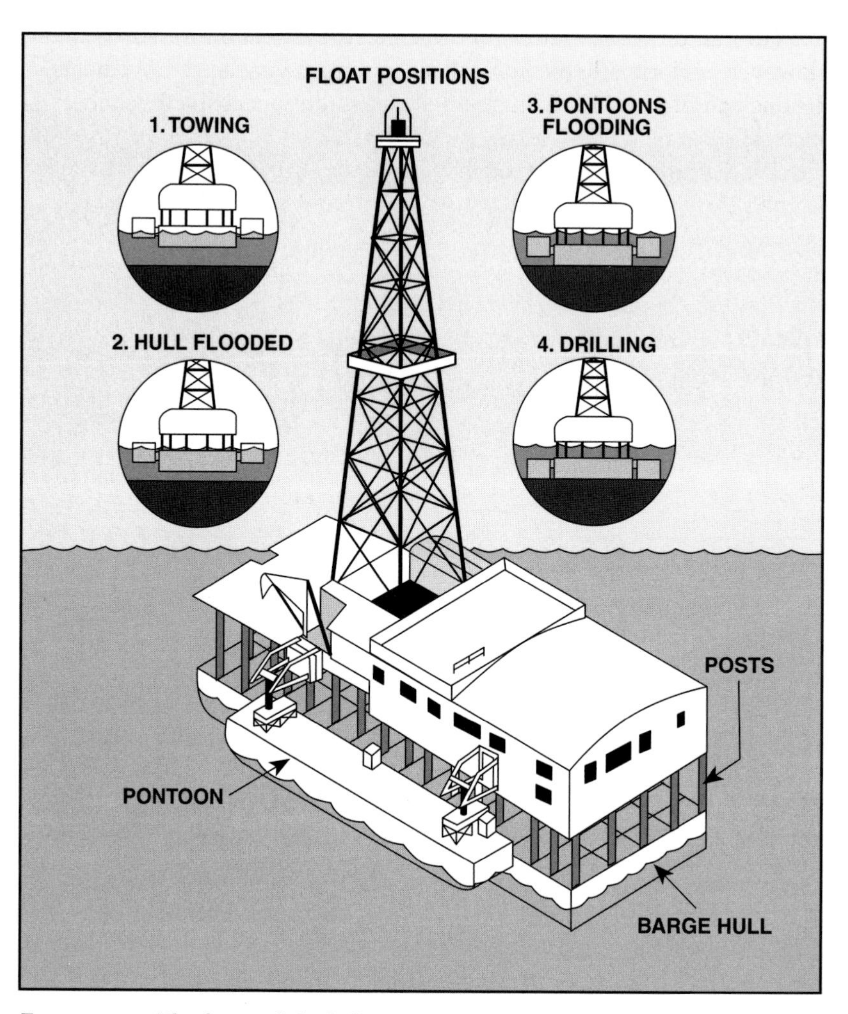

Figure 4.60 The first mobile drilling rig was a submersible barge.

The rig was floated out to the drill site near the mouth of the Mississippi River, and water was allowed to enter the barge hull at a controlled rate so that the unit slowly submerged and came to rest on the seafloor. The posts extended the drilling deck well above the waterline, and provided a stable platform for the drilling operation. After the drilling contractor successfully drilled the first well in the Mississippi River, in about 18 feet of water, they pumped out the water in the barge hull, and once again the entire unit floated on the water's surface. They then towed it to a second location, submerged it, and drilled another well. So successful was this mobile design, called a *posted barge rig*, that manufacturers are still building a few units very similar to the original for drilling in the relatively shallow waters of bays, inlets, lakes, and marshes.

Offshore Drilling Today

As offshore drilling moves into greater water depths and more hostile environments, all costs increase rapidly. One of the most striking effects is the increase in size of petroleum reserves required to justify the drilling. In some cases, oil companies must consider a reserve estimated at 100 million barrels (16 million cubic metres) of oil marginal; that is, if they estimate that they can recover only 100 million barrels from a reservoir, then that reservoir will not be profitable due to the high costs of drilling, producing, and transporting the oil to shore. Commercial reserves begin at 300 million barrels (48 million cubic metres) of anticipated production for some fields. However, exploring for offshore oil and gas is growing as improvements in drilling and production techniques reduce costs and increase the chances of success.

The type of rig used in offshore drilling depends largely on whether the company is drilling for exploration or for development. *Development wells* are the wells drilled in a reservoir that wildcat drilling has found. Oil companies almost always use mobile rigs in exploratory drilling, and often use fixed platforms with production and well maintenance facilities for development drilling. However, more and more, they are using production methods that do not require fixed platforms for development wells.

Mobile Offshore Drilling Units

To drill an offshore exploratory well, the operating company most often employs a drilling contractor who owns mobile offshore rigs, or *mobile offshore drilling units*. Several types of MODUs are available, and they can be classified in many different ways. Perhaps the most convenient is to divide them into bottom-supported units and floating units.

When a *bottom-supported unit* is drilling a well, part of its structure is in contact with the seafloor, while the remainder of it is above the water's surface. It floats on the water's surface only when it is being moved to another drill site. In contrast, *floating units* do not rest on the seafloor, but float on or slightly below the water's surface when they are on site and drilling.

Bottom-supported units include submersibles and jackups. Floating units include inland barges, drill ships and ship-shaped barges, and semisubmersibles.

Submersibles

Submersibles are bottom-supported MODUs and can be further divided into posted barges, bottle-type submersibles, and arctic submersibles.

Figure 4.61 The barge hull of this early posted barge rests on the seafloor.

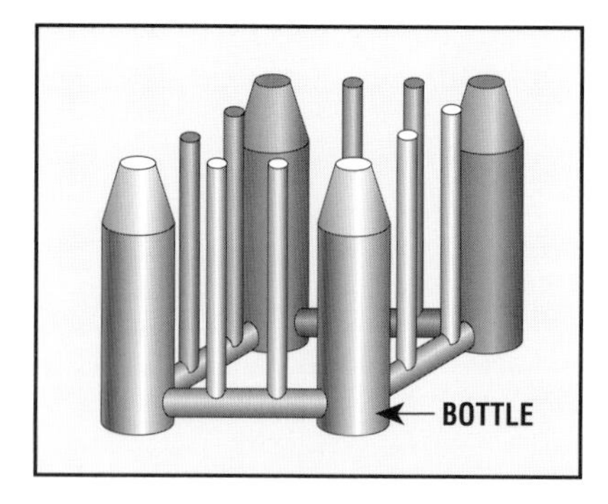

Figure 4.62 When flooded, the bottles cause a bottle-type submersible to rest on the seafloor.

Posted Barges

The earliest submersible design is the *posted barge*. Like the first posted barge described above, it consists of a barge hull with several steel posts that support an upper deck, with the drilling equipment on the deck (fig. 4.61). Posted barges are only useful in relatively shallow waters, less than about 30 feet.

Bottle-Type Submersibles

The *bottle-type submersible* is an early design that has several steel cylinders, or bottles, on top of which is a deck for the drilling equipment (fig. 4.62). When the bottles are flooded (water enters the bottles at a controlled rate), the rig submerges and comes to rest on the ocean bottom. When it is time to move the rig to its next drilling location, water is pumped out of the bottles until the unit floats again. Then, towboats move the rig to the new site. Modern bottle-type submersibles usually drill in maximum water depths of about 100 feet (30 metres).

Arctic Submersibles

Arctic submersibles are unique in that they are specifically designed for drilling in waters where moving pack ice could damage or destroy conventional submersibles. Pack ice occurs in arctic areas where winter temperatures become so low that seawater freezes. Subjected to strong currents, the ice tends to break apart and move with the currents. Moving ice can destroy a MODU unless it is specially constructed to withstand such collisions.

Several different types of arctic submersibles are available, but all share a common feature: a steel or concrete watertight chamber, a *caisson*, that rests on the seafloor when the rig is in the drilling mode (fig. 4.63). The caisson is rectangular or conical, and the drilling equipment sits on a deck, or platform, on top of it. The walls of the caisson protect the drilled wells from damage by deflecting moving ice. Also, the platform is reinforced to withstand the ravages of moving ice. Arctic submersibles can drill in water depths of up to about 150 feet (46 metres).

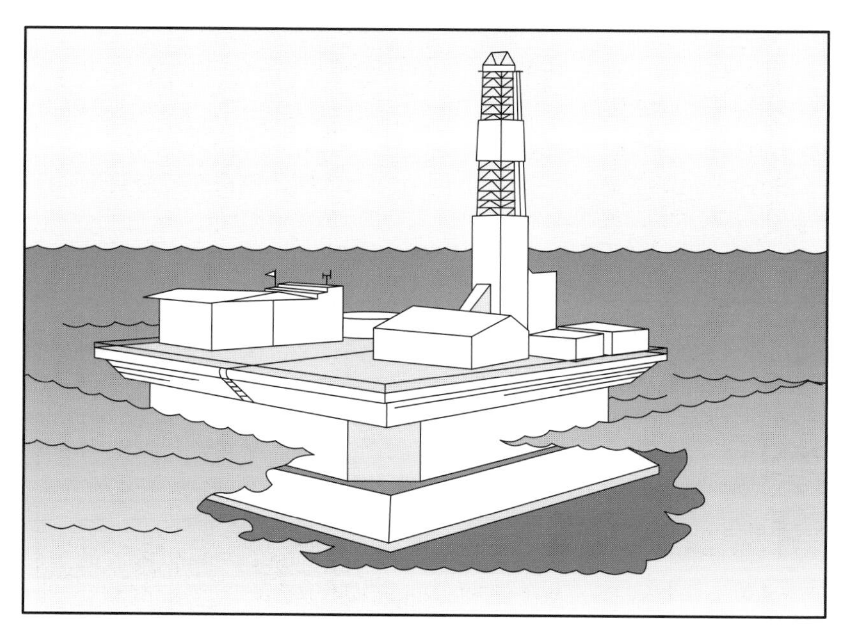

Figure 4.63 A concrete island drilling system (CIDS) features a reinforced concrete caisson with a steel base.

Jackups

The first *jackup*, or *self-elevating rig* was built in 1954 and rapidly became the most popular design in MODUs. Jackups are popular because they can drill in relatively deep water (up to about 350 feet, or 107 metres), provide a very stable drilling platform, and are fairly easy to move from one location to another. Also, they usually cost less than other types of MODUs. A variation of the jackup can drill as deep as 600 feet (214 metres) and withstand 130-mph (209-kph) winds.

A jackup rig has legs (usually three or four) that pass through its hull and can be jacked up or down (fig. 4.64). To move the rig, its legs are jacked up out of the water (or above the deck of a ship). When the rig is at the drill site, the legs are jacked down until they rest on the seabed. The hull is then jacked up above the water's surface high enough to permit the drilling crew to work unhampered by tides and waves.

Figure 4.64 The hull of a jackup is raised well above the water's surface before drilling begins.

The legs of a jackup can be either columns or trusses (fig. 4.65). Columnar legs are steel cylinders. Open-truss legs resemble high-voltage towers because they have several steel members, or trusses, that crisscross between structural corners. Columnar legs are less expensive to fabricate than truss legs but are more susceptible to twisting stresses. Because of this weakness, jackups with columnar legs are not able to operate in waters quite as deep as jackups with open-truss legs.

Both types of jackups have watertight barge hulls that can float on the surface of the water while the unit is being moved between drill sites. Usually a jackup is towed, but at least one jackup design is self-propelled; it has two engine-driven propellers, or *screws*, that power the unit as it floats. A third way to move a jackup rig is on a large ship with a flat deck. The rig is loaded onto the ship's deck, and the ship takes the rig wherever it needs to go. Used especially for long-distance moves, ships can carry a rig faster than it can be towed.

Figure 4.65 Jackups may have either open-truss legs as shown in this unit being outfitted (A) or columnar legs (B).

Inland Barges

An *inland barge rig*, considered a floating unit, is shaped very much like a traditional barge in that it is flat-bottomed and flat-sided. The drilling rig sits on the deck of the barge, and the entire unit is towed to the drill site. At the site, the barge is anchored, and drilling begins (fig. 4.66). Sometimes, however, in very shallow water, the hull is flooded and the inland barge rig is used as a submersible unit. Also called *swamp barges*, inland barges are at home in the relatively shallow waters, such as the swamps and bays of western Africa, or in large inland bodies of water, such as Lake Maracaibo in Venezuela.

Figure 4.66 An inland barge rig at work.

Ship-Shaped Barges and Drill Ships

Ship-shaped MODUs include *drill ships* and *ship-shaped barges*. Both look pretty much the same (fig. 4.67). With streamlined bows and squared-off sterns, drill ships and ship-shaped barges float on the surface of the water.

Figure 4.67 A drill ship steams down a channel on its way to a remote drill site.

For this reason they are sometimes called *surface units*. The main difference between a drill ship and a ship-shaped barge is that a drill ship is self-propelled and a barge is towed.

Both drill ships and ship-shaped barges have a derrick near the middle of the ship (*amidships* is the nautical term). Most also have a *moon pool*, a walled opening below the derrick, open to the water's surface and through which various drilling tools can pass down to the seafloor. Finally, most have a *helipad* as well, a landing pad for helicopters.

Drill ships have the advantage of being highly mobile, but their disadvantage is that they require a ship's captain and crew in addition to the usual drilling crew, so they are more expensive to operate than ship-shaped barges. However, because drill ships are so mobile, operators often use them to drill wells in remote areas, quite distant from land.

Ship-shaped MODUs are capable of drilling wells in very deep water where bottom-supported units are not practical. However, because they float on the surface, they are very susceptible to wave and wind motion and are not suitable for use in heavy seas. Fortunately, since many waters of the world are relatively calm, operators can use drill ships and ship-shaped barges all over the world.

Once a drill ship or barge is on the drill site (*on station*), it can be kept there in two ways. One is with several anchors from the vessel to the seafloor, much as any ocean-going vessel is anchored. The other way is by means of a system called *dynamic positioning* (fig. 4.68). In dynamic positioning, special propellers, or *thrusters*, are mounted on the vessel's hull below the waterline. A wind sensor and special sensors called *hydrophones* attached to the vessel send information about the wind, waves, and currents to a computer on board. The computer controls and activates the thrusters to maintain the unit precisely on station.

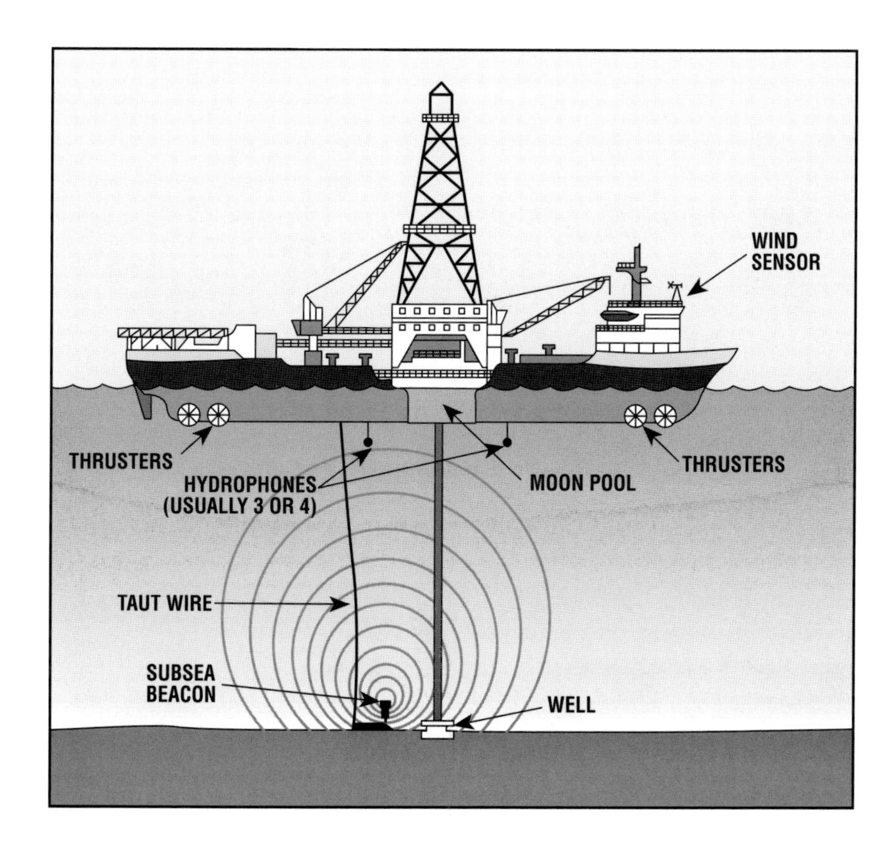

Figure 4.68 In this dynamic positioning system, hydrophones on the vessel's hull receive signals from a subsea beacon. The signals, in addition to information from a taut wire and a wind sensor, are transmitted to a computer, which activates the thrusters to keep the unit on station.

Semisubmersibles

Because drill ships and ship-shaped barges are so susceptible to wave motion, and because oil and gas have been discovered in formations that lie below rough seas, naval architects came up with a design for a floating rig that can withstand rough seas. Two designs evolved: the bottle-type and the column-stabilized semisubmersibles.

Bottle-Type Semisubmersibles

The *bottle-type semisubmersible* evolved from the bottle-type submersible. Recall that with the bottle-type submersible, the bottles are flooded so that the rig submerges and comes to rest on the seafloor. Naval architects found that if the bottles of a submersible were only partially flooded, so that some buoyancy remained, the rig settled below the water's surface but still well above the seafloor. In this semisubmerged state, wave motion does not affect the unit as much as it affects surface units. When anchored in position on the drill site, a semisubmersible provides a quite stable floating platform from which to drill wells (fig. 4.69).

Figure 4.69 A semisubmersible drilling unit floats on hulls that are flooded and submerged just below the water's surface.

At first, bottle-type submersibles were used either as submersibles or semisubmersibles. The design was so successful that naval architects soon designed units that were exclusively semisubmersible.

Column-Stabilized Semisubmersibles

One of the more popular modern semisubmersible designs is the *column-stabilized* semisubmersible (fig. 4.70). It consists of two or more pontoon-shaped hulls to which are attached several vertical cylinders, or columns.

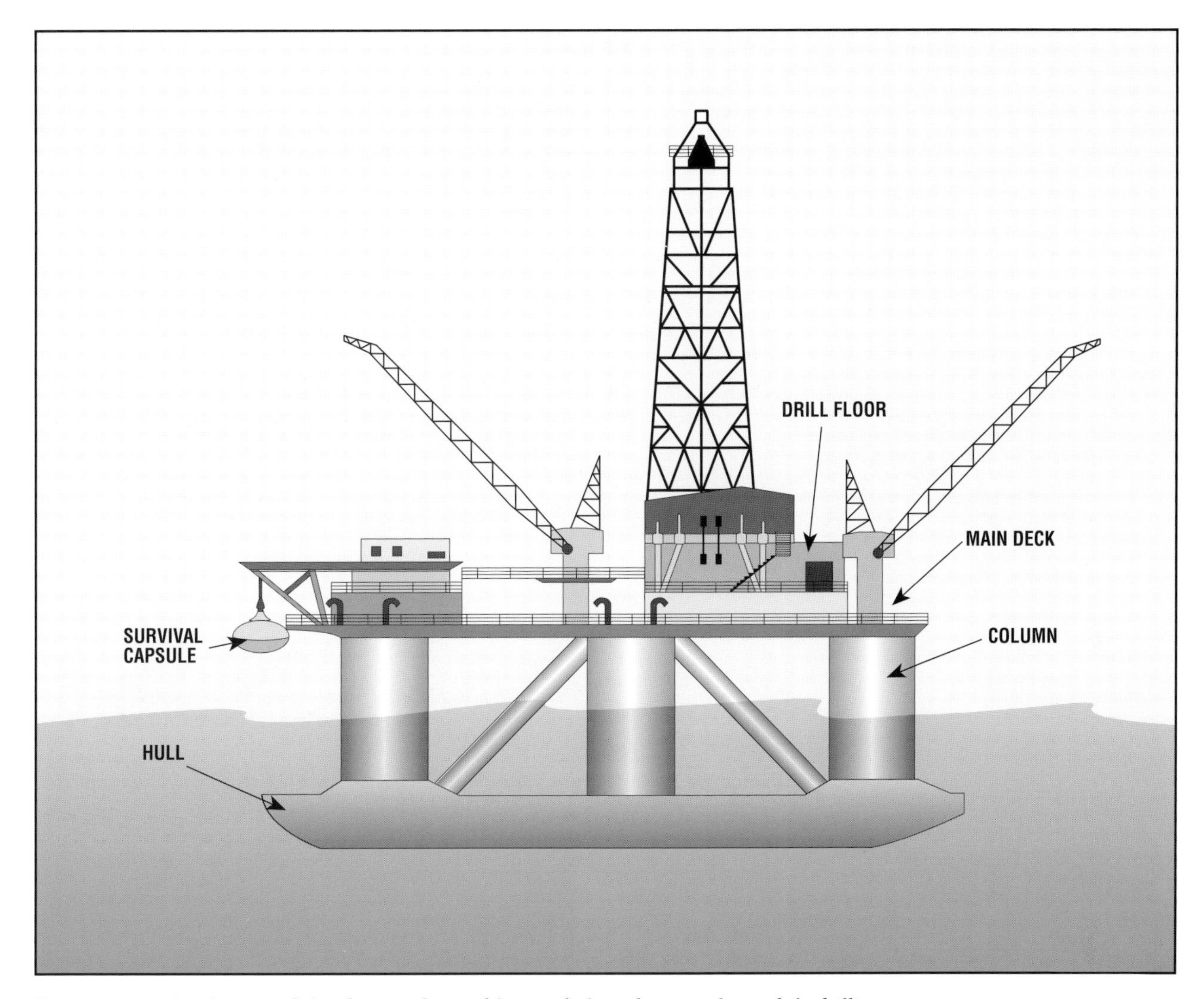

Figure 4.70 A column-stabilized semisubmersible rests below the waterline while drilling.

The main deck (the *Texas deck*) rests on top of the columns, and a derrick or mast and other drilling equipment sit on the deck.

Both the hulls and columns can be flooded with water or pumped dry. To move the unit, water is pumped out of the hulls and columns and the whole unit floats on the surface. Usually, the unit is towed to the drill site, although some semisubmersibles are self-propelled. At the drill site, the hulls and columns are flooded with enough water to submerge it to the required *drilling draft*, the depth below the surface. Anchors or dynamic positioning keep the unit on station.

Since much of the mass of a semisubmersible is below the water's surface, it does not roll or pitch much—that is, it does not move much from side to side or front to back. It does, however, tend to heave, or move up and down. As with ship-shaped MODUs, special equipment compensates for heave.

Semisubmersibles are among the largest and most expensive of all MODUs. Operating companies in all parts of the world use them in areas where the seas are rough because they are more stable than surface units. Like ship-shaped units, semisubmersibles can be used in deep water.

Figure 4.71 A self-contained platform houses all the drilling and production equipment and facilities for the crew.

Offshore Drilling Platforms

Operating companies generally use MODUs to drill exploratory wells. But once they know the offshore reservoir contains enough oil or gas to make developing it economical, they must drill several more wells. Sometimes they use MODUs to drill development wells, but usually they drill from fixed, permanent structures called *drilling platforms*. Modern platforms are large enough to house all the drilling and production equipment, crew quarters, offices, galley, and recreation rooms (fig. 4.71).

Platform rigs vary in design and appearance. Some are built of steel and some of concrete; some are rigid and some are compliant.

Rigid Platforms

Rigid platforms are the oldest design in offshore platforms and are most common in development drilling in water shallower than about 1,000 feet (about 305 metres). Rigid steel platforms are common in the Gulf of Mexico, in the Pacific Ocean off the coast of California, in the North Sea, and in other areas of the world.

Types of rigid platforms include the steel-jacket platform, the concrete gravity platform, and the caisson-type platform.

Steel-Jacket Platforms

The most common type of platform is the *steel-jacket platform*. Basically, it consists of the *jacket*, a tall vertical section made of tubular steel members that is the foundation of the platform. It is supported by piles driven into the seabed and extends upward so that the top rises above the waterline. Additional sections on top of the jacket provide space for crew quarters, a drilling rig, and the equipment needed to drill.

The height of the jacket depends on the water's depth, but at least one unit stands in over 1,000 feet (305 metres) of water. With the additional sections on top of the jacket, the entire structure reaches a height of over 1,200 feet (366 metres).

Most platform jackets are so large that they are constructed on land, launched into the water or placed onto a barge, and towed out to the site. A jacket is usually built on its side (fig. 4.72). Several of the tubular members of the jacket are sealed airtight so that when launched into the water, the jacket floats on its side, just as it was constructed.

Figure 4.72 Lying on its side, this jacket will soon be loaded onto a barge and moved to the offshore location.

At the site, large barge-mounted cranes stabilize the jacket, the tubular members are flooded, and the jacket is raised to a vertical position as the legs come to rest on the seafloor. Piles are then driven through several of the legs and deep into the ocean bottom to pin the jacket firmly to the bottom. Finally, cranes place the remaining elements of the platform on the jacket (fig. 4.73).

Figure 4.73 A large crane hoists a deck section onto the jacket. *(Courtesy of Sedco Forex)*

Concrete Gravity Platforms

Another type of rigid platform is the *concrete gravity platform,* built from steel-reinforced concrete (fig. 4.74). Concrete platforms are constructed in deep, protected waters near shore and floated or towed to the drill site in a vertical position. Tall caissons, or columns, resembling smokestacks and made of steel-reinforced concrete are the dominant feature of concrete gravity platforms. At the site, the crew floods the caissons until they rest on the seafloor.

Because the caissons are extremely heavy, the force of gravity alone is sufficient to keep them in place, so pilings are unnecessary (hence the name concrete gravity platforms). Crew quarters, drilling equipment, and other equipment sit on a deck on top of the caissons.

An advantage of concrete gravity platforms is that the operator can arrange special concrete cylinders around the base of the caissons on the seafloor to store up to a million barrels of oil. Oil storage capacity is advantageous when no pipeline exists for transporting oil to shore for refining. Another advantage of concrete gravity platforms is that they are able to withstand extremely rough seas because of their tremendous weight. For this reason, they are used extensively in the North Sea.

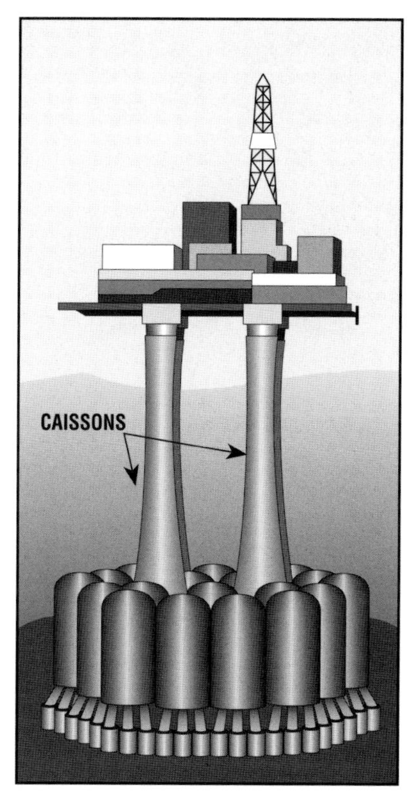

Figure 4.74 A concrete gravity platform is so heavy that gravity alone is enough to hold it in place.

Steel-Caisson Platforms

A third type of rigid platform is the *steel-caisson platform* (fig. 4.75). Engineers designed the steel-caisson platform specifically for use in the Cook Inlet of Alaska, where fast-moving tidal currents carry pack ice that can destroy conventional steel-jacket platforms.

One design has a caisson at each of the four corners of the rectangular platform, firmly affixed to the seafloor. The drilling and production decks rest on top of the caissons. The caissons are made of two layers of thick steel to prevent ice damage. Since development wells are drilled through the middle of each caisson, the drilling rig is designed so that it can be moved over the platform to each of the four caissons. The caissons also protect the top part of each well from moving ice.

Figure 4.75 This steel-caisson platform rests in the Cook Inlet of Alaska.

Compliant Platforms

Using rigid platforms in waters much over 1,000 feet deep is not practical because they become very, very expensive to build, and fabrication and installation limitations prevent their use. In deeper waters, operating companies use *compliant platforms,* which contain fewer steel parts and are lighter than rigid steel-jacket platforms. Compliant means that the platforms yield to wind and water movements, much as floating rigs do.

Types of compliant platforms include the guyed-tower platform and the tension-leg platform.

Guyed-Tower Platforms

A *guyed-tower platform* is similar to a rigid steel-jacket platform in that the bottom of the platform (the jacket) is pinned to the seafloor. However, the guyed jacket is much slimmer and does not contain as much steel as a rigid jacket, so it weighs less and is less expensive to build.

Several guy wires are attached to the jacket relatively close to the waterline and are spread out evenly around it (fig. 4.76). The guy wires are anchored to the seafloor in a unique way by means of *clump weights.* Clump weights are special weights that fit together in such a way as to behave much like a bicycle chain. The weights lie flat on the ocean floor, but as the platform moves, they lift off the floor. Then as the platform moves back, the clump weights once again lie on bottom. In short, clump weights allow the guy wires to move with the movement of the platform, yet still firmly anchor the wires.

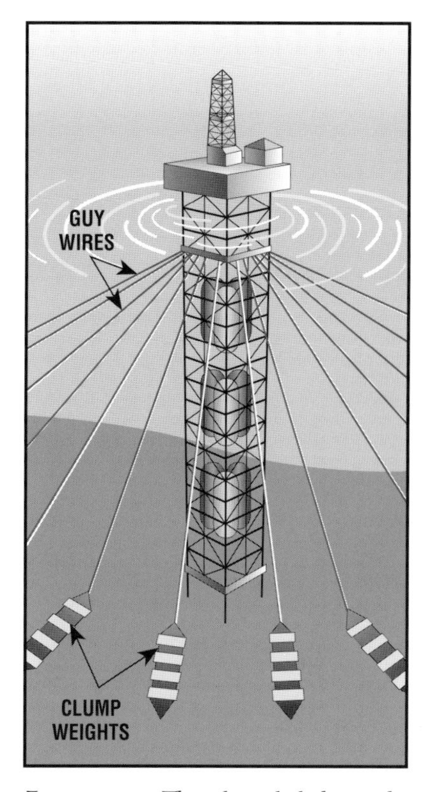

Figure 4.76 The relatively lightweight jacket of a guyed-tower platform is supported by several guy wires and clump weights.

Tension-Leg Platforms

A second type of compliant platform is the *tension-leg platform.* Like guyed towers, tension-leg platforms move with wind, waves, and currents. The top of the platform resembles a semisubmersible drilling unit; it has hulls that can be partially flooded. Unlike a semisubmersible, however, it is attached to the seafloor by several steel tubes called *tendons.* Since the tubes are firmly affixed to the platform as well as the seafloor, the buoyancy of the platform pulls and stretches the tendons, putting them under tension (fig. 4.77).

The newest tension-leg platforms designed by engineers at Shell Oil are as big as a football field and can support drilling in much deeper waters, 4,000 feet (1,220 metres) and more. Their tendons are metal pipes 28 inches (7 centimetres) in diameter and a half mile (0.8 kilometre) long attached to 375-foot-long (114-metre-long) piles driven into the ocean floor. The platform can move as much as 100 yards (90 metres) in any direction with hurricane-force winds and waves. It may soon be possible to drill in waters as deep as 1½ miles (2.4 kilometres) with them.

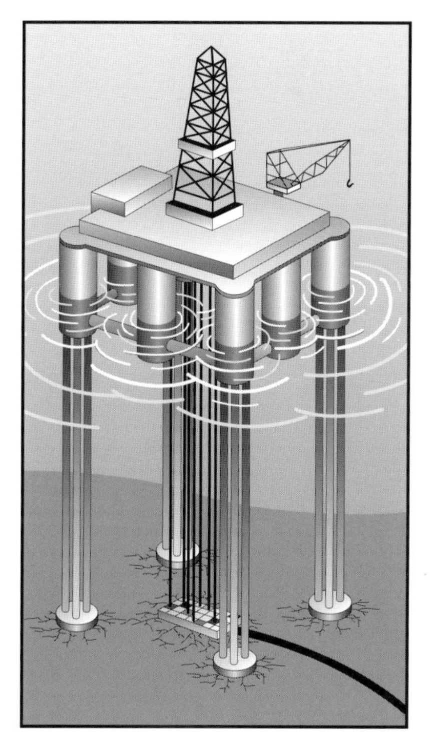

Figure 4.77 Steel tendons are kept in tension by the buoyancy of the platform on a tension-leg platform.

DIRECTIONAL DRILLING

Although wellbores are normally drilled vertically, many occasions arise that make it necessary or advantageous to drill at an angle, especially in offshore operations. This deviation from drilling a straight hole, known as *directional drilling,* makes it possible to accomplish many things that cannot be done with straight holes.

Directional wells are drilled straight to a predetermined depth, and then are gradually curved. Since the curvature of each well is gradual—only about 2 or 3 degrees per 100 feet (30.5 metres) of well depth—the straight, rigid drill stem and casing can follow the curve of the well without difficulty. Even though the curve is gradual, directional wells can be deflected off vertical to a very high degree; in fact, sometimes they run horizontally.

Uses

Improvements in directional drilling have made it possible to recover oil from reservoirs in the past several years that conventional straight-hole wells could not reach.

Offshore

Probably the most important use of directional drilling is for offshore reservoirs. Because of the huge expense of constructing a drilling platform in the water, operators cannot afford to tap offshore reservoirs unless they can drill several wells from a single platform (fig. 4.78). For example, the Gullfaks C platform in the North Sea weighs 1.5 million tons (1.36 million tonnes), covers 4 acres (1.6 hectares), and cost the US $2 billion to build. If it were not for directional drilling, an operating company would have to build several such platforms to reach the available petroleum. With directional drilling, operators can drill as many as forty or more wells from one platform.

Figure 4.78 Several directionally drilled wells tap an offshore reservoir.

Today these wells can reach an incredible distance from the platform. In January 1993, Norway's national oil company set a world record for directional drilling by drilling a well that was 23,917 feet, over 4½ miles (7,290 metres), horizontally from a platform in the North Sea.

Another way to drill an offshore well is to start drilling on land and curve to reach a pay zone offshore. It is cheaper to drill an offshore well from land—not only is there no platform to build, but also the operator needs no tankers to transport the petroleum and no service vessels to supply the offshore rig and its crew. Such a directional well is safer for the environment as well. It reduces the risk of oil spills, causes less air pollution, and does no harm to a local fishing industry.

Development Drilling on Land

As well as offshore development drilling, directional drilling has several other uses (fig. 4.79). Directional drilling allows a hole to avoid a fault line,

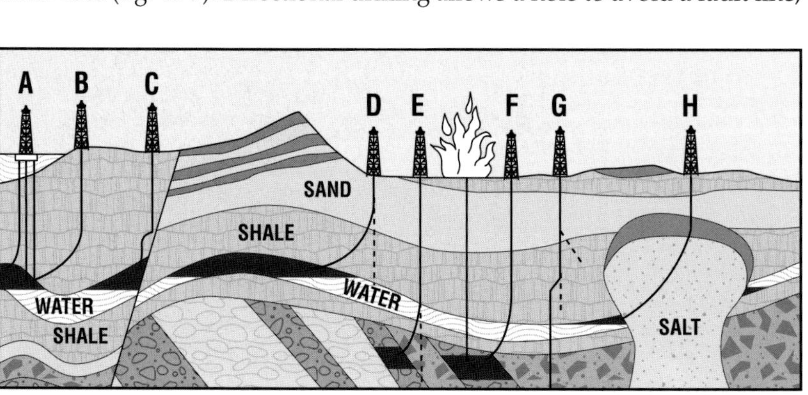

Figure 4.79 Some of the applications of directional wells

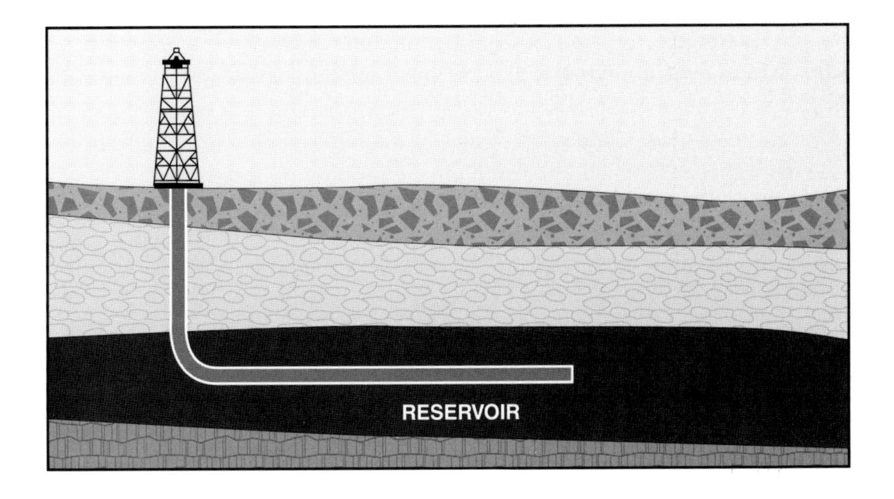

Figure 4.80 A long section of a horizontal well can pass through the oil in the reservoir.

which can cause problems in the hole. When it is impossible to locate the drilling rig over the desired spot because of a river, hill, or some other obstruction, the operator can erect the rig at one side of the obstruction and deviate the hole to the pay zone. Another reason to drill horizontally is that the orientation of the well affects how efficiently it drains the reservoir. A well drilled horizontally through a horizontal oil layer instead of vertically (fig. 4.80) produces less gas and water because the hole contacts hundreds or thousands of feet (metres) of the oil-bearing formation instead of tens of feet (metres). A horizontal well can also cross oil-containing fractures that are parallel to each other (fig. 4.81). Operators have increased production by as much as ten times in formations like this.

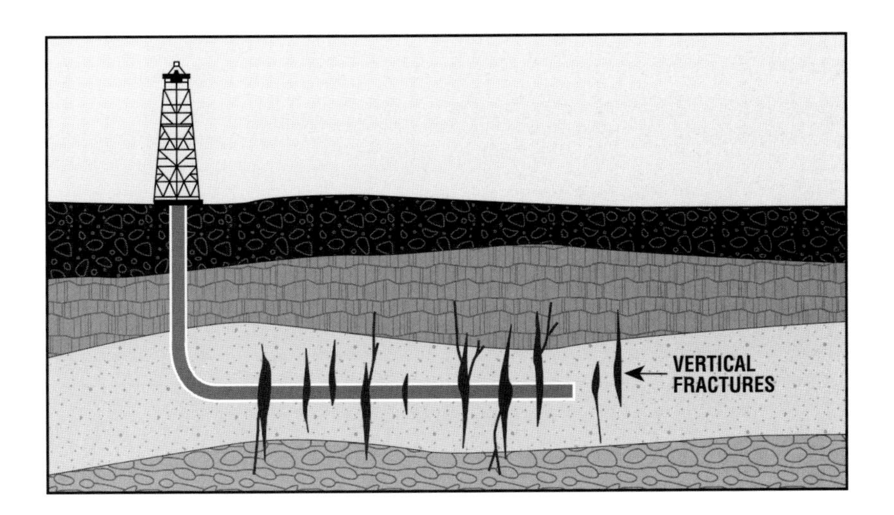

Figure 4.81 This horizontal well reaches several oil-containing vertical fractures.

Other Uses

Directional holes can also be used in exploratory drilling in cases where the straight hole misses the reservoir (see fig. 4.80). By drilling a new hole deflected from the original dry well, it may be possible to intersect a pay zone without the expense of drilling a completely new well.

Another use is in killing a *wild well,* a well that has blown out, caught on fire, and cratered (caved in). A directional well called a *relief well* is drilled so that it bottoms out near the borehole of the blown-out well. Then the crew can pump mud down the relief well to kill the wild well.

A directional well can be used to straighten a crooked hole or to sidetrack around a *fish* (an object lodged in the borehole) that cannot be removed. Finally, directional drilling is very useful when drilling into traps associated with salt domes (see Chapter 1). Typically, the traps around salt domes lie on the edges of the dome and are difficult to pinpoint accurately. A directional well makes it possible to drill into the trap if the initial straight hole misses.

Tools and Techniques

Directional drilling requires special tools and techniques. One innovation is new formulations of drilling mud that are more slippery, to reduce friction in the hole.

Drill Pipe

Because drill collars below the drill string would do nothing but lie on the bottom of a horizontal hole (fig. 4.82), drill string designers place them at the bottom of the vertical section of the string and above the lower section of drill pipe (fig. 4.83). Think of a 10,000-foot (3,048-metres) 5-inch (127-millimetre) drill string as comparable to a 1⁄32-inch (0.8-millimetre) hypodermic needle that is 60 feet (19 metres) long, and you can imagine how strong it must be to withstand the weight of the drill collars above it. Part of the solution to keeping the drill pipe in the bottom string from buckling is to use stronger pipe.

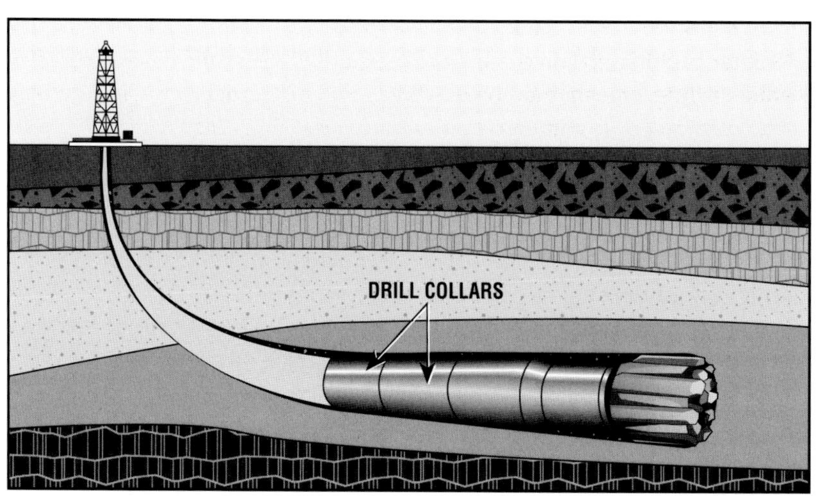

Figure 4.82 Drill collars placed at the end of the drill string in a horizontal well would not add weight to the bit.

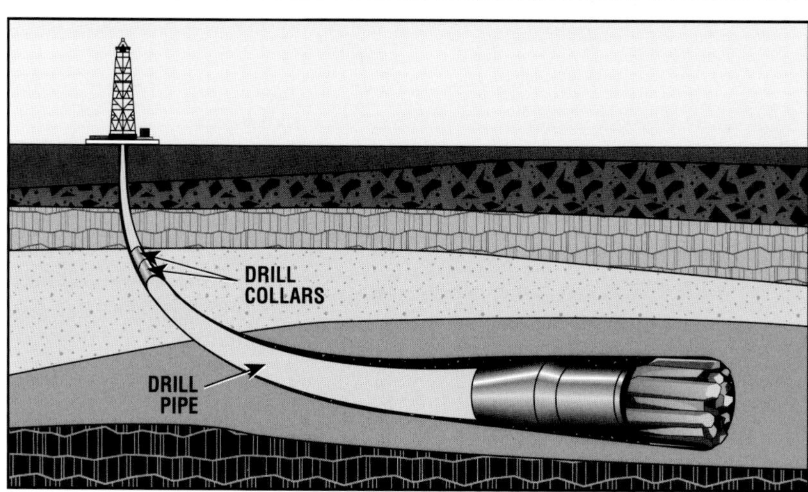

Figure 4.83 Here, the drill collars are below the vertical section of pipe and above the horizontal section.

Manufacturers have engineered new types of drill pipe especially for horizontal drilling that is light and rigid.

Bent Sub and Downhole Motor

Two other problems to overcome are how to force the bit to drill off vertical and how to rotate the bit when the drill string is straight and the hole is curved. Several ways exist to drill a directional hole, but the most common uses a bent sub and downhole motor.

A *bent sub* is a short piece of pipe with threads on both ends and bent in the middle. Typically, the bend ranges from 1 to 3 degrees off vertical. The drilling crew places it in the drill stem between the bottommost drill collar and a downhole motor. The bent sub deflects the downhole motor and bit below it off vertical, solving the first problem.

The *downhole motor* solves the second problem: it rotates only the drill bit connected to it, so that the entire drill stem need not rotate to make hole. Shaped like a piece of pipe, a downhole motor can have turbine blades, or it can have a spiral steel shaft that turns inside an elliptical opening in the housing. In the turbine type, the drilling fluid circulating inside the tool moves the turbine blades, which turns a shaft. Since the bit is attached to the shaft, it turns when the shaft turns. With the spiral type, circulating drilling mud flows through the elliptical opening and forces the spiral shaft to rotate. Again, the rotating shaft turns the bit.

Orienting the Hole

Another problem to solve is how to orient the hole in the desired direction while drilling. One method uses a directional instrument containing a magnetic or gyroscopic compass and an *inclinometer* (an instrument that measures the angle of the hole). Drilling stops while the crew runs the instrument down to the bit and then retrieves it. Another method is to use an instrument known as a *steering tool*. A steering tool sends directional information up the hole through a conductive wireline to a monitor on the rig floor. Drilling need not stop for so long with this method.

The most recent advance is the *mud pulse generator*, a wireless, self-contained instrument that transmits sonic signals up the hole, through the drilling fluid in the drill stem, to a readout device at the surface. This system, called *measurement while drilling (MWD)*, allows the driller to get information from the bottom of the hole without stopping drilling at all. MWD is useful for maintaining a straight hole as well as for directional drilling.

FISHING

Fishing is the drilling term for retrieving any object—or fish—from the wellbore. A fish can be anything from part or all of the drill stem stuck or lost in the hole to smaller pieces of equipment, called *junk,* such as bit cones, hand tools, pieces of steel, or any other item in the hole that cannot be drilled out.

Freeing Stuck Pipe

Drill pipe or drill collars can get stuck in the hole for several reasons: (1) the hole can collapse around the pipe; (2) the pipe can get stuck in a *dogleg* or *keyseat*; or (3) pressure can hold the drill collars so securely to the wall of the hole that no amount of pulling can free the pipe.

The most common reason for the wall of the hole collapsing around the pipe is that *interstitial salt water* (water contained in the interstices, or pores, of the formation rock) can sometimes attract the water in the drilling mud. For example, if the formation happens to be made of shale and the water in the mud contacts the water in the shale, the water in the mud has a tendency to transfer to the shale. The transferred water causes the shale to expand and slough off into the hole in the form of small sheets. The hole eventually fills with shale debris, and the pipe sticks.

Pipe can also get stuck in a keyseat (fig. 4.84). A keyseat is caused by a dogleg, which is a very crooked section of hole. ("It's as crooked as a dog's hind leg" is the expression that gives rise to the term.) The drill pipe tends to lean against the side of the dogleg, and as the pipe rotates, it digs out a new, smaller hole in the side of the main borehole. Then, when the drill stem is

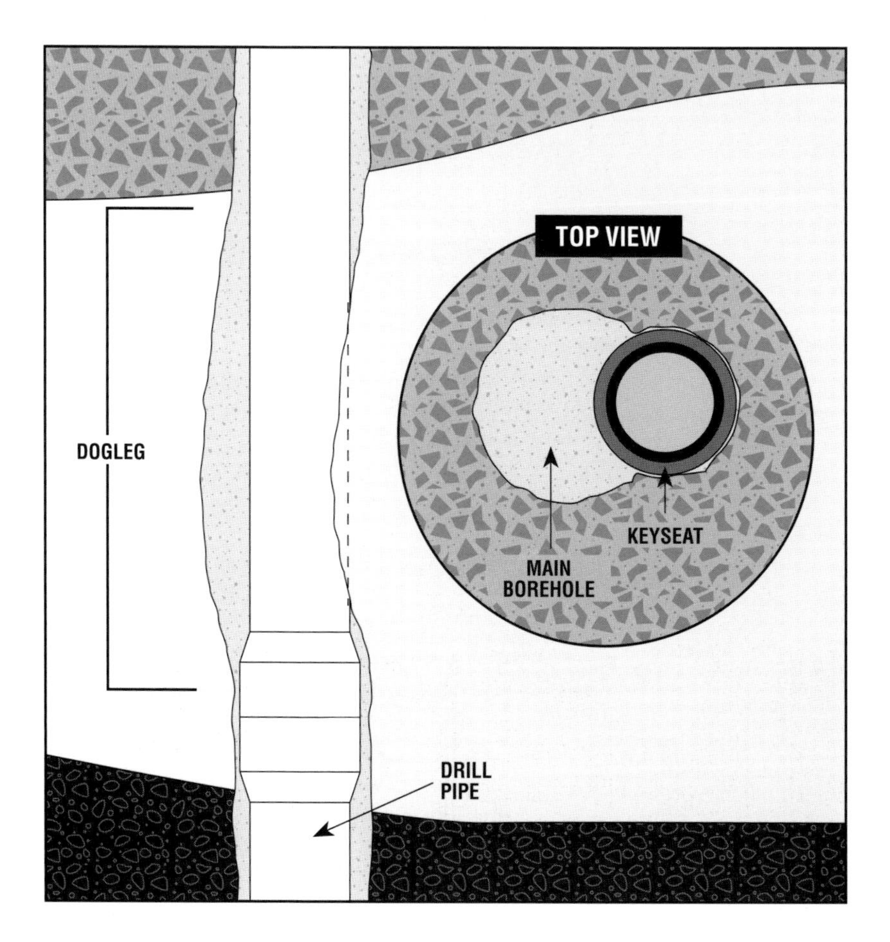

Figure 4.84 Drill pipe is stuck in a keyseat.

pulled from the hole, a tool joint of the drill pipe or the wider drill collars can jam into the keyseat so hard that they cannot be freed just by pulling on them.

Pipe can also get stuck in the hole when the pressure in the wellbore is quite a bit higher than the pressure in the formation. If the mud engineer does not adjust the properties of the mud when drilling through more permeable layers, the higher pressure of the mud in the wellbore can cause the wall cake of mud solids to build up so thickly on the inside wall of the hole that the hole becomes much smaller in diameter. Then, when the drill stem is not rotating, as when a connection is being made or during a trip, the collars may rest against one side of the hole, and the pressure from the mud can force them into the thick cake and stick them very firmly to the wall of the hole.

To free wall-stuck pipe, the crew will probably spot oil around the stuck portion and jar on the drill stem. To *spot oil* means to circulate oil or some other lubricant down the drill stem and into the annulus to the stuck portion. To *jar* on the drill stem means to install a special device—a *drilling jar* or a *bumper sub*—on top of the drill stem, which allows the driller to strike very heavy upward or downward blows on the stuck pipe. Spotting oil and jarring usually free wall-stuck pipe.

To free pipe that is stuck due to sloughing shale or to free wall-stuck pipe that remains stuck because spotting oil and jarring are not successful, the first step is to use a device called a *free-point indicator.* A free-point indicator, in spite of its name, determines the point at which the pipe is stuck (fig. 4.85).

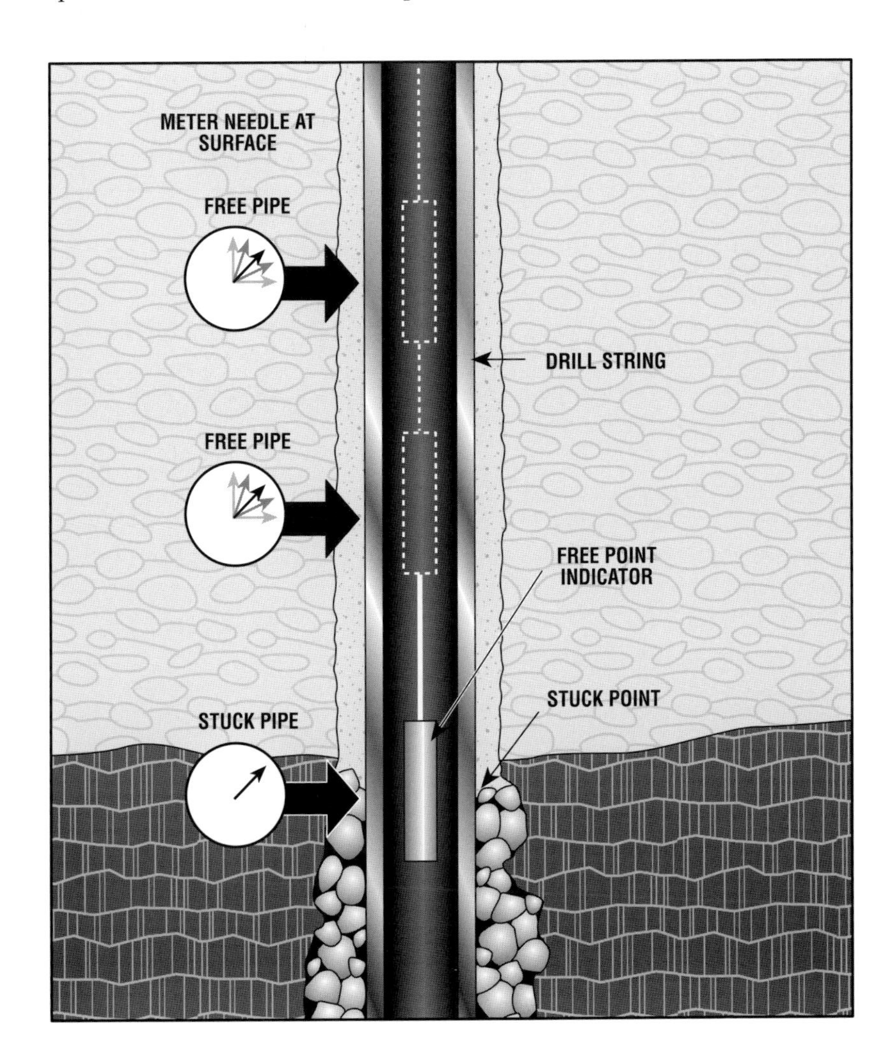

Figure 4.85 A free-point indicator locates stuck pipe.

It is lowered inside the drill stem, and then the driller stretches the drill stem by picking it up. The indicator induces a magnetic field in the pipe and sends a signal to a meter on the surface. When the indicator is opposite a point where the pipe is free, the pipe stretches a large amount, and the meter reads strong signals; therefore, the meter needle moves a great deal. But when the indicator is opposite the point where the pipe is stuck, the pipe stretches very little, and the signals are weak; therefore, the needle moves very little. By noting the depth at which the needle movement is slight, the crew can determine the stuck point.

Once they determine where the pipe is stuck, the crew positions a *string shot* opposite a tool joint several joints above the stuck point. A string shot is a long, stringlike explosive charge that is usually run below the free-point indicator. When detonated, the string shot helps loosen the threads of the pipe, just as banging on the lid of a jar that is difficult to open helps loosen it. As the string shot is set off, the driller turns the rotary to the left to back off (unscrew) the pipe. Then the crew trips the free pipe above the stuck pipe out of the hole. They usually back off the free pipe several joints above the stuck point because they must run in additional tools. The extra joints of clear pipe guide the tools to the right place.

The stuck pipe is left dangling in the hole, and a special kind of pipe called *washover pipe,* or *washpipe,* is lowered into the hole. The driller lowers the washpipe over the stuck drill stem (the fish), and begins circulation and rotation. A special cutting device, the *rotary shoe,* on the bottom of the washpipe drills the shale or wall cake that is causing the pipe to stick. However, when all the material holding the fish is removed, the fish could fall to bottom. To prevent this, the crew latches a *backoff connector* inside the washpipe (fig. 4.86).

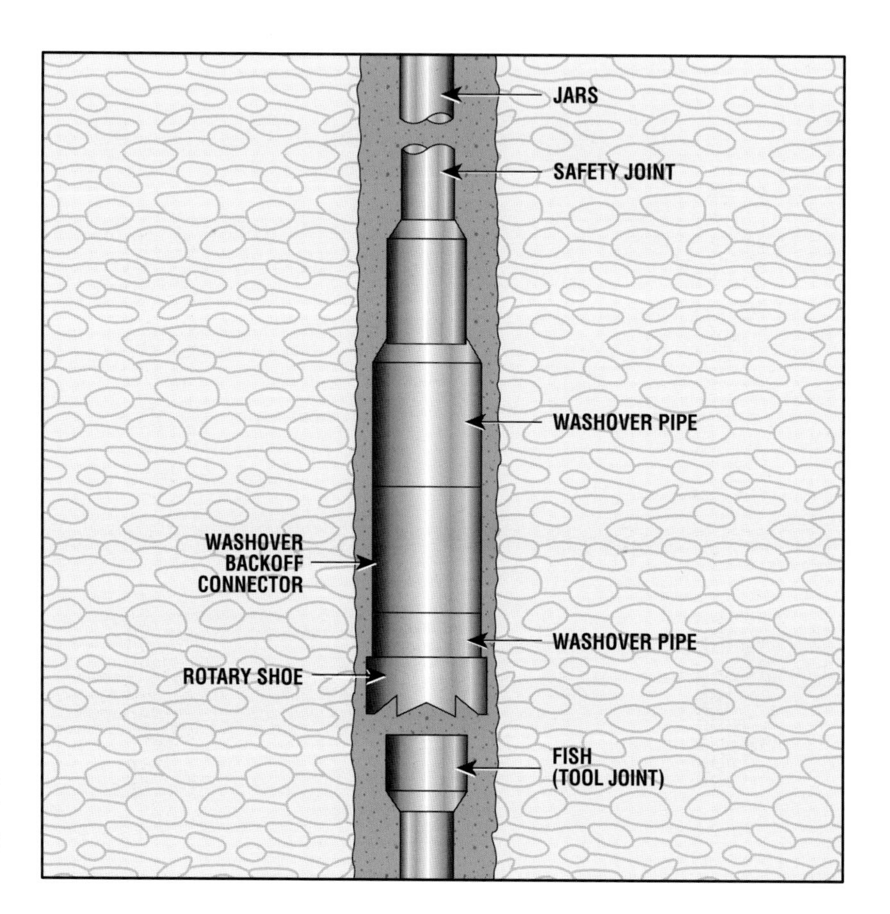

Figure 4.86 A backoff connector is used on stuck pipe inside the washover pipe to keep the drill pipe from falling when freed.

The connector is made up on the top of the fish before the washover begins; as washover proceeds, the connector is designed to remain stationary on top of the fish. When the fish is washed free and starts to fall, the crew activates the connector and it grips the inside of the washpipe and prevents the fish from falling. After all the shale or wall cake is drilled out, the crew pulls the washpipe and the drill stem (secured inside the washpipe by the backoff connector) out of the hole simultaneously.

If the drill stem is stuck in a keyseat, the fishing process is slightly different. It still starts with sending the free-point indicator and string shot down inside the pipe as before. When the crew locates the stuck point, they remove all but the last five or six joints above the stuck drill pipe or drill collar, so that the top of the fish is in the main borehole and not in the keyseat. Then they lower a bumper jar down the hole, attach it to the fish, and jar the fish loose from the keyseat. Finally, a reaming device called a *keyseat wiper* that was made up on top of the jar reams the keyseat out to normal size. In this way the tool joints and drill collars will pass through the reamed-out keyseat.

Retrieving Twisted-Off Pipe

Another fishing problem that can happen with drill pipe and drill collars is that the pipe twists off. Because of fatigue or damage, the pipe literally breaks in two; such breaks are called *twistoffs*. Fishing for twisted-off pipe can be a relatively simple operation, or it can be difficult or impossible, depending on the situation. Usually, however, retrieving a twisted-off fish is not complicated.

Two commonly used tools for retrieving twistoffs are the *overshot* and the *spear* (fig. 4.87). The crew runs an overshot or spear into the hole on drill pipe until it contacts the top of the fish. An overshot goes over the outside of the fish and grips it firmly. A spear goes inside the fish and grips the inside of

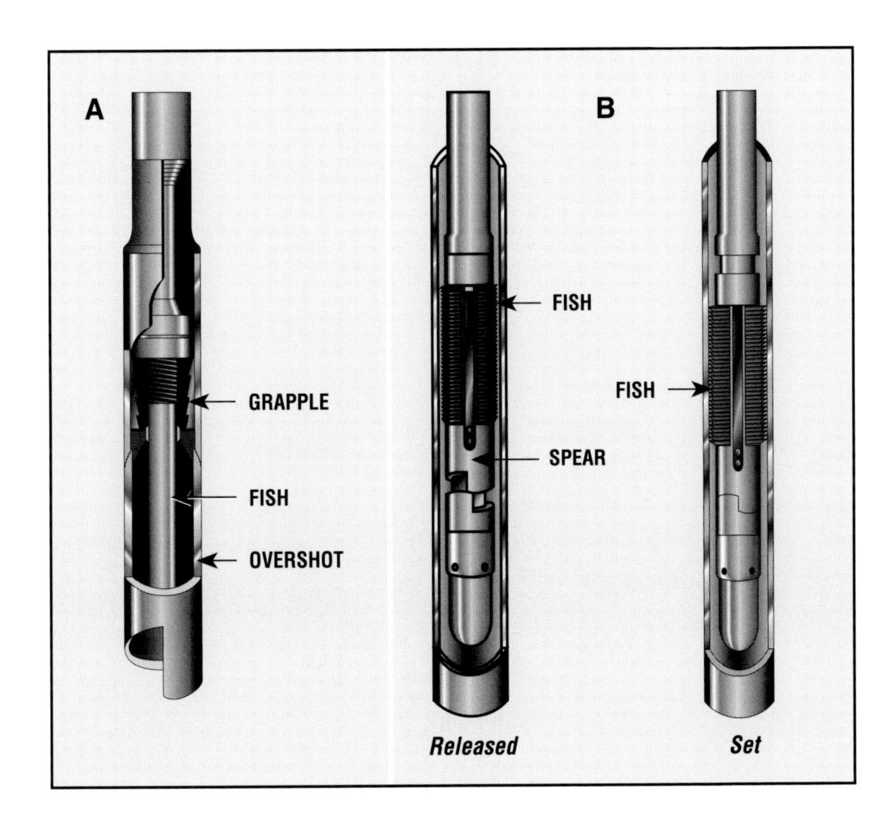

Figure 4.87 An overshot (A) and spear (B) are two commonly used tools for retrieving fish.

the pipe. A spear is used only when the diameter of the fish is close to the size of the hole, such as when a drill collar is involved. In either case, when the spear or overshot grips the fish, the crew pulls both the tool and the fish out of the hole together. Sometimes they use a special mill to smooth the top of the fish before running the overshot or spear. Milling the top of the fish makes attaching the fishing tool easier.

Fishing for Junk

Small nondrillable pieces of metal that find their way to the bottom of the hole and cause drilling to cease are known as junk. Many types of fishing tools, including powerful magnets and special baskets through which mud can be circulated, can retrieve junk. In fact, the only limit to fishing tools and methods seems to be the imagination and ingenuity of the people faced with the problem of removing fish from the hole.

AIR OR GAS DRILLING

Circulating with air or gas instead of mud as a drilling fluid is an alternate method of rotary drilling that has spectacular results but limited application. Penetration rates are higher, footage per bit is greater, and bit cost is lower with air or gas. Air or gas cleans the bottom of the hole more effectively than mud, and for that primary reason the rate of penetration is faster. Mud is denser than air or gas and tends to hold the cuttings on the bottom of the hole. As a result, the bit cannot make hole as efficiently because the bit redrills some of the old cuttings instead of being constantly exposed to fresh, undrilled formation.

Unfortunately, *air* or *gas drilling* has several disadvantages that are more significant than the advantages. The hazard of fire or explosion is always present. Because air or gas does not create significant bottomhole pressure, drilling into a high-pressure formation is dangerous. Air and gas cannot prevent formation fluids from entering the well, and most deep wells come across water-bearing formations sooner or later. If the wall of the hole tends to slough, or cave, into the hole, air or gas circulation is impossible because the drill stem may stick. Also, corrosion of the drill stem has been a problem with air or gas drilling, although chemicals have been developed to reduce this problem.

To drill with air or gas, the contractor moves large compressors and related equipment to the site. Usually, only part of the hole is drilled with air or gas; then the rig is changed over to drilling mud. Actually, air or gas is not circulated in the sense that it is used over and over again; rather, it makes one trip from the compressors, down the drill stem, out the bit, and up the annulus back to the surface, where it is blown out a *blooey line,* or vent pipe (fig. 4.88).

Figure 4.88 Skid-mounted compressors furnish the high-pressure air used on this rotary rig for drilling. The air is blowing a cloud of dust out the blooey line.

Aerated mud—that is, mud to which air or gas has been intentionally added—has been used successfully to prevent lost circulation. *Lost circulation* occurs when drilling mud leaks out of the borehole and into a formation. Thus, mud does not return to the surface but is lost downhole. Air or gas in the mud reduces the amount of pressure exerted by the mud on downhole formations and thus relieves one of the causes of lost circulation.

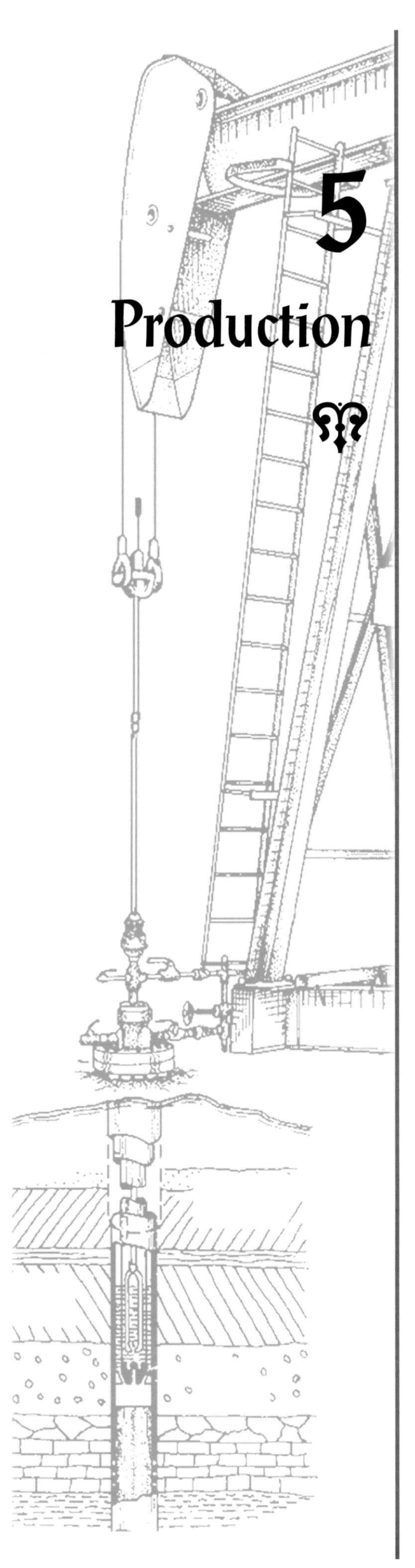

5 Production

In the petroleum industry, *production* is the phase of operation that deals with bringing well fluids to the surface and preparing them for their trip to the refinery or processing plant. Production begins after drilling is finished.

The first step is to *complete* the well—that is, to perform whatever operations are necessary to start the well fluids flowing to the surface. Routine maintenance operations, such as replacing worn or malfunctioning equipment—known as *servicing*—are standard during the well's producing life. Later in the life of the well, more extensive repairs—known as *workovers*—may also be necessary to maintain the flow of oil and gas. The fluids from a well are usually a mixture of oil, gas, and water, which must be separated after coming to the surface. Production also includes disposing of the water and installing equipment to treat, measure, and test the oil and gas before they are transported away from the wellsite.

So production is a combination of operations: bringing fluids to the surface; doing whatever is necessary to keep the well producing; and taking fluids through a series of steps to purify, measure, and test them.

EARLY PRODUCTION METHODS

The Chinese probably invented the earliest methods of well completion. Historians believe that they completed gas, water, and saltwater wells as early as 1000 B.C. Chinese technology passed down through the centuries with few changes, and it was not until the 1800s that the innovation and inventiveness that mark the petroleum industry today began.

Completion

The present production era started in 1808 with a saltwater well that two brothers, David and Joseph Ruffner, had dug in order to extract salt. They had a problem with the West Virginia well: the less concentrated sands containing salt water higher up in the well were diluting the concentrated salt water down below. To extract more of the concentrated salt water from the bottom of the well, the brothers decided that running a pipe or tube from the well bottom to the surface would seal off water in the higher zones and prevent it from entering the well. They also realized that a seal of some sort, placed just above the bottommost sand, would keep the shallower saltwater sands that migrated to the bottom from coming up through the pipe (fig. 5.1). Since no metal piping was available, they rigged up two half-tubes of wood, fitted them closely together, and wrapped twine all around them to form a long cylinder.

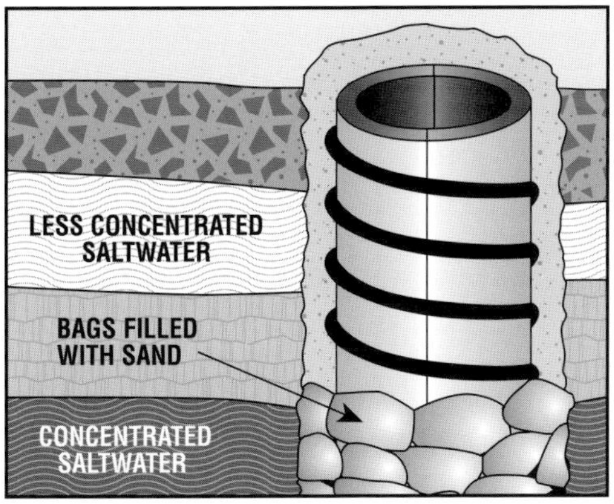

LESS CONCENTRATED SALTWATER
BAGS FILLED WITH SAND
CONCENTRATED SALTWATER

Figure 5.1 Running a pipe to line the sides of a well and sealing off the outside of the pipe near the bottom prevented fluids from the formations near the surface from entering the well.

After fitting a watertight bag snugly near the lower end as a seal, the brothers placed the homemade tube carefully in the well. With this invention, concentrated brine flowed freely through the tube, and the bag kept out the less desirable water. This West Virginia experiment was the first step toward the completion techniques that are now so vital to the oil industry.

Pumping

Another important operation borrowed from the water-well industry was the use of the reciprocating walking beam to operate a pump. A reciprocating walking beam moves up and down. In the decade after Drake's discovery well in 1859, a cable-tool drilling rig remained on a well. The walking beam on the rig that the driller used to raise and drop the bit could later operate a pump. The cable-tool rig also had spools with cable wrapped around them, similar to the drawworks on a rotary rig, and tools attached to the cables could be lowered into and raised from the well. This equipment was known as the "standard rig front" and was mostly made of wood and driven by steam.

Central pumping power, developed by the oil industry around 1880, was the first major innovation in production equipment. Central pumping consisted of using a single *prime mover*, or power source, to pump several wells. More developments followed at the turn of the century. Rotary drilling evolved, and *internal-combustion engines* began to replace steam engines. All of these inventions laid the groundwork for continuing improvements in equipment and procedures. Adoption of new devices was slow, however, with operators using old equipment as long as possible. By the early 1920s, though, the demand for a method to replace the on-site rig led the oil industry to focus on the *beam pumping system*, a self-contained unit, mounted on each well at the surface, that operated a pump in the hole.

After a slow start, the transition was quite rapid. Problems with beam pumping fostered the search for alternate ways to lift oil artificially. The electric submersible pump was developed in 1930 as an acceptable alternative, followed by marketable gas-lift devices a few years later. But even today, most wells on artificial lift are using beam pumping systems.

Storage and Handling

The industry also had a problem with what to do with the oil after it came out of the well. Often, wooden barrels were the only containers available, and early operators used these for collecting, storing, and shipping petroleum (fig. 5.2). A barrel even became the standard for measuring petroleum, as it remains today. A *barrel* of crude equals 42 gallons, or 0.15899 cubic metres (1 cubic metre is 6.2897 barrels).

Figure 5.2 Wooden barrels were the first containers for produced oil.

Sometimes, operators dug earthen pits into the ground and routed the well fluids into them. Handling and separation were difficult in any case and became even more difficult when considerable amounts of water and sand came mixed with the produced oil. Tanks slowly replaced wooden barrels, first made of wood, then riveted iron, and finally bolted or welded steel.

After 1920, large advances were made in the operation of lease facilities. Petroleum engineers improved methods for oil, gas, and water separation, chemical and electrical emulsion treatment, and water handling. Although producing oil and gas has taken advantage of many new technologies, modern petroleum-producing facilities still retain some similarity to old-time designs. The emphasis now, however, is on using efficient techniques and methods to improve lease operation.

WELL COMPLETION

After a well has been drilled and the company has determined that the reservoir will be economical to produce, the work of setting the final string of casing, preparing the well for production, and bringing in the oil or gas begins.

Completion equipment and the methods employed are quite varied, and operators make the decisions for an individual well based on the type of oil or gas accumulations involved, the requirements that may develop during the life of the well, and the economic circumstances at the time the work is done. An operator may use low-pressure pipe, sometimes secondhand, if the oil accumulation has a marginal payout, and other expenditures will be scaled down accordingly. If the operator anticipates high pressure and a long well life, however, the best grade of pipe will be necessary.

Production Casing and Liners

Many oil and gas wells require four concentric strings of large pipe, each one reaching to the surface: *conductor pipe, surface casing, intermediate casing,* and *production casing* (fig. 5.3). The production casing is often called the *oil string* or the *long string* in the oil patch and is the final casing for most wells. Usually, the production casing completely seals off the producing formation, but in rare instances the production casing stops near or just on top of the potential pay zone.

Another type of pipe that is not uncommon in wells over 10,000 feet (3,048 metres) is called a *liner*. Liners are really just like casing—that is, they serve the same purpose—but they do not extend all the way to the surface (fig. 5.4). Instead, a liner hangs from the end of larger casing above it by means of a *liner hanger*. A liner can function as production casing, in which case it is called a *production liner*. Since it does not go to the surface, which is sometimes a considerable distance, the operator has a lower pipe cost.

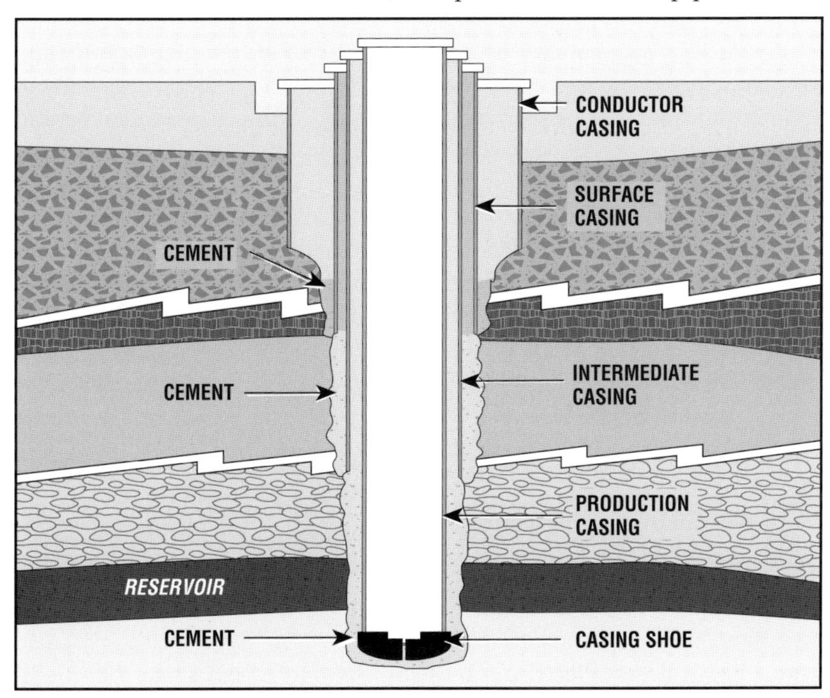

Figure 5.3 Conductor, surface, intermediate, and production casing are cemented in the well. Note that the production casing is set through the producing zone and seals it off.

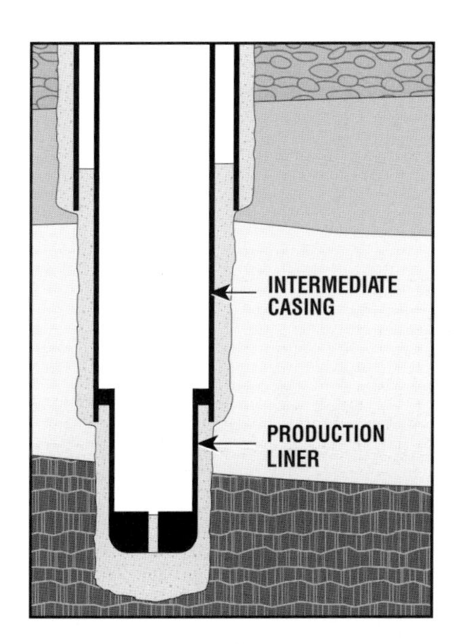

Figure 5.4 A production liner is cemented in place but hangs from the bottom of the intermediate casing rather than extending to the surface.

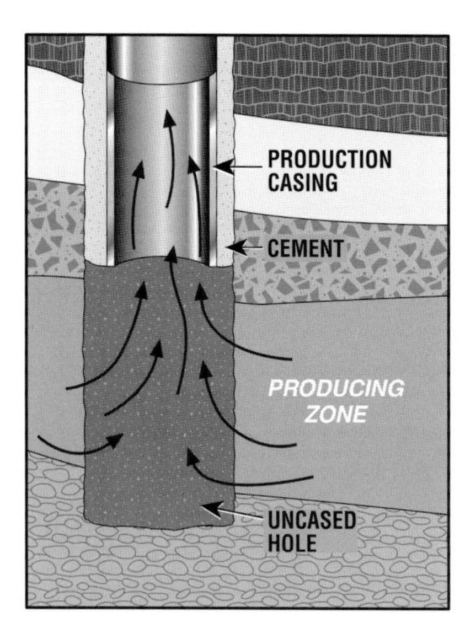

Figure 5.5 An open-hole completion allows reservoir fluids to flow into the uncased hole.

Open-Hole, Perforated, and Wire-Wrapped Screen Completions

The production casing either runs to the bottom of the hole and blocks off the production zone completely or, rarely, stops just above the production zone. If the casing stops above the production zone and leaves the hole open, it is an *open-hole* or *barefoot completion*. The reservoir fluids simply flow from the formation into the uncased hole. Usually, however, the casing runs to the bottom of the hole. The hydrocarbons cannot get out of the reservoir and into the casing because the cement and casing walls seal it off. In this case the crew perforates the casing, or shoots holes in it, so that the fluid can flow into the well. This is a *perforated completion*.

To complete a well, a well servicing contractor may move in a small rig, or the drilling contractor may do the work before moving to another site. The operator determines the type of completion method to use by looking at the characteristics of the reservoir and its economic potential.

Open-Hole Completion

An open-hole, or barefoot, completion has no production casing or liner set opposite the producing formation (fig. 5.5). Instead, reservoir fluids flow unrestricted into the open wellbore. This type of completion, which is rare and is generally restricted to limestone reservoirs, is useful in wells with only one productive zone and low-pressure formations. In an open-hole completion, intermediate casing is set just above the pay zone, and drilling proceeds into the productive zone as far as necessary to complete the well.

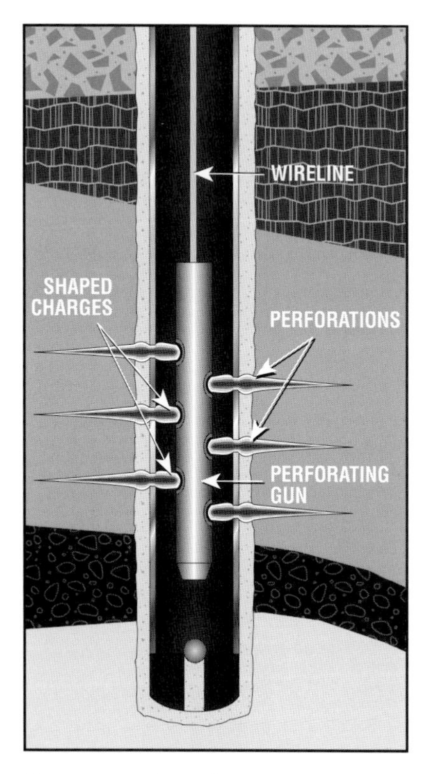

Figure 5.6 A perforating gun creates holes in the production casing or liner, allowing fluids from the formation to enter the well.

Perforated Completion

The perforated completion—by far the most popular method of completing a well—requires a good cementing job and the proper perforating method. Perforating is the process of piercing the casing wall and the cement behind it to provide openings through which formation fluids may enter the wellbore.

A *perforating gun*, or *perforator*, is the tool that makes these openings. The completion crew lowers the long cylindrical gun down the production casing or liner until it is opposite the reservoir zone. The perforator carries several bullets or special explosive charges called *shaped charges* (fig. 5.6) that are aimed at the walls of the casing or liner. The shaped charges or bullets shoot smooth, round holes in the casing or liner and penetrate the rock as well. Because the explosion of a shaped charge is actually a jet of high-energy gases and particles, it is called *jet perforating*.

Wire-Wrapped Screen Completion

Some reservoirs produce not only oil and gas, but also sand. A formation produces sand when the individual grains making up the reservoir rock are *unconsolidated*. In other words, the rock grains do not adhere to each other. Sand can flow into the well along with the fluids and can clog it up to reduce or stop production. Sand also damages the equipment in the hole.

In reservoirs with sand, then, a completion sometimes includes a *wire-wrapped screen*, or *screen liner*, plus a *gravel pack* to keep sand out (fig. 5.7).

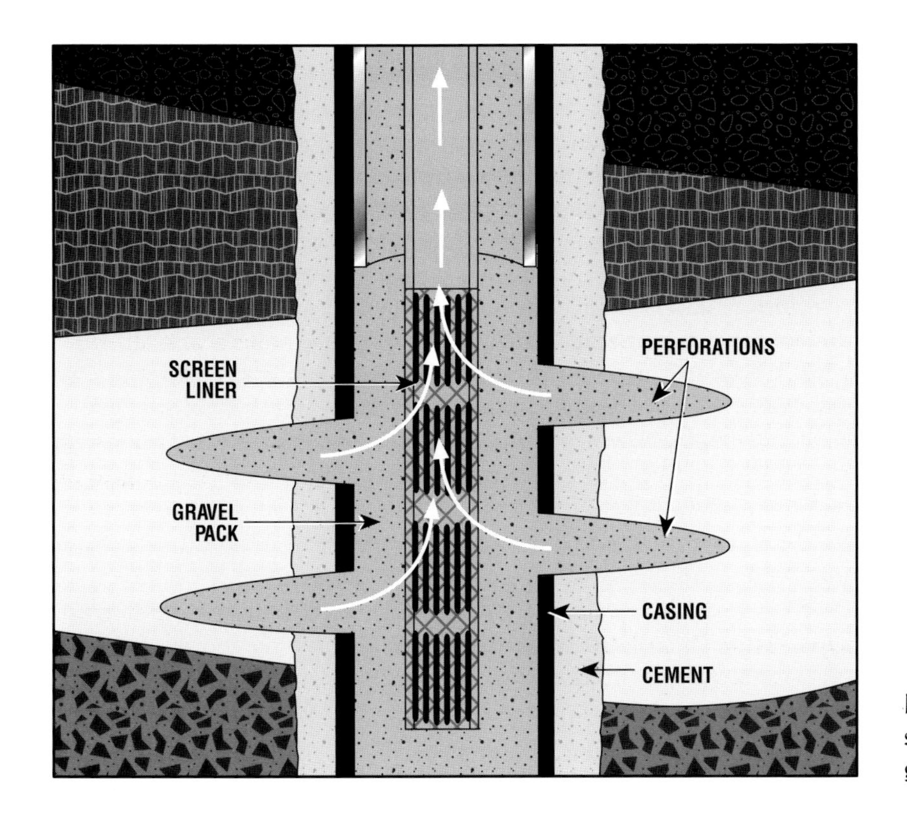

Figure 5.7 A wire-wrapped screen, or screen liner, is often combined with a gravel pack inside perforated casing.

Wire-wrapped screen is a relatively short length of pipe with holes or slots in its sides (the *slotted liner*) and a specially shaped wire wrapped around it (fig. 5.8). A gravel pack is made up of a fine gravel, which is actually coarse sand, that filters out the finer sand in the formation. To place the gravel pack, first the crew circulates a fluid into the hole to wash out sand at the bottom. Then the crew pumps the gravel down the tubing in a viscous (thick) fluid. Finally, the crew runs in the wire-wrapped screen.

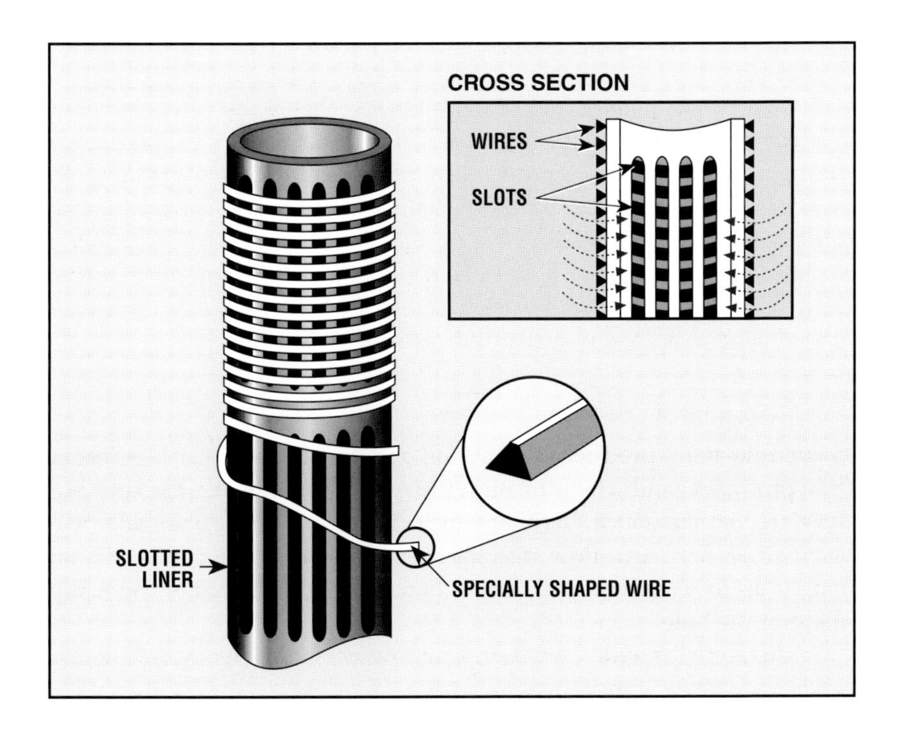

Figure 5.8 The slotted liner is a pipe with holes in it with specially shaped wire wrapped around it. Together the slotted liner and wire make up the wire-wrapped screen.

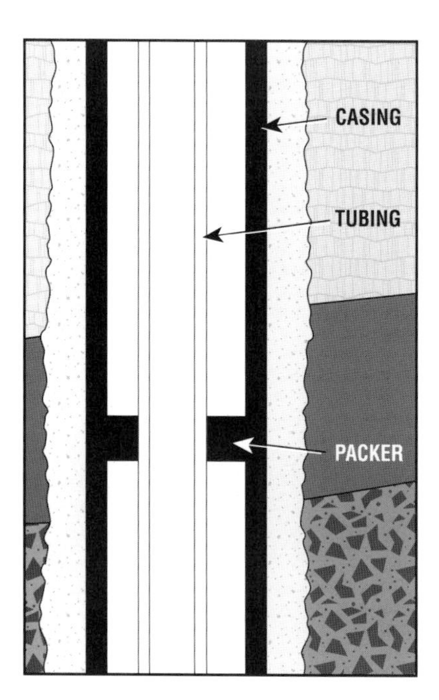

Figure 5.9 Tubing is smaller in diameter than casing.

Tubing and Packers

After cementing the production casing, the completion crew runs a final string of pipe called the *tubing* (fig. 5.9). The well fluids flow from the reservoir to the surface through the tubing. Unlike casing, tubing is not cemented in but hangs from the wellhead at the surface. Tubing is smaller in diameter than casing—the outside diameter ranges from about 1 to 4½ inches (about 25 to 114 millimetres).

Rig operators nearly always run tubing in a well because oil and gas produce better through small-diameter tubing than through large-diameter casing, the way a river flows faster through a narrow channel than through a wide bed. Also, servicing crews can remove tubing if it becomes plugged or damaged. They cannot easily remove casing because it is cemented in the well.

Packers

A *packer* is a ring made of metal and rubber that fits around the tubing. It provides a secure seal between everything above and below where it is set (fig. 5.10). It keeps well fluids and pressure away from the casing above it. Since the packer seals off the space between the tubing and the casing, it forces the formation fluids into and up the tubing.

One reason to isolate the casing is to keep sand and sediment in the produced fluids from eroding it, in the same way sandblasting etches glass. Sulfur and other naturally occurring chemicals in the reservoir fluids are another enemy of casing. They cause the metal to corrode. Tubing can erode and corrode too, but it can be pulled and repaired or replaced, since it is not cemented in the hole. Also, corrosion-resistant tubing is cheaper than corrosion-resistant casing because tubing is smaller.

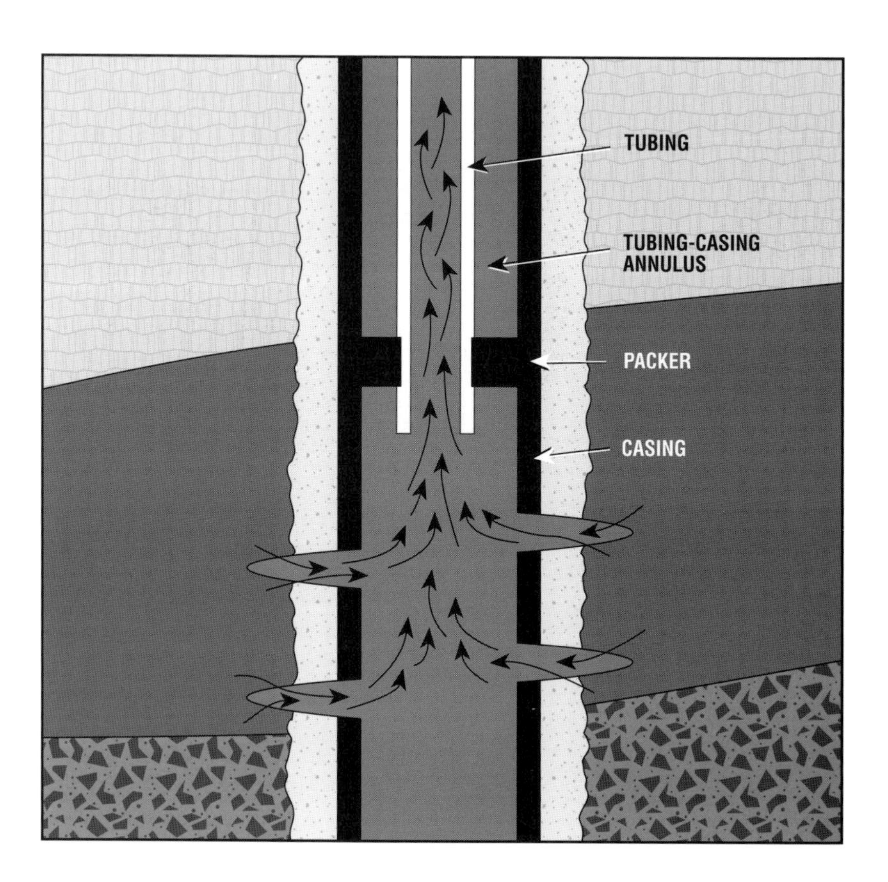

Figure 5.10 A packer between the casing and tubing keeps well fluids out of the tubing-casing annulus.

The sealing element of a packer, the *packing element*, is a dense synthetic rubber ring that expands against the side of the casing. A packer may have one packing element or several separated by metal rings. *Slips* hold the packers in place. A slip is a serrated piece of metal that grips the side of the casing (fig. 5.11). Most packers also have a circulation valve that allows fluids to pass through it. Although most of the time a packer must seal fluids off from the annulus, sometimes the crew opens the valve and uses a circulating fluid to set or retrieve the packer. Retrievable and permanent packers are the two main types.

Figure 5.11 The slips grip the walls of the casing to hold the packer in place.

Subsurface Safety Valve

Another device frequently installed in the tubing string near the surface is a *subsurface safety valve*. The valve remains open as long as fluid flow is normal. When the valve senses something amiss with the surface equipment of the well, it closes, preventing the flow of fluids.

Tubingless Completion

Although an operator will usually complete a well with tubing, a small-diameter well that uses small-diameter casing may be completed without the tubing; this is a *tubingless completion*. The casing is cemented and perforated opposite the producing zones (fig. 5.12). Tubingless completions are used mostly in small gas reservoirs that produce few, if any, liquids and are low in pressure.

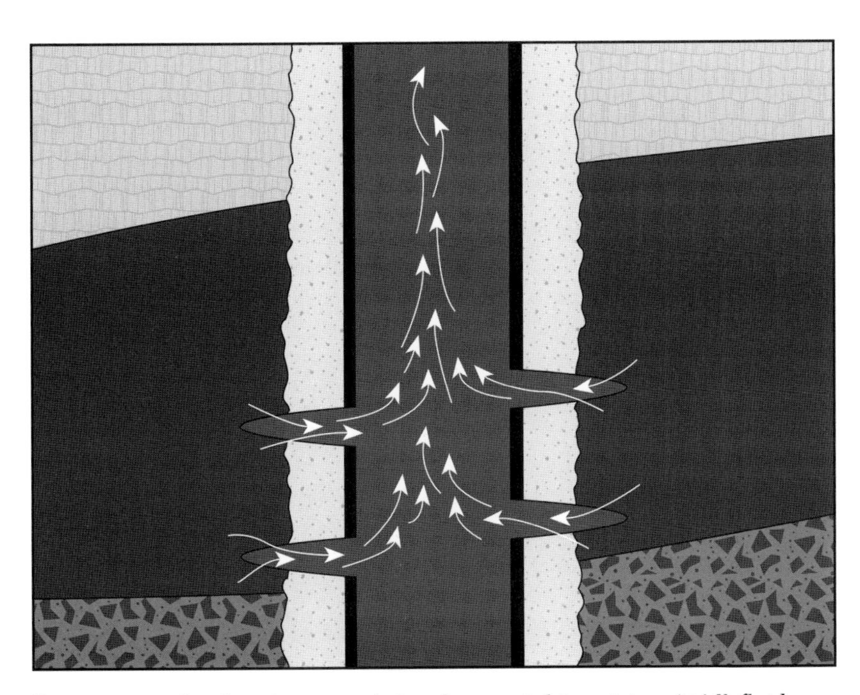

Figure 5.12 A tubingless completion has no tubing string. Well fluids are produced through small-diameter casing.

Multiple Completions

The operator uses a *multiple completion* when one wellbore passes through two or more zones with oil and gas in them (fig. 5.13). Usually, a separate tubing string with packers is run in for each producing zone. For example, a triple completion has three tubing strings and three packers, and each zone produces independently of the others.

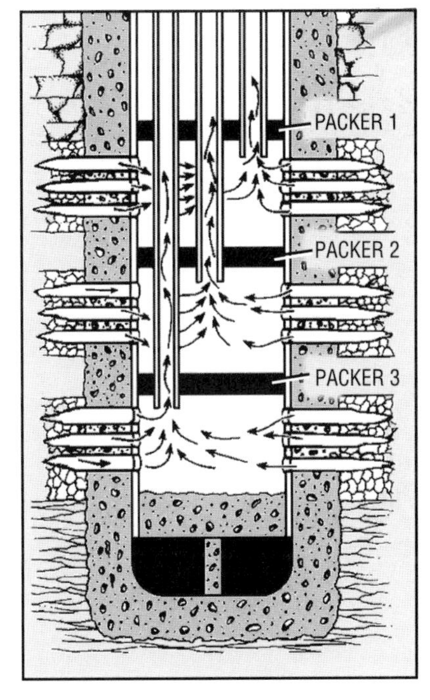

Figure 5.13 A multiple completion, such as this triple, usually has a separate tubing string and packer for each producing zone. Also, the casing is perforated opposite each producing zone.

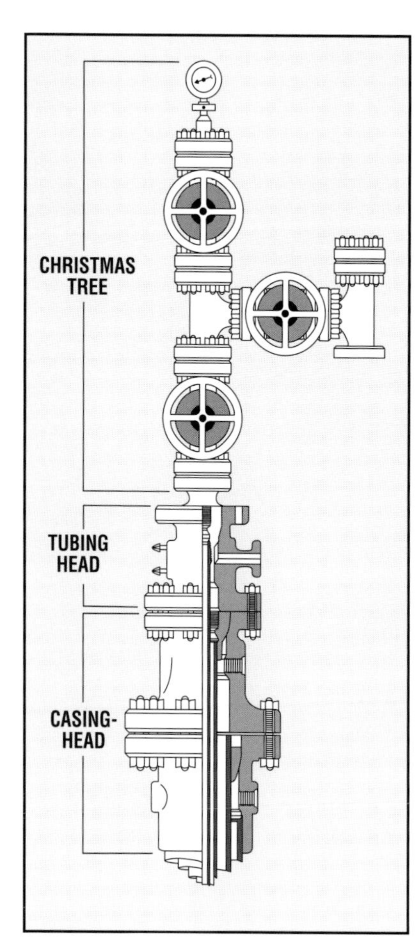

Figure 5.14 A wellhead usually has one or more casingheads, a tubing head, and a Christmas tree.

Wellhead

The *wellhead* includes all equipment on the surface that supports the various pipe strings, seals off the well, and controls the paths and flow rates of reservoir fluids. The operator determines the type of wellhead according to the conditions in the individual well. Sometimes, a simple assembly to support the weight of the tubing in the well is enough. Other wells may need a high-pressure wellhead to control formation pressure. Pressures greater than 20,000 pounds per square inch (140 megapascals) have been found in some fields.

All wellheads have at least one casinghead and casing hanger, usually a tubing head and tubing hanger, and a Christmas tree (fig. 5.14).

Casinghead

Each string of casing usually hangs from a *casinghead*, a heavy steel fitting at the surface. Metal and rubber seals in the casinghead prevent fluids from moving within the wellhead or escaping to the atmosphere. Each casinghead also has a place for a pressure gauge to warn of leaks.

Tubing Head

Similar in design and use to the casinghead, the *tubing head* supports the tubing string, seals off pressure between the casing and the inside of tubing, and provides connections at the surface to control the flowing liquid or gas. The tubing head often stacks above the uppermost casinghead. Like the casingheads, it has outlets to allow access to the annulus for gauging pressure or connecting valves and fittings to control the flow of fluids.

Christmas Tree

High-pressure wells usually have a group of valves and fittings called a *Christmas tree* (fig. 5.15), a name derived from its treelike appearance.

Figure 5.15 Valves on the Christmas tree control the flow of fluids from the well.

Low-pressure wells may also have Christmas trees. These trees are less complex than high-pressure ones but serve the same purpose: to regulate, measure, and direct the flow of fluids from the well.

Pressure gauges on the Christmas trees measure pressure both in the casing and in the tubing. Rig operators have more control over the well if they know the pressures under various operating conditions. For example, pressure gauges can warn of leaks.

Valves on the tree work like those on an outdoor water faucet. They can be opened and closed to control the flow of fluids from the well. The main valve is the *master valve* just above the tubing head. Opening the master valve allows the fluids to flow to the rest of the tree, and closing it shuts off the flow of reservoir fluids entirely.

Another important part of a Christmas tree is the *choke*. The choke restricts the line through which well fluids enter the tree. The size of the choke's opening may be fixed or adjustable. A fixed choke must be replaced to change the well's producing rate. The opening of an adjustable choke changes size when a worker turns a control handle.

Starting the Flow

As a well is drilled, pressure in the production zone is offset by the pressure of the drilling fluid in the hole. The drilling crew leaves this heavy mud inside the casing after drilling is complete so that its pressure will continue to offset the formation pressure after the casing is perforated. The completion crew usually runs production tubing into the well while the drilling fluid is still in place.

Washing In

To start the flow of formation fluids, the crew displaces the mud in the tubing by pumping salt water, or brine, into it, called *washing in*. They then set the packer. This results in a column of water holding back the formation pressure. Water exerts less pressure than drilling mud, so there may be enough pressure in the formation to overcome it and start the oil or gas flowing through the perforations into the well and up the tubing. Generally, almost pure gas or oil will follow soon after all the water is out. Until the production is pure, the mixture of oil, gas, and water goes into a tank for disposal later.

Swabbing

If reservoir pressure is not high enough to push the hydrocarbons up and out of the well, *swabbing* may be necessary to *unload* the well. Unloading a well means to remove some of the brine pumped in. Less brine means less pressure for the formation pressure to overcome, so often the well will begin flowing after it is swabbed.

Other Means

Another means of starting production flow is to force high-pressure gas into the tubing before setting the packer. The gas pushes the brine in the tubing down, out into the annulus between the tubing and casing, then up. The packer is then set, leaving only a short column of fluid in the tubing holding pressure on the formation. The well will often start flowing immediately.

Sometimes the well does not flow at a fast enough rate, given the known amount of hydrocarbons in the reservoir, after these completion operations. The company representative may decide to stimulate the reservoir to make the well produce at an acceptable rate. *Stimulation* is one of several processes that enlarge or create channels in the reservoir rock so that the oil and gas can move through it and into the well. A section later in this chapter describes these processes.

If a well does not flow on its own after washing in, swabbing, and stimulation, it needs an artificial means of lifting oil to the surface, usually some sort of pump. A later section describes the various means of artificial lift.

When hydrocarbons do flow out of the well, they go through a pipe called the *flow line* to a storage tank and gas separators. From there separate gas and oil pipelines transport them, sometimes many miles or kilometres for refining and sale.

Completing Gas Wells

Gas wells are generally completed in the same way as oil-producing wells except that they never require artificial lift—natural gas always flows without help. Pressure at the surface is usually higher in a gas well than in an oil producer, but if the pressure is low, swabbing off water in the tubing may be necessary.

RESERVOIR DRIVE MECHANISMS

After the well has been completed, gas and oil begin their journey from the reservoir to the surface. This first period in the producing life of a reservoir is called *primary recovery*, or *primary production*. During this stage, natural energy in the reservoir often displaces the hydrocarbons from the pores of a formation and drives it toward the wells and up to the surface. *Reservoir drive mechanisms*, as the natural energies are called, include dissolved-gas drive, gas-cap drive, water drive, combination drive, and gravity drainage.

In order of importance, the three natural forces that move the fluids in a reservoir are water drive, gas drives, and gravity drainage.

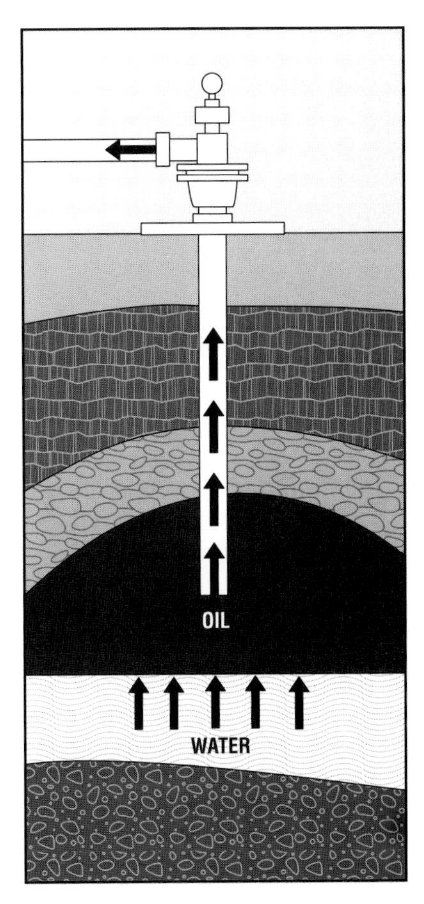

Figure 5.16 In a water drive reservoir, salt water under the oil pushes the oil to the surface.

Water Drive

Water drive occurs when there is enough energy available from free water in the reservoir to move hydrocarbons out of the reservoir, into the wellbore, and up to the surface (fig. 5.16). Water located beneath the oil in a reservoir is under pressure proportional to the depth beneath the surface; in other words, the deeper the water, the higher the pressure. As soon as there is an opening (the well) for the oil to move into, normal water pressure forces the oil into it. Water is quite efficient at displacing oil from reservoir rock, moving into the pores where oil has moved out. The process is similar to displacing oil in a tank by adding water at the bottom and letting the oil spill over the top.

The water pressure remains high as long as an equal volume of water replaces the volume of oil withdrawn. Eventually, however, the water level rises to the point where the well is producing mostly water (fig. 5.17). When this happens, one solution is for a workover crew to seal off the bottom of the well and perforate the casing higher up so that the rest of the oil will flow out.

Water-drive reservoirs can have bottom-water drive or edgewater drive. In a *bottom-water drive* reservoir, all of the oil accumulation has water under it (see fig. 5.17). A well drilled anywhere through a reservoir of this type penetrates oil first and then water. In an *edgewater drive* reservoir, oil almost completely fills the reservoir. Water occurs only on the edges of the reservoir, so only wells drilled along the edges penetrate water (fig. 5.18). Wells drilled near the top of the structure penetrate oil only.

Water drive is the most efficient natural drive. Sometimes 50 percent or more of the oil in the reservoir will flow out due only to water drive.

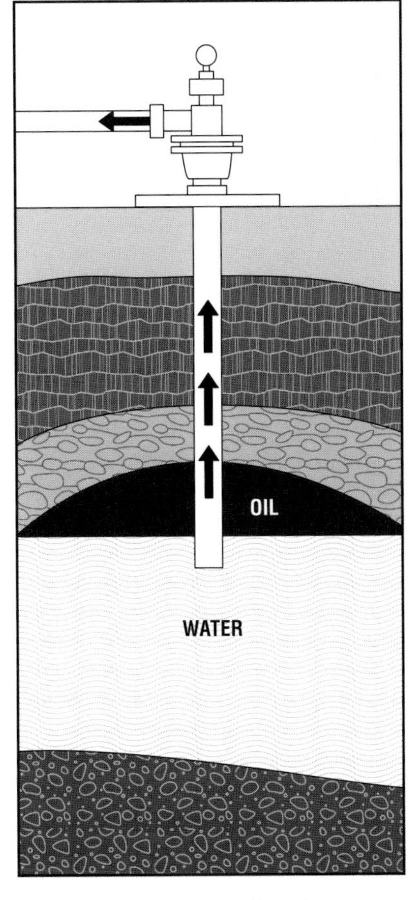

Figure 5.17 Eventually, the water level rises to the bottom of the well in a water drive reservoir.

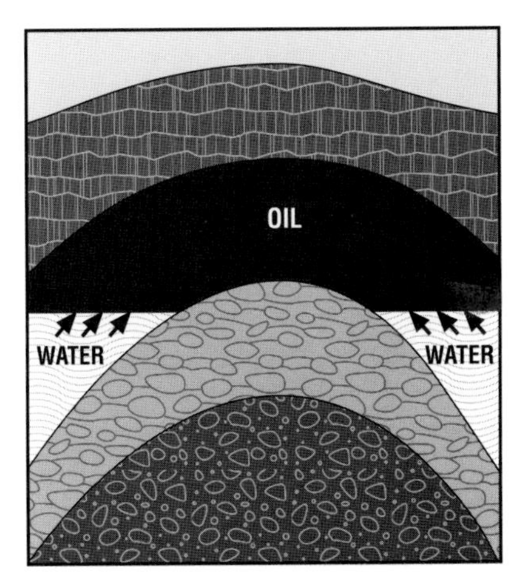

Figure 5.18 Water at the edges of the reservoir helps drive the oil into the wellbore.

Figure 5.19 In a dissolved-gas drive reservoir, gas comes out of the oil, expands, and lifts it to the surface.

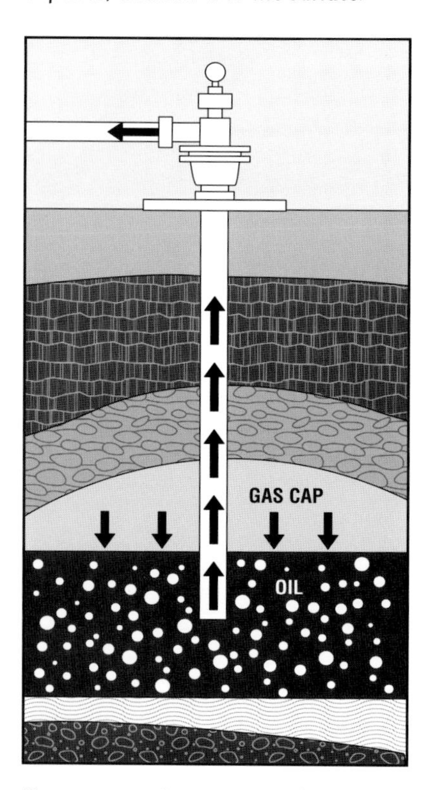

Figure 5.20 In a gas-cap drive reservoir, the free gas in the cap and the gas dissolved in the oil expand to move oil up to the surface.

Gas Drives

The two types of gas drives are dissolved-gas drives and gas-cap drives. They are also called *depletion drives* because when the gas is gone, or depleted, the pressure that drives the oil out is gone. When the drive gives out, a service crew must add a pump or some other sort of artificial lift to the well to supplement the natural drive.

Both types of gas drives work because the pressure of a gas is related to its volume, or the space it fills. A compressed gas, like the air in a tire, has a high pressure. In other words, the air presses out against the walls of the tire with greater and greater strength as more air is forced into the space inside the tire. When the air is released to the outside of the tire, it expands into the atmosphere and loses pressure; the tire goes flat.

Dissolved-Gas Drive

A *dissolved-gas drive* works because some of the hydrocarbons in the oil are light enough that they become gaseous when the well releases pressure from the reservoir (fig. 5.19). This gas is like the bubbles in a soft drink. If you shake up a bottle of soda with the cap on, nothing happens. But when you uncap it, the carbon dioxide dissolved in the liquid becomes gaseous and the liquid and gas foam up out of the bottle. Opening up the wellbore is like uncapping a soda bottle. The lighter hydrocarbons turn into gases and the oil and gas flow up to the surface. Just as the carbon dioxide gas in a soda bottle goes away quickly after you uncap it, a dissolved-gas drive usually depletes quickly. For this reason such a well will require some form of artificial lift at an early stage.

The amount of oil recovered varies from 5 to 30 percent. Usually, dissolved gas-driven wells produce little or no water.

Gas-Cap Drive

In some reservoirs, not all gas is dissolved in the oil. Instead, it forms a cap on top of the oil. A reservoir that has a gas cap will also have *gas-cap drive.* When the wellbore opens an escape route for the oil in the reservoir, the pressure of the compressed natural gas in the gas cap pushes the oil into it. As the level of oil in the reservoir drops, the gas cap expands and continues to push the oil into the well and up to the surface (fig. 5.20). The more space the oil leaves empty in the porous reservoir rock, the more the gas expands to take its place. Gradually, the gas loses pressure. This drive works like a can of spray paint. A compressed, or pressurized, gas in the can sprays out a gas and paint mixture when the valve on top is opened. Eventually, the pressure of the gas inside the can is no greater than the pressure of the air outside it, and it can no longer push out any paint left in the can.

The pressure of a gas-cap drive depletes more slowly than a dissolved-gas drive. From 20 to 40 percent of the oil in the reservoir may flow out before a gas-cap drive fails.

Combination Drives

More than one drive can work in a reservoir at the same time, and this is called a *combination drive.* One type of combination drive occurs when the oil has a gas cap above it and water below it. Both the gas cap and water drive the oil into the well and up to the surface. Another combination is dissolved-gas drive plus water drive.

Gravity Drainage

The least common type of reservoir drive is *gravity drainage*. The force of gravity is, of course, always at work in a reservoir. Usually, gravity causes oil to migrate upward, until it runs into an impenetrable formation, by pulling the water down beneath it. This happens because water is heavier than oil. However, in shallow, highly permeable, steeply dipping reservoirs and in some deeper, nearly depleted reservoirs, the oil may flow downhill to the wellbore (fig. 5.21).

Figure 5.21 In a gravity drainage reservoir, the oil flows downhill to the well.

ARTIFICIAL LIFT

After the tubing has been run in, the packer set, and the well perforated, the hydrocarbons usually flow to the surface immediately or after swabbing. When the pressure from natural reservoir drive falls to the point where the well cannot produce on its own, an artificial method of lifting the hydrocarbons is necessary. The most common ways of providing artificial lift are some sort of pump and a method that involves injecting a gas into the well.

Beam Pumping

By far the most common method of pumping oil from the formation to the surface in land-based wells is beam pumping. Beam pumping consists of surface and subsurface equipment. The *beam pumping unit* sits on the surface (fig. 5.22). It sends an up-and-down motion, called *reciprocating action*, to a string of rods called *sucker rods* (fig. 5.23). Sucker rods are solid, high-strength steel (or sometimes fiberglass) rods connected together. The top of the sucker rod string is attached to the front of the pumping unit, usually to a *walking beam*, and hangs down inside the tubing. At the end of the string, near the bottom of the well, is a *sucker rod pump*. The walking beam's reciprocating action moves the rod string up and down to operate the pump.

Figure 5.22 Beam pumping units like this one are a common sight in oil country.

Electric Submersible Pumps

In many fields, older wells that have been using beam pumping units gradually begin producing much more water than oil. In order to recover enough oil to be profitable, tremendous volumes of fluid have to be lifted from the well. For this reason, electric *submersible pumps* have become popular because the rig operator can stack as many pumps as needed in the well. Both the pump motors and the pumps are downhole, submerged in the well fluid, on the end of the tubing string (fig. 5.24). A protector between the pump and motor seals the well fluids away from the motor and buffers the pump

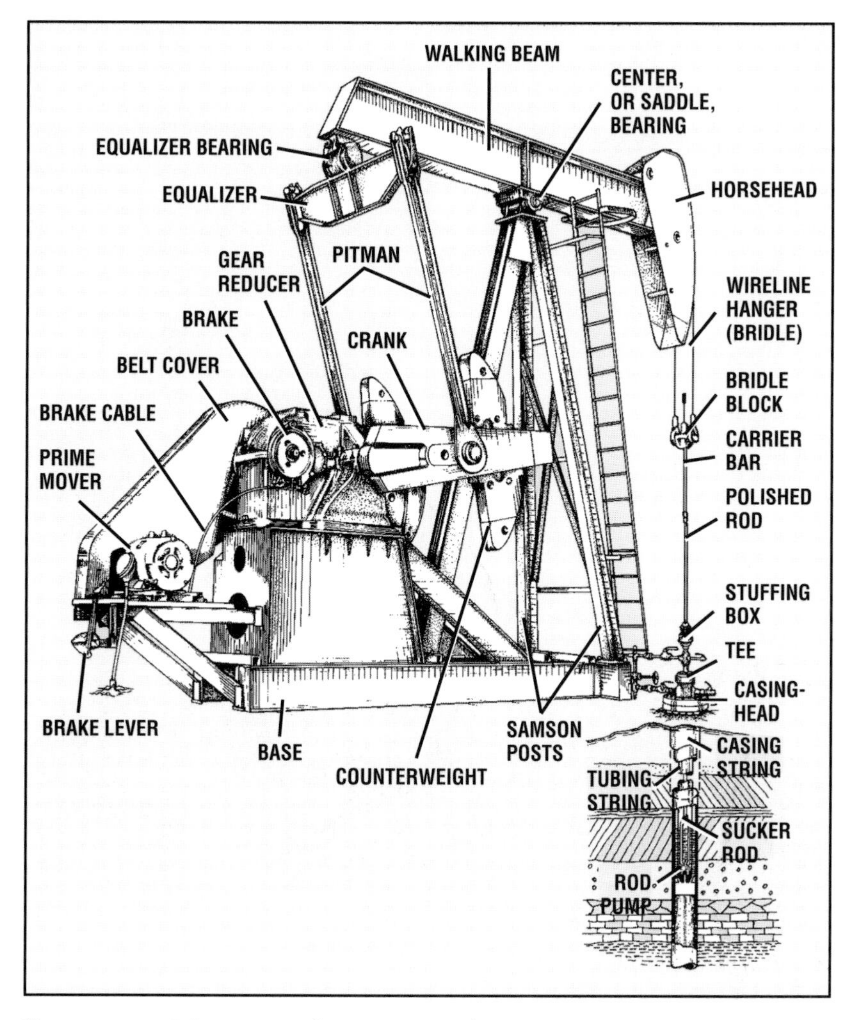

Figure 5.23 A beam, or rod pumping unit has many components.

from the movement of the motor. A special heavy-duty armored cable supplies electricity to the unit. The main disadvantage of submersible pumps is that they are sensitive to sand and gas.

Subsurface Hydraulic Pumps

A third pumping system is hydraulic pumping. *Hydraulic pumping* is similar to beam pumping in that an engine on the surface powers a pump in the hole, but it uses hydraulic energy, or the energy of a flowing liquid, to make the pump work instead of sucker rods. Because the hydraulic pump does not use sucker rods, it is less complicated to service than the mechanical system, but the pump wears out easily due to erosion. Another advantage is that hydraulic pumps can work in deeper wells than sucker rod pumps.

A hydraulic system actually has two pumps: an ordinary electric or engine-driven pump on the surface to force a liquid, called the *power fluid*, down the hole, and a hydraulic pump at the bottom of the well that the power fluid runs (fig. 5.25).

One type of hydraulic system recycles crude oil pumped from the reservoir into a settling tank on the surface as the power fluid. Another type is a closed system that uses treated water as the power fluid, which is kept separated from the produced oil.

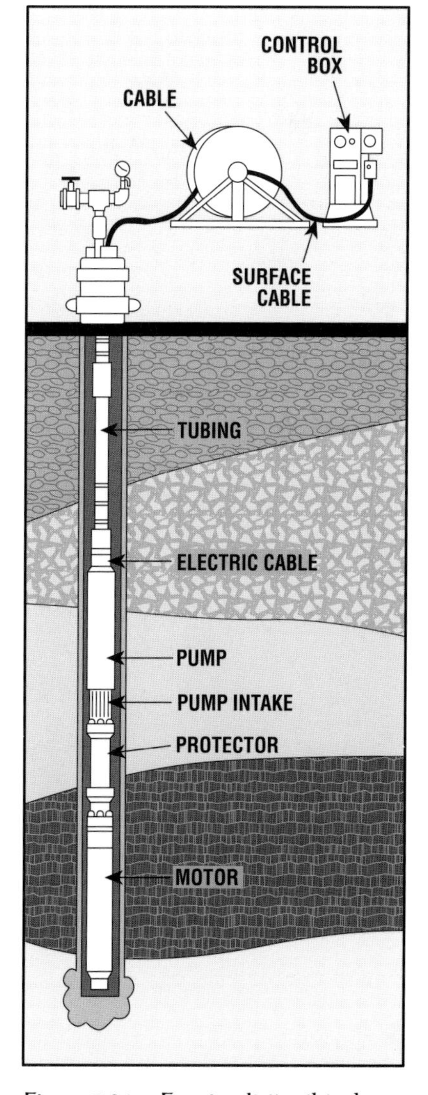

Figure 5.24 For simplicity, this drawing shows one electric submersible pump, although a single well usually has at least 15 to 20 pumps stacked together.

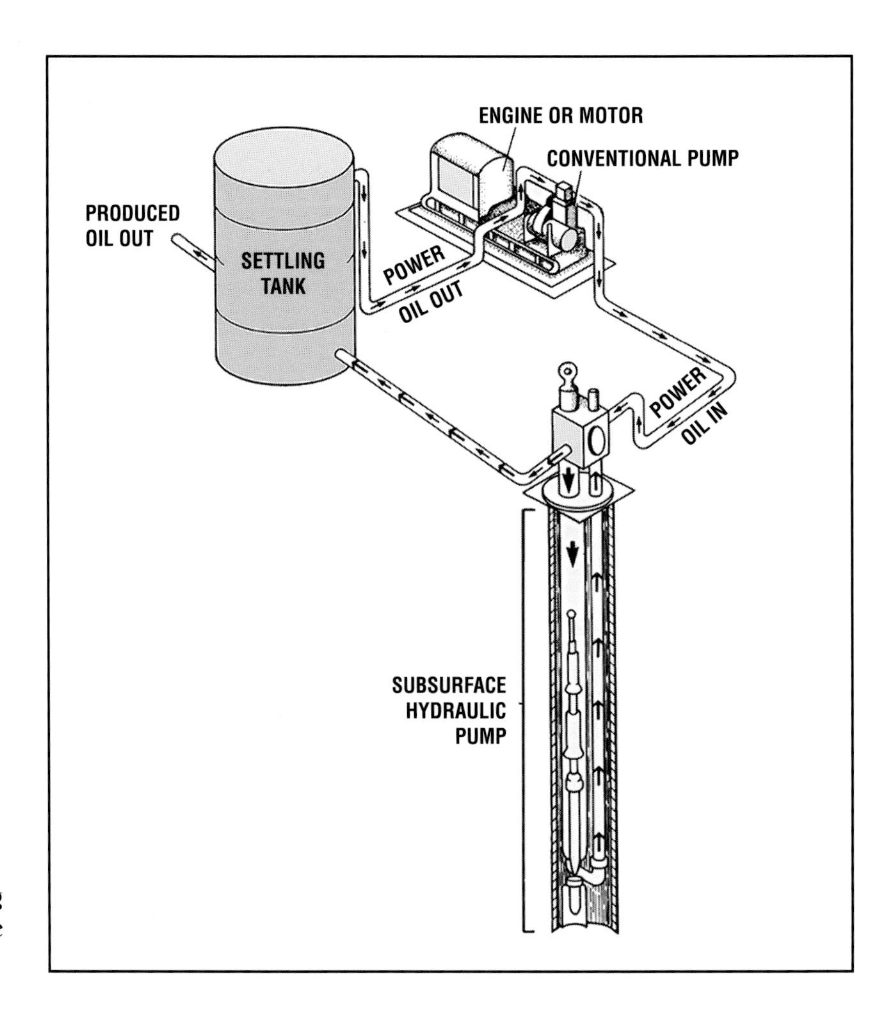

Figure 5.25 A hydraulic pumping system has components on the surface and in the well.

Gas Lift

The term *gas lift* covers a variety of methods by which a gas is used to increase the production of a well or to restore production to a well that has stopped flowing—a *dead well*. When a well flows on its own, it may be using a natural form of gas lift—dissolved-gas drive or gas-cap drive. Natural gas can also be injected into the well to lift the oil artificially on the same principle. The completion crew installs *gas-lift valves* on the production tubing. The valves allow a gas to be forced into the annulus to pass into the tubing and into the liquid there.

To understand the principle of gas lift, recall that gas is lighter than oil and water. For example, 1 gallon (1 cubic metre) of water weighs a little over 8 pounds (960 kilograms), while 1 gallon (1 cubic metre) of natural gas weighs only a fraction of an ounce. When the natural gas enters the liquid in the tubing, the gas makes this liquid column much lighter in weight. Since the liquid column is lighter, it exerts less pressure on the bottom of the well. With the pressure lower at the bottom, the pressure remaining in the reservoir becomes sufficient to push reservoir fluids to the surface through the tubing. Gas may be injected continuously or intermittently, depending on the producing characteristics of the well and the arrangement of the gas-lift equipment.

Gas lift is common when a supply of gas is economical and available and when the amount of petroleum it will lift justifies the expense. Gas lift is especially suitable for offshore use because platform space is limited and gas-lift equipment is largely downhole.

WELL TESTING

Production tests help determine how much and how fast a well will produce. The company representative may run production tests before or after setting casing, cementing, and perforating. Each test reveals certain information about a particular well and its reservoir. Accuracy is, of course, very important, and the data from these tests form the case history of a well.

Potential Test

One of the most frequently performed production tests after completion is the *potential test.* A potential test measures the largest amount of gas and oil a well can produce over a 24-hour period under certain fixed conditions. Basically, the test involves allowing the well to produce for a given period of time and accurately measuring the production. Lease operators perform potential tests both when first producing the well and again several times during its life. State regulatory agencies usually require potential test information to establish the allowable production for the well.

Bottomhole Pressure Test

A *bottomhole pressure test* uses a gauge to measure the pressure at or near the bottom of the well. The test is usually conducted after the well has been shut in for 24 to 48 hours. When an operator schedules this test at certain intervals, it reveals valuable information about the decline or depletion of the zone in which the well is producing. For example, pressure measurements can tell the company representative the most efficient rate of flow for the well.

Productivity Test

A *productivity test* is a combination of a potential test and a bottomhole pressure test. It determines the effects of different flow rates on the pressure in the well. This helps the company representative decide the rate at which the well should be produced for the best results. Producing at the maximum flow rate may deplete the drive too quickly or damage the well.

The procedure for the test is first to measure the closed-in bottomhole pressure of the well and then to measure the flowing bottomhole pressure at several stabilized rates of flow. This type of testing is done on both oil and gas wells and is the most widely accepted method of determining the capacity of gas wells. In many states, the regulation of gas production is based upon a productivity test.

Wireline Formation Test

A *wireline formation test* measures the pressures at specific depths. The tester is a tool that the crew runs into the well on a conductor line. The tool perforates one or two holes in the casing. A valve opens a testing chamber to let in fluid and record its pressure. Finally, a sample chamber opens and draws in a few gallons or litres of formation fluid to bring to the surface. This test is good for quick readings of pressures, confirming porosity and permeability data from other logs, and predicting general productivity.

WELL STIMULATION

The term *well stimulation* encompasses several techniques used to enlarge old channels or to create new ones in the producing formation. Since oil usually exists in the pores of sandstone or the cracks of limestone formations, enlarging or creating new channels causes the oil or gas to move more readily to a well. Sometimes the problem is low permeability. In this case, the well will be stimulated immediately after completion. In other cases, the natural permeability of the rock may be adequate, but the formation near the wellbore may be damaged in a way that restricts the flow channels in porous rock. Formation damage can occur during drilling, completion, workover, production, or injection.

There are three ways to do this. The first and oldest method is to use *explosive fracturing*. During the 1930s, *acid stimulation*, or *acidizing*, became commercially available. *Hydraulic fracturing*, the third stimulation method, was introduced in 1948.

Explosives

As early as the 1860s, crews exploded nitroglycerin inside wells to improve their productivity. They simply lowered a nitro charge into the open hole on a conductor line and detonated it to fracture the formation. *Nitro shooting* became fairly routine until the advent of acidizing and hydraulic fracturing. For a time in the 1960s, lease operators experimented with nuclear explosives in a limited number of gas wells. While this method increased production somewhat, the cost was prohibitive.

Oil companies are still interested in explosive techniques because certain kinds of tight formations do not respond readily to either acidizing or hydraulic fracturing. Research continues today to find other techniques that might increase production, but currently fracturing and acidizing are the most effective well stimulation methods.

Hydraulic Fracturing

Hydraulic fracturing is all about pressure. Several powerful pumps (fig. 5.26) inject a liquid, the *fracturing fluid*, into the well at a fast rate. The fluid develops a high pressure that actually splits, or fractures, the rock. To visualize this, imagine splitting a log with an axe. The axe head is a wedge.

Figure 5.26 Several powerful, truck-mounted pumps are arranged at the well site for a fracturing job.

A wedge first cuts a tiny crack in the log that the force of the blow enlarges into a wider cut until the log splits. In fracturing, the fluid acts as a wedge and its high pressure is the force that pushes it into the rock. Hydraulic fracturing splits the rock instead of the casing because the casing is stronger than the rock.

Hydraulic fracturing improves the productivity of a well by either creating new fractures that act as flow channels or extending existing flow channels farther into the formation. Fracturing is a usual part of completion, and refracturing to restore productivity of an old well is a regular procedure. Workover people commonly shorten the word fracturing to *frac*, as in frac job and frac unit. Hydraulic fracturing works well in sandstone reservoirs.

Proppants

During early experimental work, engineers discovered that a hydraulically formed fracture tends to heal, or lose its fluid-carrying capacity, after the parting pressure is released unless the fracture is propped open in some manner. *Proppants,* or propping agents, hold the fractures open. Sand, nutshells, and beads of aluminum, glass, and plastic may be used as proppants (fig. 5.27). Spacer materials are used between the particles of the proppant to ensure its optimum distribution.

Figure 5.27 Sand is one proppant used to hold fractures open.

Fracturing Fluid

The fracturing fluid may be either oil based or water based. In reality, the fluid is nearly always brine because it is safe, available, and cheap. Some fracturing fluids are gels, which suspend the proppants better. Polymer additives reduce friction between the fluid and the walls of the tubing. Although this may not sound as if it would be a factor, any slowing of the fluid due to friction requires larger pumps to keep the injection rate high enough. Finally, additives reduce fluid loss into the formation.

Acidizing

In acid stimulation, or acidizing, an acid reacts chemically with the rock to dissolve it (fig. 5.28). As in hydraulic fracturing, this enlarges existing flow channels and opens new ones to the wellbore. Well servicing crews stimulate both new and old wells with acid. Reservoir rocks most commonly acidized are limestone (calcium carbonate) and dolomite (a mixture of calcium and magnesium carbonates), or *carbonate reservoirs.*

Figure 5.28 Acid enlarges existing channels or makes new ones.

Types of Acids

Acids that are strong enough to dissolve rock are often strong enough to eat away the metal of the pipes and equipment in the well. Acidizing, therefore, always involves a compromise between acid strength and additives to prevent damage to the equipment. Oilfield acids must create reaction products that are soluble; otherwise, solids would precipitate and plug the pore spaces just opened up. Since acidizing uses large volumes of acid, it must be fairly inexpensive.

Workers on acidizing jobs must be trained to handle the acids they use, many of which have dangerous fumes and can burn the skin. Acidizing contractors choose the type of acid based on the formation and the conditions in the well. The choices include hydrochloric, hydrofluoric, acetic, and formic acids.

Additives

Additives are used with oilfield acids for many reasons, but one of the most important is to prevent or delay corrosion—that is, to inhibit the acid from attacking the steel tubing or casing in the well. A *surfactant,* or surface active agent, is another type of additive. It is mixed in small amounts with an acid to make it easier to pump the mixture into the rock formation and to prevent spent acid and oil from forming emulsions. An *emulsion* is a thick mixture like mayonnaise.

Other common additives are *sequestering agents,* which prevent the precipitation of ferric iron during acidizing, and *antisludge agents,* which prevent an acid from reacting with certain types of crude and forming an insoluble sludge that blocks channels or reduces permeability.

Types of Acidizing Treatments

There are two basic kinds of acid stimulation treatments: acid fracturing and matrix acidizing.

Acid fracturing, or *fracture acidizing,* is similar to hydraulic fracturing, with acid as the fluid. Acid fracturing does not require proppants, however, because it does not just force the rock apart, but also eats it away. It is the more widely used treatment for well stimulation with acid. Since most limestone and dolomite formations have very low permeabilities, injecting acid into these formations, even at a moderate pumping rate, usually results in fracturing.

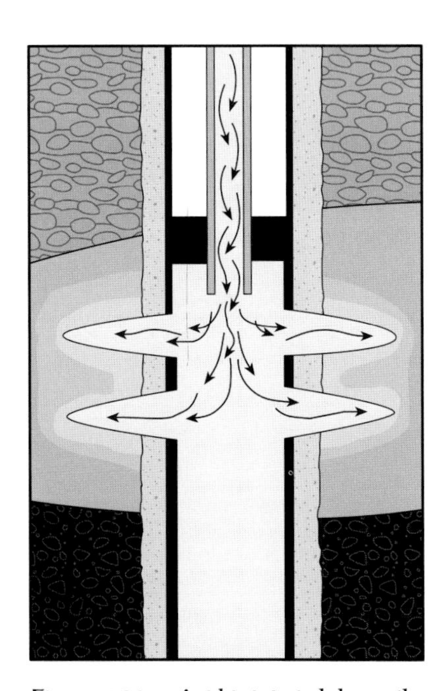

Figure 5.29 Acid is injected down the tubing and into the formation through perforations to remove formation damage without fracturing the formation.

Matrix acidizing can be subdivided into two types. The first is *wellbore cleanup,* or *wellbore soak.* In wellbore soak, the crew fills up the wellbore with acid without any pressure and allows it to react merely by soaking. It is a relatively slow process because little acid actually comes in contact with the formation. The second matrix acidizing method is a low-pressure treatment that does not fracture the formation, but allows the acid to work through the natural pores (fig. 5.29). This second process is what people in the oil patch are usually referring to when they speak of matrix acidizing. Operators generally use matrix acidizing when the formation is damaged or when a water zone or gas cap is nearby and fracturing might result in excessive water or gas production.

IMPROVED RECOVERY TECHNIQUES

After a well has used up the reservoir's natural drives and gas lift or pumps have recovered all the hydrocarbons possible, statistics show that 25 to 95 percent of the oil in the reservoir may remain there. This amount of oil can be worth recovering if prices are high enough. A reservoir may approach the end of its primary production life having produced only a small fraction of the oil in place for many reasons. The operator may have begun production of the reservoir before improved development and production practices were known. Unknown problems such as a casing leak could have resulted in wasted reservoir energy, or the owner might not have been willing to invest more money for maintenance while wells were producing profitably. The petroleum industry has developed several techniques to produce at least part of this remaining oil.

The major methods of improved oil recovery are waterflooding, gas injection, chemical flooding, and thermal recovery. Table 5.1 summarizes each method's process and its main use.

Waterflooding

When the wells drilled into one reservoir stop flowing, the company representative may hire a workover contractor to pump, or inject, water into some of them (fig. 5.30). The wells into which water is pumped become *injection wells*. This water kills the wells and then sweeps into the reservoir and moves some of the oil that remains in the rock toward other wells in the same reservoir, the *producing wells*. The producing wells then pump up the oil and water, often by means of a beam pumping unit. Several injection wells surround each producing well. This workover is called *waterflooding*. Waterflooding is the least expensive and most widely used secondary recovery method.

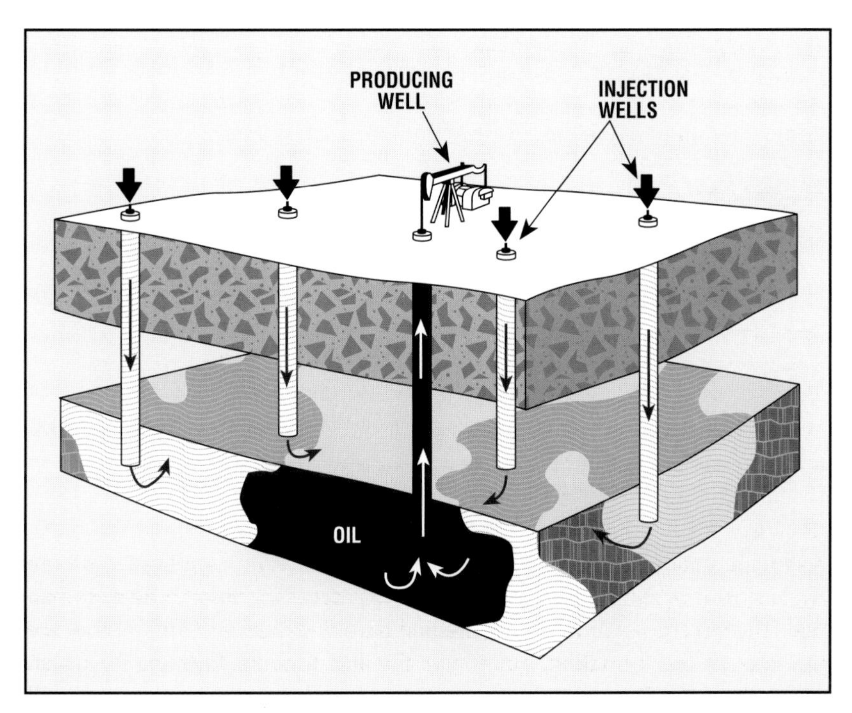

Figure 5.30 In waterflooding, water is injected into wells around the producing well. This is a five-spot pattern—four injection wells and one producer—but many other patterns can be used.

TABLE 5.1
IMPROVED RECOVERY METHODS

Method of Recovery		Process	Use
Waterflooding	Water	Water is pumped into the reservoir through injection wells to force oil toward production wells.	Method most widely used in secondary recovery.
Immiscible Gas Injection	Natural gas, flue gas, nitrogen	Gas is injected to maintain formation pressure, to slow the rate of decline of natural reservoir drives, and sometimes to enhance gravity drainage.	Secondary recovery.
Miscible Gas Injection	Carbon dioxide	Under pressure, carbon dioxide becomes miscible with oil, vaporizes hydrocarbons, and enables oil to flow more freely. Often followed by injection of water.	Secondary recovery or tertiary recovery following waterflooding. Considered especially applicable to West Texas reserves because of carbon dioxide supplies located within a feasible distance.
	Hydrocarbons (propane, high-pressure methane, enriched methane)	Either naturally or under pressure, hydrocarbons are miscible with oil. May be followed by injection of gas or water.	Secondary or tertiary recovery. Supply is limited and price is high because of market demand.
	Nitrogen	Under high pressure, nitrogen can be used to displace oil.	Secondary or tertiary recovery.
Chemical Flooding	Polymer	Water thickened with polymers is used to aid waterflooding by improving fluid-flow patterns.	Used during secondary recovery to aid other processes during tertiary recovery.
	Micellar-polymer (surfactant polymer)	A solution of detergent-like chemicals miscible with oil is injected into the reservoir. Water thickened with polymers may be used to move the solution through the reservoir.	Almost always used during tertiary recovery after secondary recovery by waterflooding.
	Alkaline (caustic)	Less expensive alkaline chemicals are injected and react with certain types of crude oil to form a chemical miscible with oil.	May be used with polymer. Has been used for tertiary recovery after secondary recovery by waterflooding or polymer flooding.
Thermal Recovery	Steam drive	Steam is injected continuously into heavy-oil reservoirs to drive the oil toward production wells.	Primary recovery. Secondary recovery when oil is too viscous for waterflooding. Tertiary recovery after secondary recovery by waterflooding or steam soak.
	Steam soak	Steam is injected into the production well and allowed to spread during a shut-in soak period. The steam heats heavy oil in the surrounding formation and allows it to flow into the well.	Used during primary or secondary production.
	In situ combustion	Part of the oil in the reservoir is set on fire, and compressed air is injected to keep it burning. Gases and heat advance through the formation, moving the oil toward the production wells.	Used with heavy-oil reservoirs during primary recovery when oil is too viscous to flow under normal reservoir conditions. Used with thinner oils during tertiary recovery.

Injection Water

The injected water must be clear, stable, and similar to the water in the formation where it is being injected. It also must not be severely corrosive and must be free of materials that may plug the formation. If the water is severely corrosive, it may contain additives to lessen corrosion, or corrosion-resistant equipment may be used. Deaerating, softening, filtering, chemical treating, stabilizing, and testing are common water-treatment processes.

Waterflooding in the Arctic

The use of waterflooding as an improved recovery method in the Arctic has posed many technological challenges to the industry. The harsh environment, unfriendly to man and machine, has made the development of new methods necessary.

As an example, treated seawater has been injected into the Sadlerochit formation in the Prudhoe Bay of Alaska's North Slope at rates of up to 2 million barrels (318,000 cubic metres) per day. The waterflooding system designed for this region incorporates several special features, including the building of huge water injection pumps and a structure for seawater intake specifically adapted to the shallow Beaufort Sea to minimize the loss of marine life. Such a system must be reliable under harsh conditions, since problems can shut down the system for as long as a year. Equipment breakdown of even a minor sort is expensive in this part of the world because of high labor costs. Also, the presence of permafrost means that water cannot be distributed through underground pipes, the usual practice in most waterfloods operating today. Instead, the Prudhoe Bay system employs specially insulated and protected aboveground pipes for water movement. Special emergency circulating and heating, disposal, evacuation, filling, and cleaning systems keep the water and oil flowing.

Immiscible Gas Injection

Sometimes the crew injects an immiscible gas—one that will not mix with oil—into injection wells in alternating steps with water to improve recovery. Immiscible gases include natural gas produced with the oil, nitrogen, or flue gas. This process is called *immiscible gas injection.*

Immiscible gas injected into the well behaves in a manner similar to that in a gas-cap drive: the gas expands to force additional quantities of oil to the surface. Gas injection requires the use of compressors to raise the pressure of the gas so that it will enter the formation pores.

Miscible Gas Injection

A second type of gas injection uses a gas that is miscible (mixable) with oil to move the oil to the well. When a miscible gas is injected, some of its molecules mix with the oil molecules. It works similarly to the natural dissolved-gas drive. As injection continues, the gas moves part of the oil to the producing well (fig. 5.31).

The petroleum industry first began to use miscible gas injection projects in the 1950s. The injection gases include liquefied petroleum gases (LPGs) such as propane, methane under high pressure, methane enriched with light hydrocarbons, nitrogen under high pressure, and carbon dioxide used alone or followed by water. LPGs are appropriate for use in many reservoirs because they are miscible with crude oil on first contact. However, LPGs are in

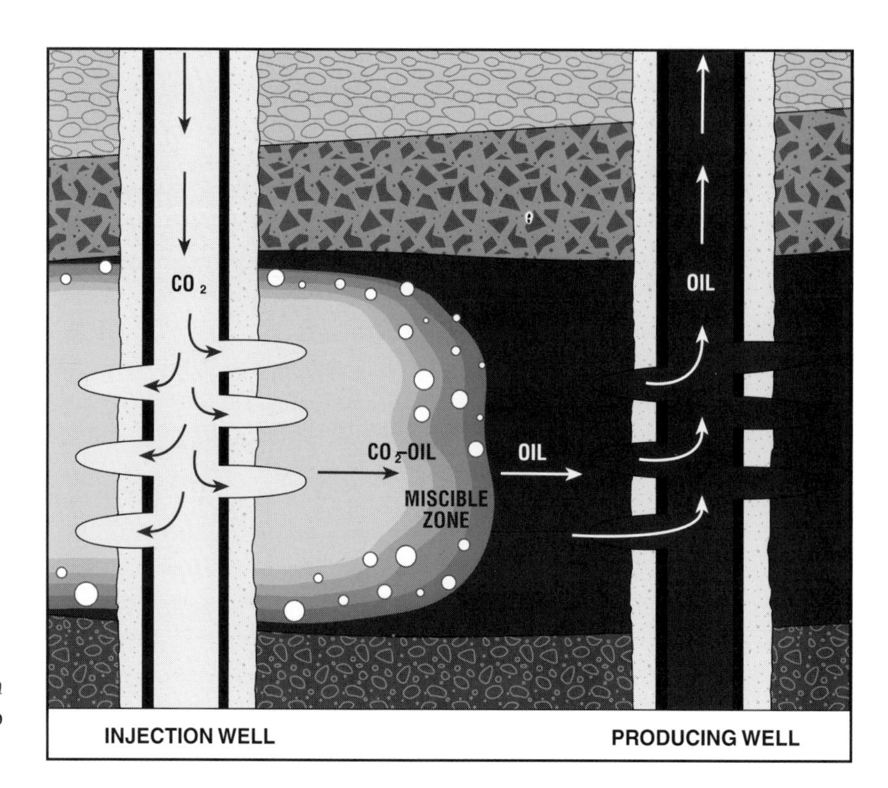

Figure 5.31 One type of gas injection involves injecting carbon dioxide into the reservoir.

such demand as a marketable commodity in their own right that their use in improved recovery is limited. In the 1970s carbon dioxide began to be used frequently as an injection gas. Carbon dioxide has a greater viscosity under pressure than many other gases and displaces oil at low pressures. In areas with a ready supply of carbon dioxide, it is a very attractive choice because it costs less than the LPGs and methane.

Chemical Flooding

Chemical flooding is a general term for injection processes that use special chemicals in water to push oil out of the formation. The principle is to add a surfactant to the injection water. Ordinary dishwashing detergent contains surfactants. A surfactant molecule is attracted to both oil and water, so it reduces surface tension and breaks up oil into tiny droplets that can be drawn from rock pores by water—in other words, the surfactant causes the oil and water to mix.

In one type of chemical flooding, *micellar-polymer flooding*, the workover crew injects a batch, or *slug*, of water containing the surfactant (fig. 5.32). Then the crew mixes a special chemical called a *polymer* with the water and injects it into the well behind the surfactant. Polymers increase water viscosity, decrease effective rock permeability, and are able to change their viscosity with the flow rate. Small amounts of polymers dissolved in water increase the viscosity of water. This higher viscosity slows the progress of the water through a reservoir and makes it less likely to bypass the oil in low-permeability rock.

In *alkaline*, or *caustic, flooding*, the crew injects an alkaline, or caustic, solution. The alkaline solution reacts with natural acids present in certain crude oils to form surfactants within the reservoir. The surfactant formed in the reservoir works in the same way as an injected surfactant to move additional amounts of oil to the producing well.

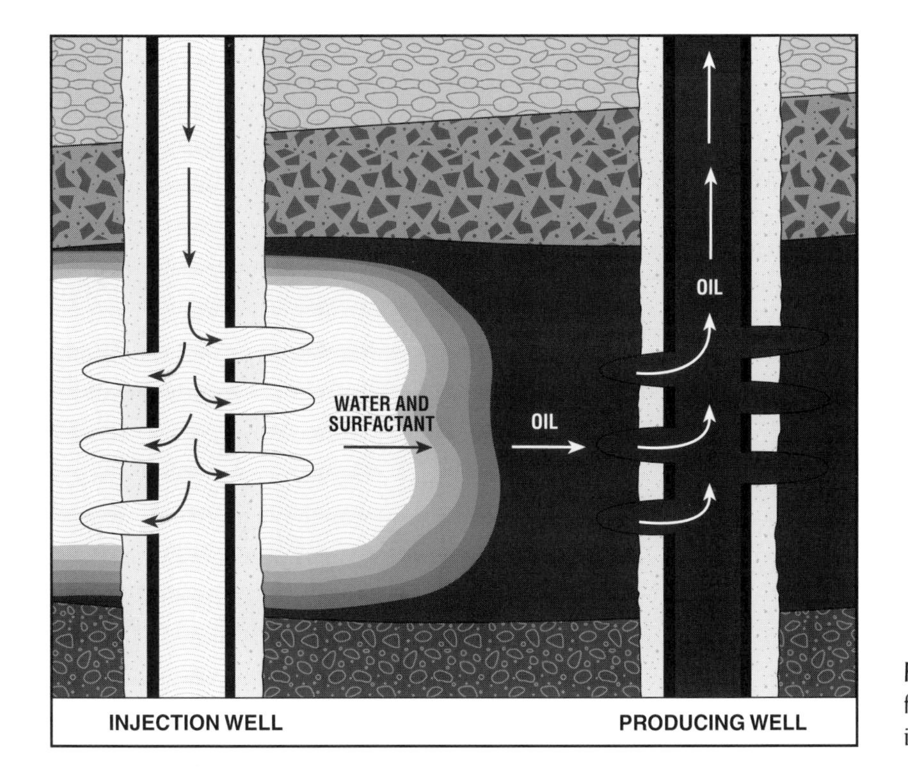

Figure 5.32 In one type of chemical flooding, water containing a surfactant is injected into the reservoir.

Thermal Recovery

Oil in some reservoirs is so viscous, or thick, that it cannot flow through the reservoir and into a well. Just as tar or other solid materials can be made to flow by heating them, so can some viscous oils. Recovery techniques that use heat are called *thermal processes* or *thermal recovery*. A hot fluid is pumped into the injection wells and moved toward the producing wells.

Steam Drive

Steam injection, also known as *steam drive* or *continuous steam injection,* involves generating steam on the surface and forcing this steam down injection wells and into the reservoir (fig. 5.33). When the steam enters the

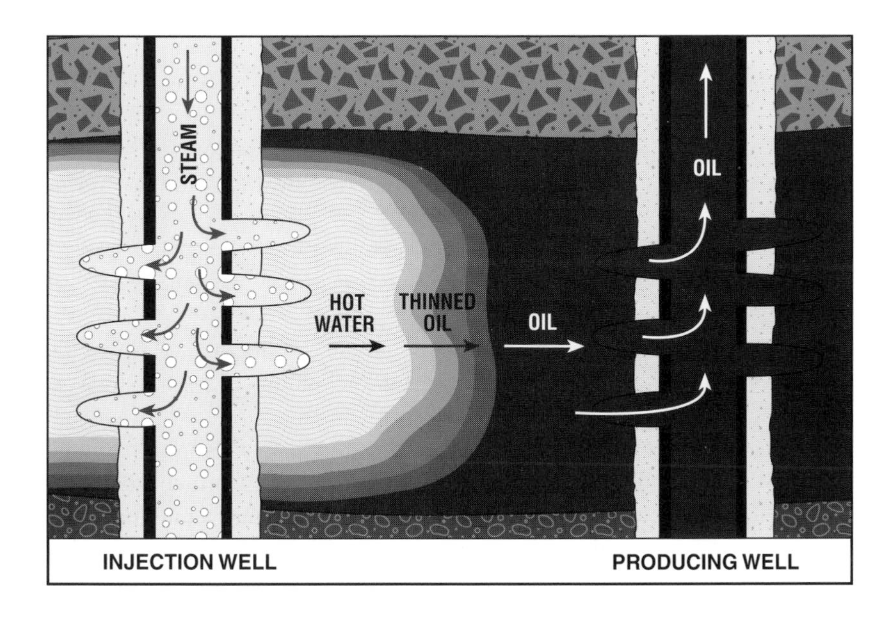

Figure 5.33 In steam flooding, steam and hot water thin the oil and move it to producing wells.

reservoir, it heats up the oil and reduces its viscosity. The steam flows through the reservoir, cools a little, and condenses (forms hot water). The heat from the steam and hot water vaporizes lighter hydrocarbons, or turns them into gases. These gases move ahead of the steam, cool, and condense back into liquids that dissolve in the oil. In this way, the gases and steam provide additional gas drive. The hot water also moves the thinned oil to production wells, where oil and water are produced.

Cyclic Steam Injection

Another method is *cyclic steam injection*, or *huff and puff*. Each huff-and-puff operation involves only one well, but the workover crew may install huff-and-puff equipment on several wells in the oilfield. As in steam drive, the method injects hot steam down the well and into the reservoir to heat the oil. Then steam injection stops, and the operator closes in the well and lets the reservoir soak for several days. In the reservoir, the steam condenses, and a zone of hot water and less viscous oil forms. Finally, the crew reopens the well, and the hot water and thinned oil flow out. This process of steam injection, soaking, and production can be repeated until oil recovery stops.

Fire Flooding

Still another way to use heat in a reservoir is *fire flooding*, or *in situ combustion* (fig. 5.34). In situ (pronounced "in SIT-choo") means in place. In fire flooding, the crew ignites a fire in place in the reservoir. To do this, they first inject compressed air down an injection well and into the reservoir, because oil cannot burn without air. A special heater in the well ignites the oil in the reservoir and starts a fire. As the fire burns, it begins moving through the reservoir toward production wells. Heat from the fire thins out the oil around it, causes gas to vaporize from it, and changes water in the reservoir to steam. Steam, hot water, and gas all act to drive oil in front of the fire to production wells.

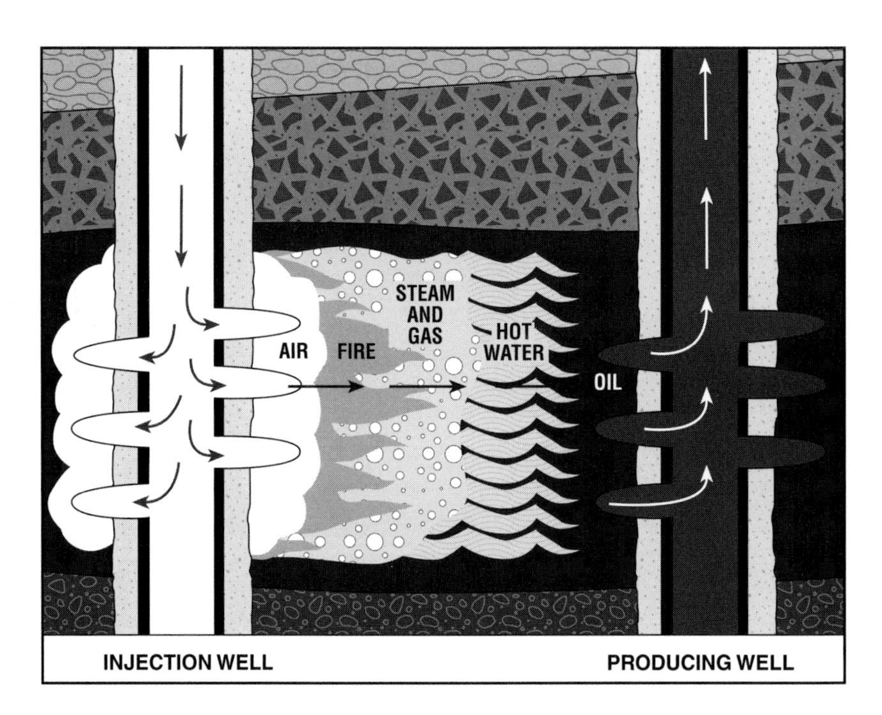

Figure 5.34 In the fire flooding method, compressed air is injected into the reservoir, and the oil is ignited.

SURFACE HANDLING OF WELL FLUIDS

Oil and gas are not usually salable as they come from the wellhead. Typically, a well stream is a high-velocity, turbulent, constantly expanding mixture of hydrocarbon liquids and gases mixed with water and water vapor, solids such as sand and shale sediments, and sometimes contaminants such as carbon dioxide and hydrogen sulfide. Several steps are necessary to get oil or gas ready to transport to its next stop (fig. 5.35).

The well stream is first passed through a series of separating and treating devices to remove the sediments and water, to separate the liquids from the gases, and to treat the emulsions for further removal of water, solids, and undesirable contaminants. The oil is then stabilized, stored, and tested for purity. The gas is tested for hydrocarbon content and impurities, and gas pressure is adjusted to pipeline or other transport specifications.

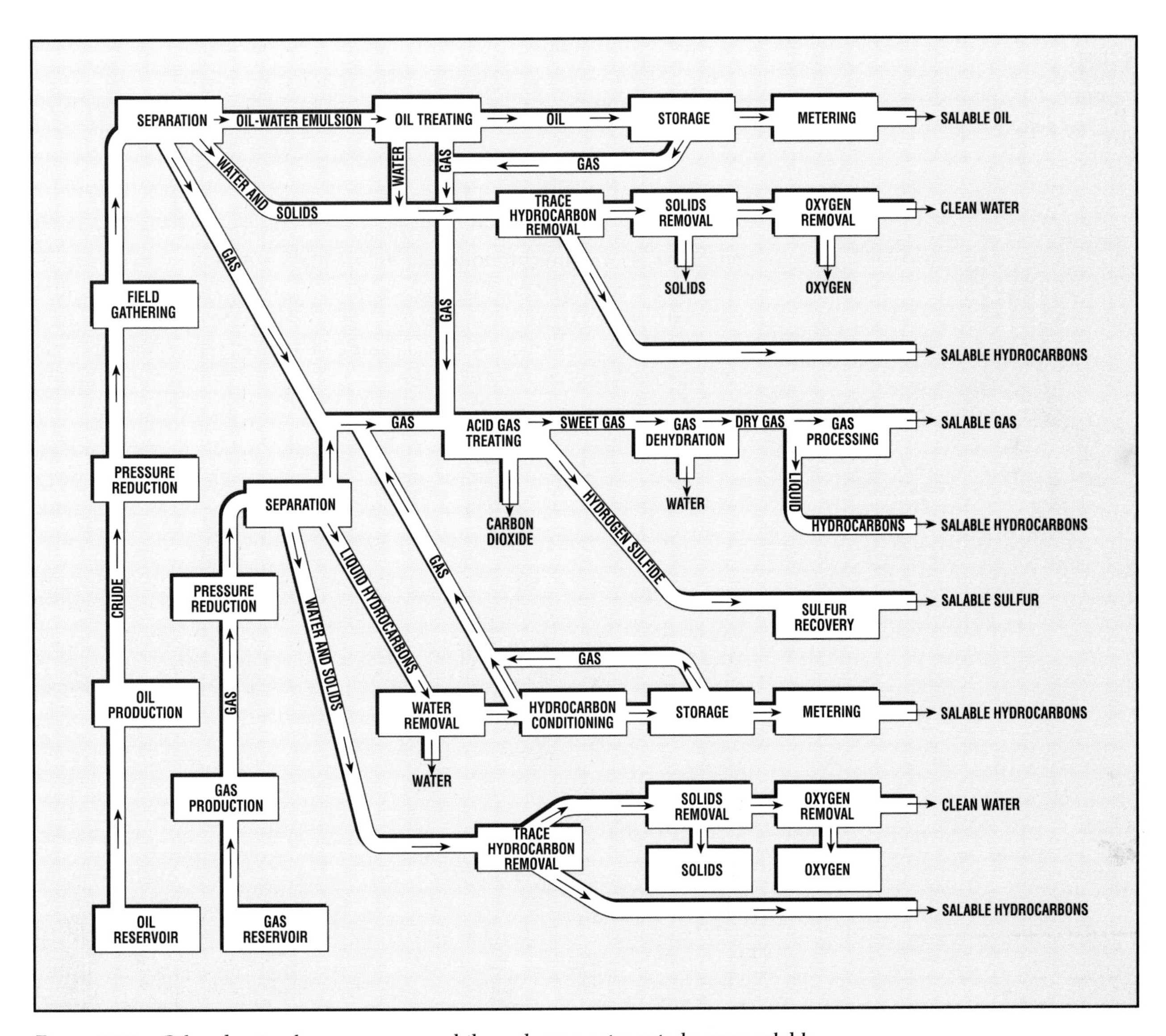

Figure 5.35 Oil and natural gas are processed through many stages to become salable.

Removing Free Water

Some of the water produced with the oil may not be mixed with it; this water is known as *free water.* When given the opportunity, this free water will readily separate from the oil by the force of gravity alone, since the water is heavier than the oil. A *free-water knockout* (sometimes abbreviated as FWKO) is a vertical or horizontal vessel that provides a space for free water to settle out of the well stream (fig. 5.36).

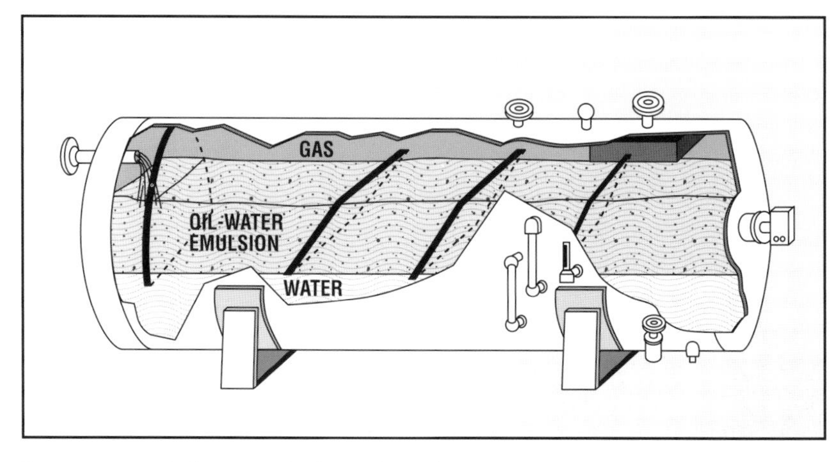

Figure 5.36 A horizontal, three-phase free-water knockout provides space for the free water to settle out and the gas to rise.

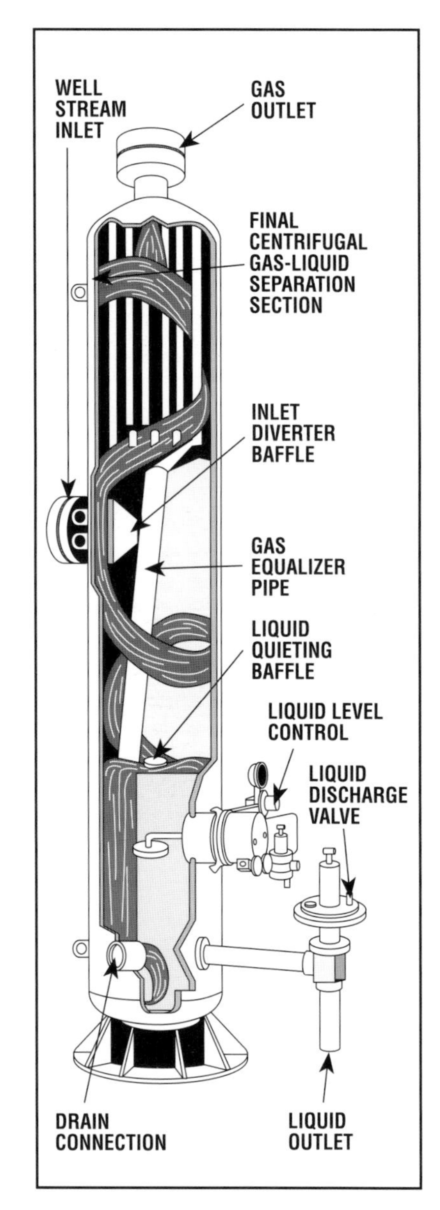

Figure 5.37 Both gravity and centrifugal force are used to separate well fluids in a vertical separator.

Separating Liquids from Gases

The simplest type of equipment used to separate liquids from gases is a tank in which the force of gravity separates them. Oil, which is heavier than gas, sinks to the bottom of the tank and is then removed for additional treatment or sent to the storage tanks. The lighter element, gas, is removed from the top of the tank and enters the gas gathering system.

Separators

Vertical and horizontal *separators* use centrifugal force in addition to gravity to separate well fluids. Separators may be two-phase or three-phase devices. A two-phase vessel separates the well fluids into liquids and gas. A three-phase vessel separates the fluids into oil, gas, and water; however, the oil and gas may still contain some water at this stage of treatment. Well fluids that contain relatively small amounts of free water may go through a three-phase separator.

Vertical Separators

A *vertical separator* is a long vertical cylinder (fig. 5.37). Well fluids enter the vertical separator in about the middle of the cylinder through an inlet with a diverter baffle attached to it. The diverter baffle causes the fluids to swirl. Here, gravity and centrifugal force work to separate the oil and gas. Gravity causes the heavier oil to sink to the lower part of the separator, and centrifugal force from the swirling causes the heavy oil particles to collect on the walls of the separator. The gas, which still contains some liquid particles, rises through the chamber where it is again swirled, and the same forces remove more liquid. Finally, the gas flows out through the gas outlet near the top of the separator, and the heavier oil leaves through an outlet at the bottom of the separator.

Horizontal Separators

A *horizontal separator* operates in much the same way as a vertical separator, although, as the name implies, the unit is in a horizontal position instead of upright (fig. 5.38). The same gravity and centrifugal forces are used, and the oil leaves from the bottom of the unit and the gas from the top.

Horizontal separators may be of either the single-tube or the double-tube design. The double-tube horizontal separator has two horizontal units mounted one above the other and joined by flow channels near the ends of the units. The well fluids enter at one end of the upper unit and are swirled, and the liquids fall through the flow pipe into the liquid reservoir in the lower portion of the bottom unit. The separating process continues in both the upper and the lower units. Gas removed from the liquid in the lower unit rises through the flow channel and joins the gas stream leaving the upper separator at the gas outlet. Oil is discharged through a connection at the lower part of the bottom tube.

Figure 5.38 This horizontal separator has a single tube, although some separators have more.

Multistage Separation

Production fluids often pass through more than one separator to separate more gas from the liquids. In two-stage separation, the well fluids pass through either a vertical or a horizontal separator for the first stage of separation. In the first stage, the fluids are often under pressure as they leave the well. The liquid resulting from the first stage has lost some of its pressure, and it is then sent into a second separator. The second separator removes more gas from the liquid because the lower pressure allows more of the hydrocarbons to vaporize. Any number of separators may be used in stage separation as long as each stage operates at a lower pressure than the previous one.

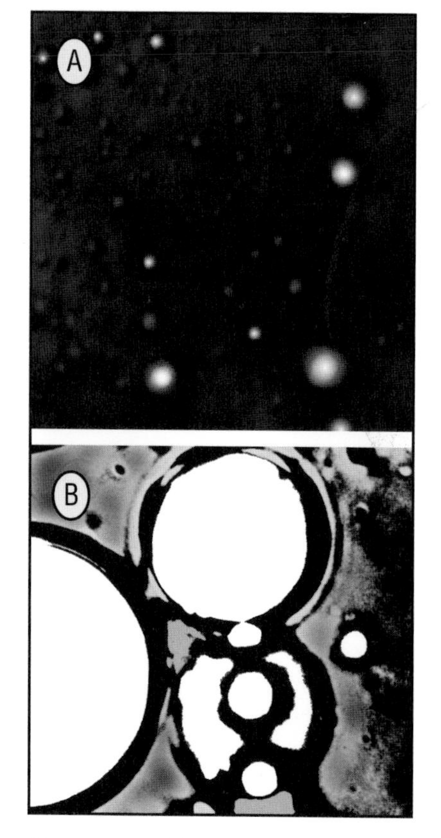

Figure 5.39 This photomicrograph (A) of a water-oil emulsion shows water droplets (light-colored circles) dispersed in oil. A closeup (B) shows that the water droplets touch but are unable to merge because of the oil film around them.

Controlling Paraffin

Hydrocarbons from a well troubled with *paraffin* (a white hydrocarbon wax sometimes found in petroleum) can also mean problems for the surface equipment used to treat the well fluids. Paraffin may not only reduce the efficiency of oil and gas separators but also make the equipment inoperable by partially filling the vessel or blocking fluid passages. While the best place to treat paraffin problems is downhole (see the section "Well Servicing and Repair" in this chapter), paraffin blockage can also be handled in separators. Steaming and using solvents are effective. Another method involves coating all internal surfaces of the separator with a plastic to which the paraffin will not stick and so preventing it from building up to a harmful thickness.

Treating Oilfield Emulsions

Even in multistage separation, the liquid that remains is not pure oil. Separators commonly remove only free water and gas, leaving an emulsion as the liquid. An *emulsion* is a mixture in which one liquid is uniformly distributed (usually as tiny globules) in another liquid. In a *water-in-oil emulsion*, the water is dispersed in the oil; in an *oil-in-water emulsion*, the reverse is true. Milk is an example of an oil-in-water emulsion, and butter is a water-in-oil emulsion.

The two liquids that form an emulsion, oil and water in this case, are *immiscible* liquids; that is, they will not mix together under normal conditions (fig. 5.39). They form an emulsion only when they are agitated to disperse one liquid as droplets in the other and when the mixture contains an emulsifying agent, or *emulsifier*. Emulsifiers commonly found in petroleum emulsions include asphalt, resinous substances, and oil-soluble organic acids.

To break down a stable emulsion into its components, some form of treating is necessary. An emulsion is *tight* (difficult to break) or *loose* (easy to break). Whether an emulsion is tight or loose depends on the properties of the oil and water, the percentage of each found in the emulsion, and the type and amount of emulsifier present.

Treating facilities may use a single process or a combination of processes to break down an emulsion, depending upon the emulsion being treated. To break down a water-in-oil emulsion (the most common type), the properties of the emulsifying agent must be neutralized or destroyed so that the droplets of water can come together. Treatments that do this use chemicals, heat, or electricity, along with gravity.

Chemical Treatment

The emulsion that remains after separation from the gas and free water is often piped into special tanks, or vessels, for chemical treatment. Chemicals called *demulsifiers* are added to the emulsion in order to make the droplets of water merge, or *coalesce*. When droplets merge, they get bigger, and big, heavy water drops settle out faster than small, light ones. A *bottle test* helps determine which chemical is the most efficient demulsifier for a particular emulsion (fig. 5.40). Results from a bottle test also indicate the smallest amount of the treating chemical necessary to break the quantity of emulsion being produced.

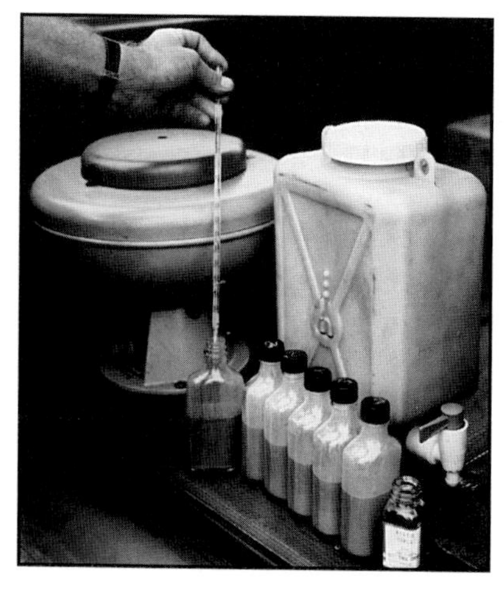

Figure 5.40 A bottle test helps determine the best chemical to use as a demulsifier for treating an emulsion.

Heat Treatment

When an emulsion is heated, it becomes more viscous, and the water and oil molecules move about rapidly, causing the water droplets to strike each other. When the force and frequency of the collision are great enough, the film of emulsifier that surrounds each droplet breaks, and the water drops merge and separate from the oil. Heat alone may not cause an emulsion to break down. Usually the application of heat is an auxiliary process to speed up separation. If possible, the treating process uses little or no heat because heat allows valuable lighter hydrocarbons to boil off from the oil.

Treatment with Electricity

Electricity is also used to treat emulsions, usually in conjunction with heat and chemicals. The film around the water droplets formed by the emulsifier is composed of molecules that have a positive and a negative end, very much like a bar magnet. When an electric current disturbs this film of polar molecules, the molecules rearrange themselves. The film is no longer stable, and adjacent water droplets coalesce freely until large drops form and settle out by gravity.

Types of Emulsion Treaters

To apply any or a combination of these treatments, the emulsion is sent into a special holding tank where the water can settle out, or through a treating tank or vessel called a *flow-line treater* (fig. 5.41). A flow-line treater is a tank in which chemicals, heat, and often electricity are applied to an emulsion. A treater may combine an oil-gas separator, a freewater knockout, a heater, and a filter. Different treaters emphasize different functions and are available in a number of sizes and shapes to handle different volumes of well fluids. Some treaters are designed for extremely cold climates; other models are especially designed to treat foaming oil, for example.

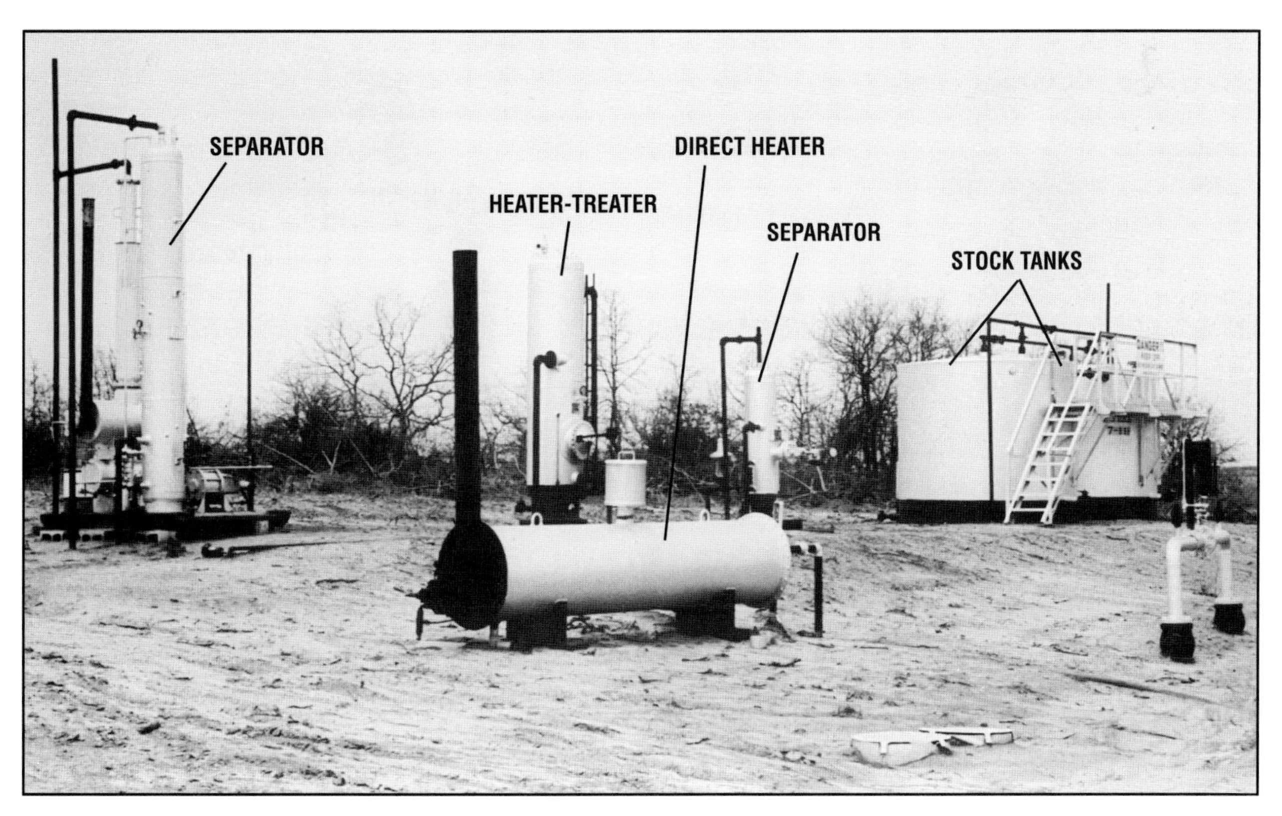

Figure 5.41 Well fluids pass through various vessels on the lease site for separation and treatment.

Treaters can operate at atmospheric pressure, but they often operate under low pressure. For this reason, the treater can be a low-pressure, second-stage separator as well as a treating unit. Where the pressure in the flow line (a pipe from the well to the processing equipment) is low, the treater can operate as a primary separator, which eliminates the need for a regular separator.

Flow-line treaters are either vertical or horizontal cylinders constructed so that the emulsion enters the vessel with the treating chemical already added. Commonly used emulsion treaters are heater-treaters and electrostatic treaters. These devices combine various pieces of equipment to treat the emulsion in one vessel.

Heater-Treater

A heater-treater is a combination of a heater, free-water knockout, and oil and gas separator. In a *vertical heater-treater* (fig. 5.42), the chemically treated emulsion first flows through a heat exchanger, a tube outside the main tank. Cold, incoming emulsion flows up to the top of the treater and is preheated by warm, already treated crude oil flowing down and out of the heat exchanger.

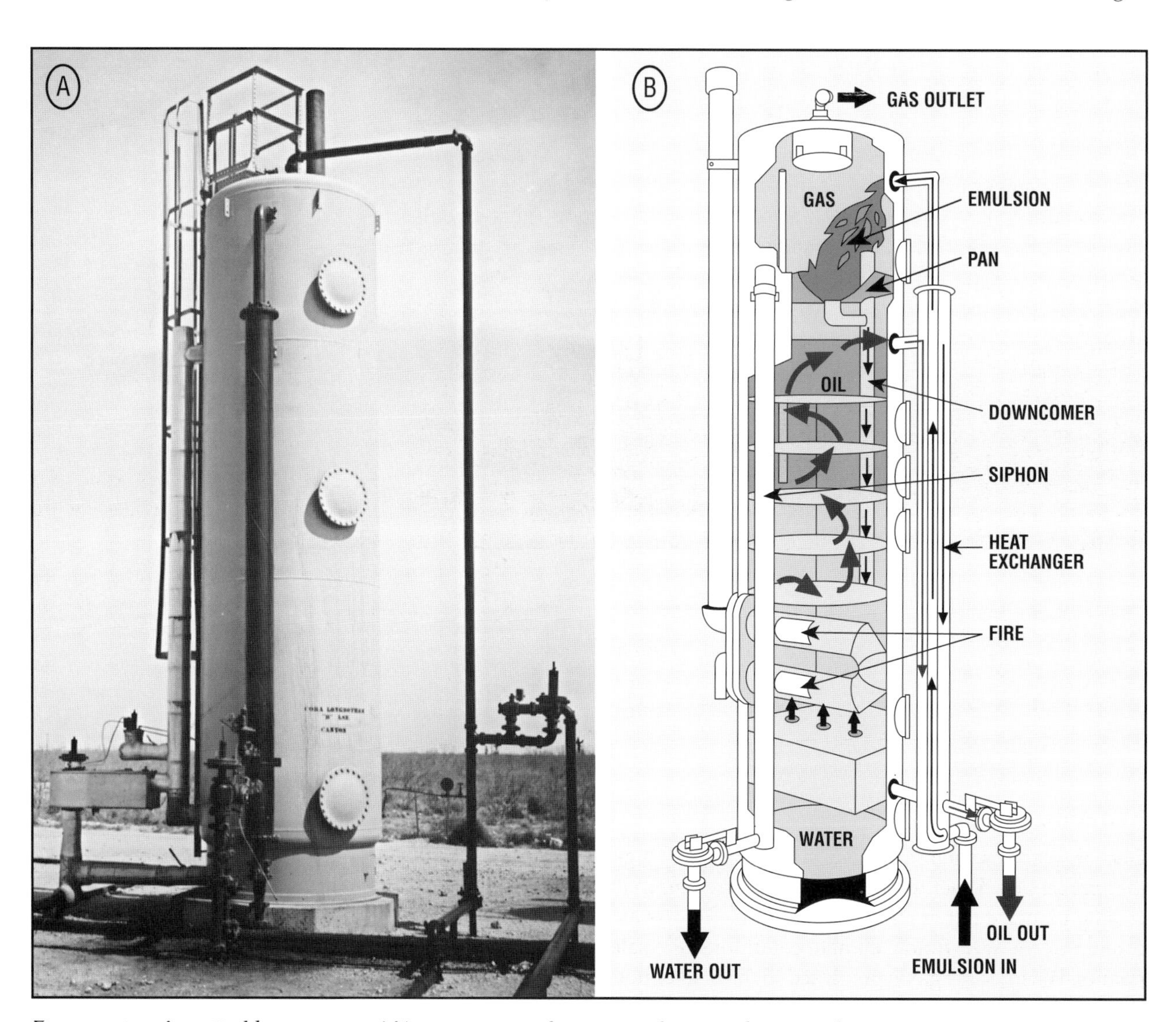

Figure 5.42 A vertical heater-treater (A) separates an oil-water emulsion, as shown in the cutaway view (B).

The emulsion then splashes over a pan at the top of the treater and falls downward through a tube called a *downcomer*. Gas breaks out of the emulsion as it splashes over the pan. When the emulsion reaches the lower section of the treater, free water and sediment fall out. The emulsion rises through the water, which is heated by a *fire tube*, a pipe with steam or hot gases flowing through it. The emulsion settles midway in the vessel. There, with the help of heat and the chemical already in it, the oil-water emulsion separates. The water falls to the bottom, while the crude oil rises upward, passes through an oil outlet to the heat exchanger, and warms incoming emulsion. As the clean crude exits the heater-treater, it flows through a pipe to a *stock tank* for storage.

A *horizontal heater-treater* (fig. 5.43) is similar in operation to a vertical one but has the advantage of a larger treating section and can handle larger volumes of fluid. However, a horizontal treater requires more space and cannot handle sediment as well as a vertical treater.

Figure 5.43 The treating action of a horizontal heater-treater is like that of a vertical heater-treater.

Electrostatic Treater

An *electrostatic treater*, similar in design and operation to a horizontal heater-treater, features a high-voltage electric grid instead of a fire tube. The emulsion rises to the grid, and the water droplets receive a charge that causes them to collide, merge, and settle out.

Handling Natural Gas

Once the gas has been separated from the liquids, it generally requires further processing. Whether this processing happens in the field, in a gas processing plant, or not at all depends on such factors as the type of well, the content of the well fluids, climate conditions, the sales contract, and economics. Whether the gas is processed in the field or not, its pressure must be regulated before sale. The operator installs a regulator to reduce pressure or a compressor to raise pressure, as needed.

Preventing Hydrate Formation

Most natural gas contains substantial amounts of water vapor. If the water vapor cools below a certain temperature, hydrates can form. A *hydrate* is a solid crystalline compound formed by hydrocarbons and water under reduced temperature and pressure. It often resembles dirty snow. When gas comes from the wellhead, it is usually at a high temperature and pressure. As the gas cools, water condenses out, so hydrate formation becomes a real danger. Hydrates may pack solidly in gathering lines and equipment, blocking the flow of gas. Heating the gas and injecting chemical inhibitors are two ways of preventing the water vapor in the gas from forming hydrates.

The most common equipment for heating gas is an *indirect heater* because it is simple, economical, and relatively trouble-free. It consists of a heater shell, a removable fire tube and burner assembly, and a removable coil assembly (fig. 5.44). The heater shell is usually filled with water, which completely covers the fire tube and the coil assembly. The fire tube heats the water bath, the water bath heats the coil assembly, and the gas is heated as it passes through the coil assembly.

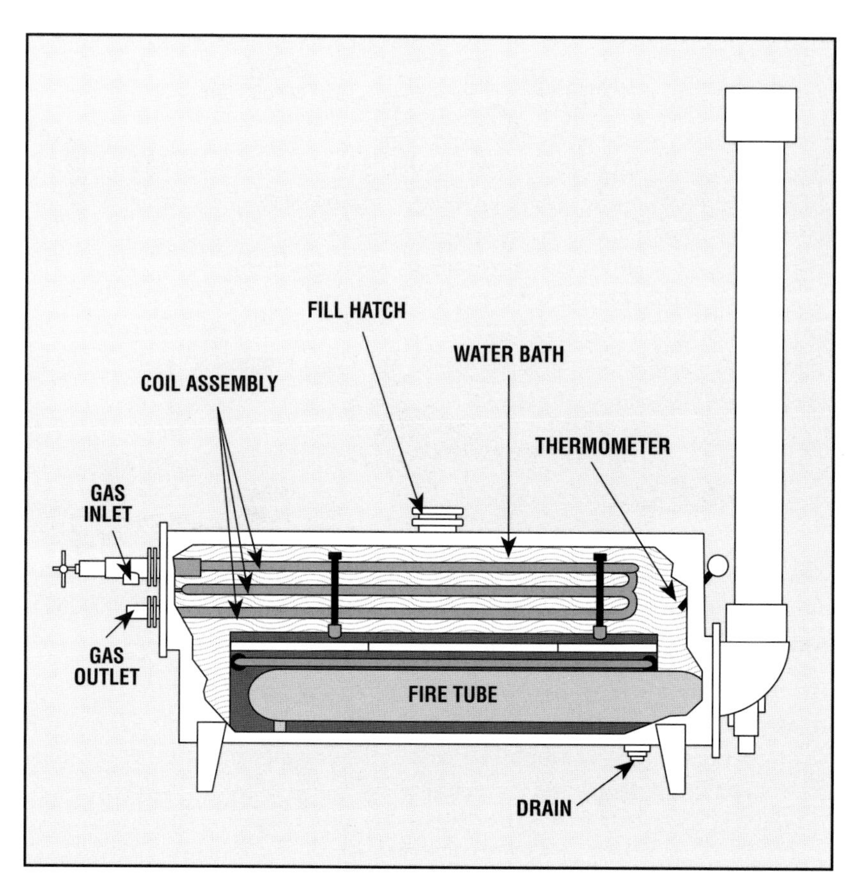

Figure 5.44 An indirect heater warms the water bath, which in turn warms the gas flowing through the coil assembly.

Figure 5.45 The arrows show where glycol or methanol is injected into the gas stream to prevent hydrate formation.

Chemicals inhibit the production of hydrates by lowering the freezing point of water vapor so that it does not condense as easily. Examples of such chemicals are ammonia, brine, glycol, and methanol. They are injected into the flowing gas at strategic points (fig. 5.45).

Dehydrating

Besides forming hydrates, water accelerates corrosion of pipelines and equipment. Also, too much water is undesirable from the buyer's standpoint. Many pipeline companies will not buy gas containing more than 7 pounds of water per million cubic feet of gas. So to prevent operating problems and to meet contract specifications, the gas is usually dehydrated. *Dehydration* is the removal of water.

Two methods are used: absorption and adsorption. In *absorption*, the gas is bubbled through a liquid desiccant, often a glycol solution, that absorbs the water vapor (fig. 5.46). A *desiccant* is a substance that has a special attraction or affinity for water vapor. The liquid and gas can then be separated (fig. 5.47). In *adsorption*, the vapor-laden gas is passed over a solid desiccant in an *adsorption tower* (fig. 5.48). The water adheres to the solid desiccant, and the dry gas flows out of the adsorption tower. The desiccant bed can be reused after it is dried by heating and vaporizing the water.

Each of the processes has its own special advantages and disadvantages. Dry-bed adsorption is more likely to be used in a gas processing plant than in the field.

Figure 5.46 This glycol liquid-desiccant dehydrator unit is skid-mounted.

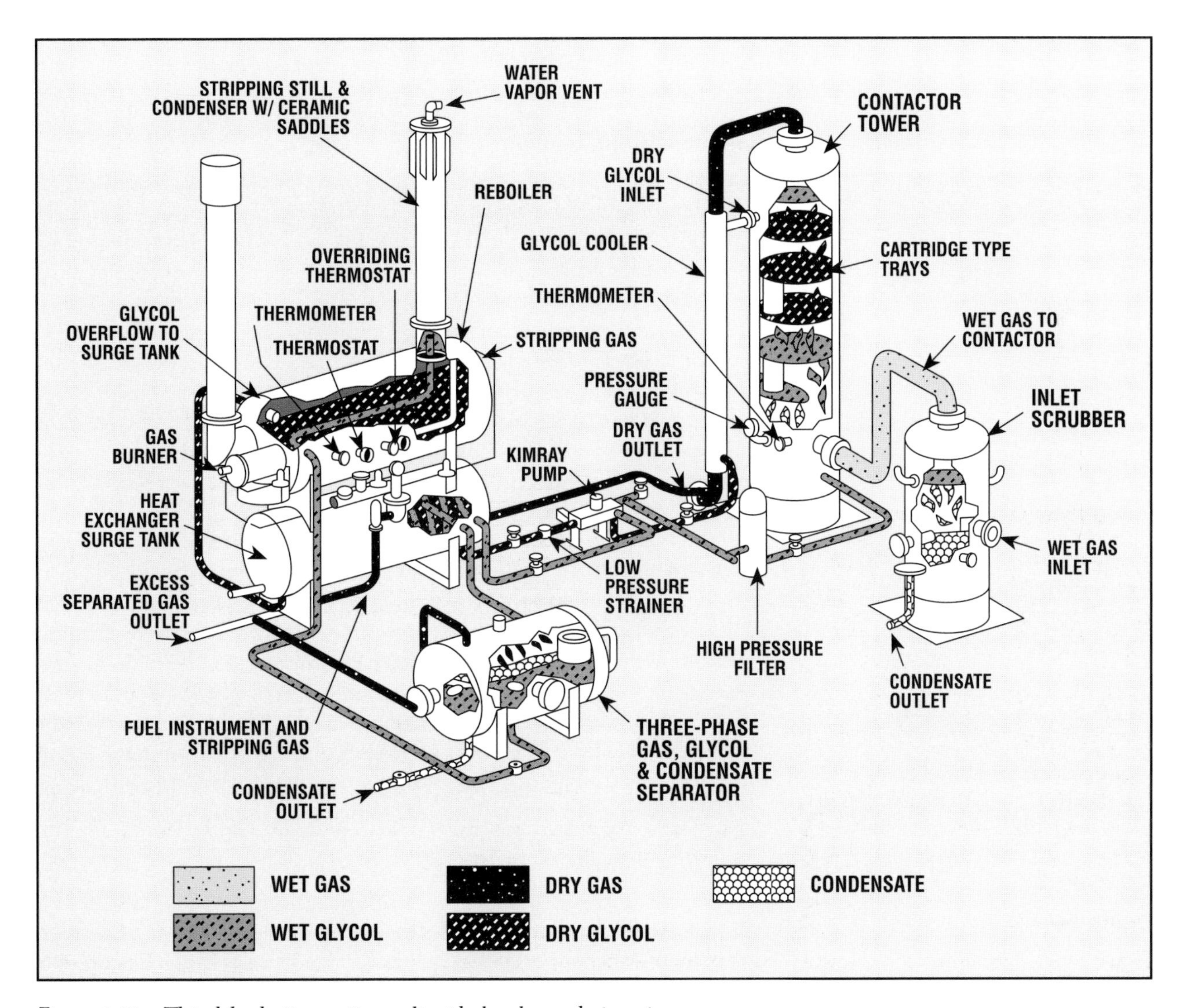

Figure 5.47 This dehydration unit uses liquid glycol as a desiccant.

Figure 5.48 This dry-bed dehydration system uses a solid desiccant.

Removing Contaminants

Natural gas often contains carbon dioxide and compounds of sulfur such as hydrogen sulfide. These gases are called *acid gases* because they form acids or acidic solutions in the presence of water. Gases with high concentrations of sulfur are also called *sour gases.* If present in significant quantities, acid gases must be removed from natural gas. Both carbon dioxide and hydrogen sulfide are corrosive when water is present, and hydrogen sulfide is very poisonous in relatively small concentrations. Removing these contaminants from natural gas is often called *sweetening* the gas.

The processes for sweetening natural gas are similar to the ones for removing water vapor: absorption and adsorption. In the two types of absorption processes, the acid gas stream contacts a liquid absorbent that selectively removes certain acid gases from the natural gas.

In *chemical absorption,* the liquid absorbent reacts chemically with the acid gases but not with the natural gas. The purified gas flows out, and heat and/or low pressure is used to separate the acid gases from the absorbent so that the absorbent can be used again. The most widely used sweetening processes in the industry are the amine processes (fig. 5.49). An amine process is a continuous operation that uses a solution of water and a chemical known as an amine to remove carbon dioxide and several sulfur compounds.

In *physical absorption,* the acid gases physically dissolve in the liquid absorbent, and the natural gas does not. Again, heat and low pressure separate the acid gases from the absorbent so that it can be reused. Commercial processes of this type include the Selexol, sulfinol, Rectisol, and fluor solvent processes.

In *adsorption,* or dry-bed, processes, the acid gas stream flows over a solid adsorbent, as in dehydrating. Rather than removing water, however, this adsorbent removes sulfur compounds and/or carbon dioxide. Heat or lowered pressure vaporizes the acid gas and removes it from the adsorbent bed.

Figure 5.49 This sweetening unit uses an amine to remove carbon dioxide and sulfur compounds. *(Courtesy of Sivalls, Inc.)*

Removing Natural Gas Liquids

Natural gas is a mixture of methane, ethane, propane, and some heavier hydrocarbons such as butane and natural gasoline (ranging from pentanes through nonanes or decanes). In field separators, some of the heavier hydrocarbons liquefy and are removed. For economic reasons, the lease operator often wants to liquefy some or all of the remaining ethane and heavier hydrocarbons and remove them from the methane gas. The liquefied hydrocarbons, known as *natural gas liquids* (NGLs), are more valuable as liquid fuel or refinery feedstocks than as components of natural gas. Also, if butanes and heavier hydrocarbons are allowed to remain in the natural gas stream, they may condense in the transmission lines during cold weather.

When gas is produced from high-pressure gas wells, the NGLs are usually separated in field facilities. It is generally more economical to process gas coming from an oilwell in a gas processing plant.

The usual equipment for processing gas from a gas well is the *low-temperature separation (LTX) unit* (fig. 5.50). This equipment partially dehydrates the gas stream as well as removing NGLs. It takes advantage of the fact that the gas stream coming from the wellhead is under high pressure.

Figure 5.50 A low-temperature separation unit separates NGLs produced from a gas well.

After passing through a free-water knockout, the compressed gas is cooled and allowed to expand to pass through a choke into a separator vessel. The expansion further cools the gas and condenses the heavier hydrocarbons into a liquid. The hydrocarbon liquid and the hydrates formed from water vapor drop to the bottom of the separator, and the dry gas passes out the top. Then the gas is warmed and allowed to enter the gas gathering system. The hydrates melt, the water goes to a disposal system, and the NGL is stored in a low-pressure tank.

Storing Crude Oil

Oil, water-cut oil, and water produced by the well move from the wellhead or separator through the treating facilities and finally into a group of *stock tanks* for storage, or a *tank battery* (fig. 5.51). The number of tanks in a battery will vary, as will their size, depending on the daily production of the well or wells and the frequency of pipeline runs. The total storage capacity of a tank battery is usually three to seven days' production. Also, since a battery has two or more tanks, one tank can be filling while oil is being run from another.

Figure 5.51 This central treating station has a free-water knockout, electrostatic treaters, and stock tanks.

Measuring Oil in the Tank

Before a tank battery is put into operation, each tank is *strapped*, or calibrated. A special strapping contractor measures the dimensions of the tank and then computes the amount of oil that it contains for each height increment of the tank. The contractor prepares a table showing the capacity in barrels according to the height of the liquid in the tank, the *tank capacity table.* The tank capacity table commonly shows the capacity in increments of ¼ inch (6 millimetres) from the bottom to the top.

To measure the amount of oil in a tank, a technician called a *gauger* lowers a steel tape with a weight on the end into the tank until it just touches the bottom. The highest point at which oil wets the tape shows the level or height of the oil in the tank. By referring to the tank table, the gauger can determine the amount of oil in the tank. A gauger may work for either the lease operator or the pipeline company that will buy the oil.

An *automatic tank gauge*, a newer method of measuring, consists of a steel gauge line contained in a housing. A float on the end of the line rests on the surface of the oil in the tank. The other end of the line, which is coiled and counterbalanced on the outside of the tank, runs through a reading box that shows the height of oil in the tank.

Tank Construction

Most tanks are constructed of either bolted or welded steel. They have a bottom drain outlet for draining off *sediment and water (S&W),* still commonly referred to as BS&W (basic sediment and water). Sometimes a worker must enter an empty tank to clean out paraffin and sediment that cannot be removed through the drain outlet.

Oil enters the tank at the top at an inlet opening. The pipeline outlet is usually 1 foot (30 centimetres) above the bottom of the tank; the space below this outlet provides room for the collection of S&W. A metal seal closes the pipeline outlet valve when the tank is being filled and is similarly locked in the open position when the tank is being emptied. The seals guarantee that the buyer receives the quantity and quality of oil mutually agreed upon by the buyer and the lease operator, or *producer*.

MEASURING AND TESTING OIL AND GAS

The lease operator owns the oil or gas as it comes from the well. The operator then sells it, and the pipeline company, often shortened to *pipeline*, transports it to the buyer. The operator, pipeline company, and buyer all measure and test the oil and gas at different times.

Oil Sampling

Pipeline companies may test the oil they receive at any time. Therefore, to assure that the pipeline company will accept the oil, the operator should sample and test the oil in the same manner as the pipeline company. Because the procedures for taking samples and making water and sediment tests vary from field to field and company to company, both the operator and the pipeline must agree on them.

Sampling Methods

Many methods exist for sampling oil. Broadly, sampling can be done either automatically or manually. Frequently, a sampler manually samples the oil in lease stock tanks. Perhaps the most common way is by *thief*, or *core, sampling*. In thief sampling, the sampler lowers a *thief*, a round tube about 15 inches (38 centimetres) long, into a tank to any level within a half inch (13 millimetres) of the bottom of the tank (fig. 5.52). The thief has a spring-operated, sliding valve that can be tripped, thus trapping the sample.

A better but more difficult manual sampling method is *bottle sampling*. One method of bottle sampling uses a 1-quart (about 1-litre) bottle or beaker with a stopper and cord (fig. 5.53). The sampler lowers the sealed bottle to the desired depth, removes the stopper, and pulls the bottle or beaker to the top at a uniform speed. If pulled at the proper speed, the bottle should be about three-fourths full of liquid; if not, the sampler must start over.

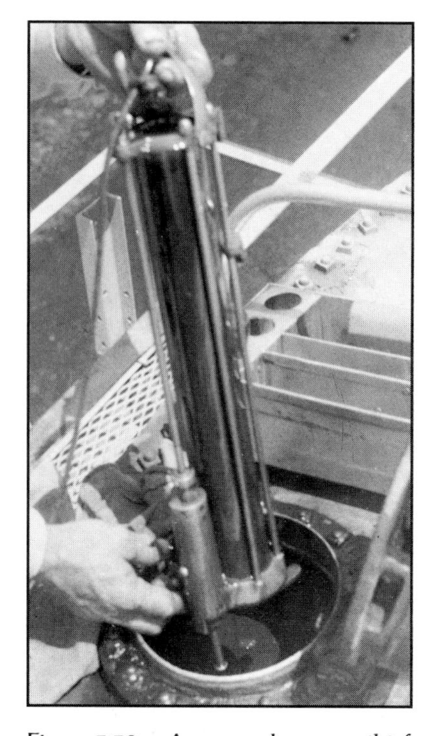

Figure 5.52 A person lowers a thief to obtain an oil sample from a storage tank.

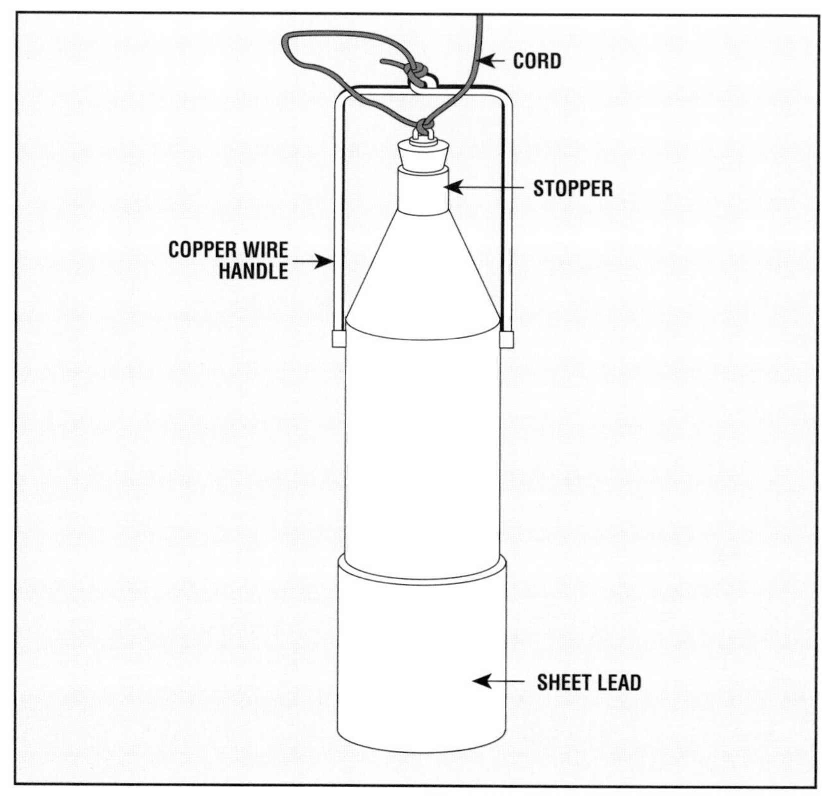

Figure 5.53 A 1-quart (1-litre) beaker is used in bottle sampling.

Types of Samples

A sample can come from different parts of the tank, and all parties concerned should agree on where to take it from. An *average sample* is one that consists of proportionate parts from all sections of the tank. A *running sample* is a bottle sample, taken while the sample bottle is moving. A *spot sample* is one obtained at some specific location in the tank by means of a thief, bottle, or beaker.

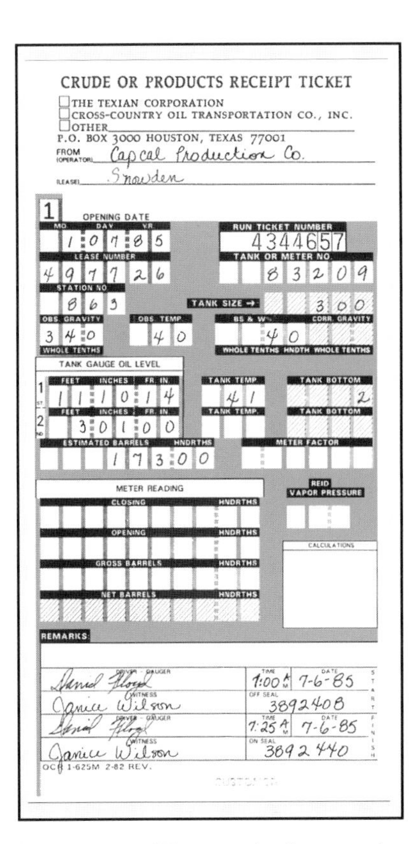

CRUDE OR PRODUCTS RECEIPT TICKET

☐ THE TEXIAN CORPORATION
☐ CROSS-COUNTRY OIL TRANSPORTATION CO., INC.
☐ OTHER
P.O. BOX 3000 HOUSTON, TEXAS 77001
FROM (OPERATOR) Capcal Production Co.
(LEASE) Snowden

1 OPENING DATE MO. DAY YR. 1 07 85
RUN TICKET NUMBER 4344657
LEASE NUMBER 497726
TANK OR METER NO. 83209
STATION NO. 863
TANK SIZE → 300
OBS. GRAVITY 34.0 (WHOLE TENTHS)
OBS. TEMP. 40
BS & W% .40 (WHOLE TENTHS HNDTH)
CORR. GRAVITY (WHOLE TENTHS)

TANK GAUGE OIL LEVEL
1ST FEET 11 INCHES 10 FR. IN. 1/4 — TANK TEMP. 41 — TANK BOTTOM 2
2ND FEET 3 INCHES 01 FR. IN. 00 — TANK TEMP. — TANK BOTTOM
ESTIMATED BARRELS 173 HNDRTHS 00
METER FACTOR

METER READING
CLOSING HNDRTHS
OPENING HNDRTHS
GROSS BARRELS HNDRTHS
NET BARRELS HNDRTHS
REID VAPOR PRESSURE
CALCULATIONS

REMARKS:

DRIVER - GAUGER: Daniel Floyd	TIME 1:00 AM	DATE 7-6-85
WITNESS: Janice Wilson	OFF SEAL 3892408	
DRIVER - GAUGER: Daniel Floyd	TIME 7:25 AM	DATE 7-6-85
WITNESS: Janice Wilson	ON SEAL 3892440	

START / FINISH

OCR 1-625M 2-82 REV.

Figure 5.54 The run ticket becomes the basis for all payments for oil delivered from the lease.

Oil Measurement and Testing

Crude oil is bought, sold, and regulated by volume. The operator usually measures the volumes of oil, gas, and salt water produced by each lease every 24 hours. This is important because in many fields the government regulates how much oil or gas may be produced each day. Regulations base the amounts on the market demand for oil or gas and the most efficient rates of production for the particular fields.

When measuring crude oil, the volume must be corrected for any S&W present. Since the volume varies according to temperature, the volume must also be corrected to a standard base temperature of 60°F (15.5°C). These requirements call for a series of tests for temperature, density, and S&W content. Technicians run these tests and measurements in the presence of witnesses representing both the lease operator and the pipeline company. The information they get is written on the pipeline *run ticket* (fig. 5.54).

Temperature Measurement

The temperature of oil in lease stock tanks generally is close to that of the air surrounding the tanks unless the oil has just recently been produced and still has a higher temperature or if the oil has been heated in a treater to separate it from S&W or salt water. Temperature is usually measured with a special thermometer that is lowered into the oil on a line and then withdrawn to observe the reading.

Gravity and S&W Content Measurement

The S&W content and the API gravity (specific gravity, or density, measurement) of oil in stock tanks are measured from samples taken from the tanks. A sampler takes samples by using a thief or similar device or by withdrawing oil through sample cocks installed at various levels in the tank. Different buyers require different levels of cleanliness, but the maximum S&W content in most states is 1 percent. Most pipeline companies accept all oil from 4 inches (10 centimetres) below the bottom of the pipeline connection on up. A *centrifuge test*, also called a *shake-out test*, determines the S&W content of the samples (fig. 5.55). The test uses a glass container that is graduated so that the percentage of S&W can be read directly.

An instrument called a *hydrometer* measures the API gravity of the oil. The technician pours the crude into a cylinder with accurate levels marked on it and drops the hydrometer into it. It floats at a certain level in the crude (the higher it floats, the lighter the oil). The markings on the cylinder show the API gravity in °API at 60°F (15.5°C). Since the gravity, or density, of oil varies with its temperature, the technician must correct gravity readings of oil at temperatures other than 60°F (15.5°C). Tables are available that make these corrections easy.

Figure 5.55 A high-speed centrifuge determines the precise S&W content.

Figure 5.56 A skid-mounted LACT unit is an efficient system for measuring and testing oil.

Figure 5.57 A sample container on a LACT unit holds oil samples automatically taken from the oil stream flowing through the unit.

LACT Units

The development of *lease automatic custody transfer (LACT) units* has changed the processes of measuring, sampling, testing, and transferring oil from a tedious, time-consuming business that required many hours and was error-prone into an efficient measuring and recording system that leaves lease personnel free to do other operations (fig. 5.56). Despite the complexities of gauging and testing crude oil in the field, automatic equipment can perform the following tasks:

- Measure and record the volume of the oil.
- Detect the presence of water in the oil stream and calculate the percentage.
- In cases of excessive water, divert the flow into special storage where it is held for treatment.
- Isolate individual wells on a lease and test them separately from the combined flow.
- Determine and record the temperature of the oil.
- Verify the accuracy of flowmeters and provide for calibration when necessary.
- Take samples from the stream in proportion to the rate of flow and hold these for verification by conventional test procedures (fig. 5.57).
- Measure and record the API gravities of the oil produced.
- Switch well production from a full tank to an empty one and switch the oil from the full tank to the pipeline system.
- Shut in the wells and relay an alarm signal to a remote point in case of any malfunction.

Of course, not all LACT installations are as elaborate as the system just described. Lease operators use a system that best meets their individual needs and budgets.

Gas Sampling

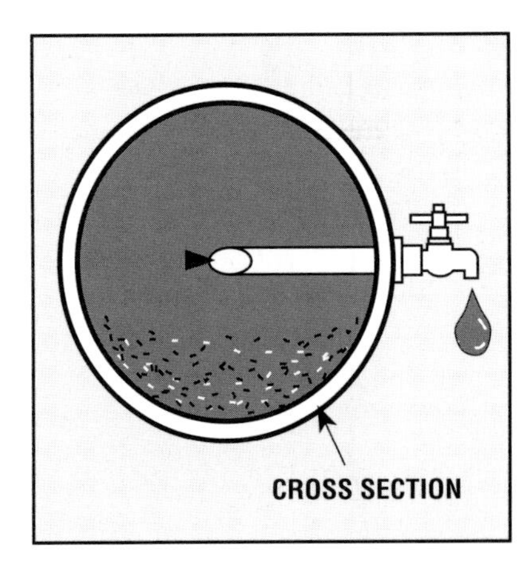

Figure 5.58 A sample should be taken near the center of the pipe where water vapor has not condensed on the walls.

Like oil, gas must be sampled before it can be sold. Sampling, which is always done in the field, is the way the operator determines the sales potential, composition, heating value, specific gravity, and other physical properties of the gas being produced from a specific well. It also reveals information about the performance of the field and the processing equipment. Because each well has its own unique characteristics—age, equipment downhole and on the surface, production rates, and so forth—sampling procedures vary from well to well. Some wells may require that gas samples be taken frequently from several different points on the surface equipment; other wells may have samples taken less frequently and from perhaps only one point. Regardless, sampling is vitally important to both the seller and the buyer.

The best place on the pipeline to sample gas is wherever the sample will accurately represent the composition of the whole stream. A sample taken at an elbow in the pipe or some other location where impurities collect will not be representative. Similarly, a sample taken near the side of a pipe, where water vapor may have condensed on the walls, is not accurate. The probe should extend into the center of the pipe (fig. 5.58).

Several manual methods are used to sample gas: purging at reduced pressure, fluid displacement, air displacement, and the use of a floating piston and vacuums. Each has its advantages and disadvantages, but they all require that sampling or testing equipment be leak-free and that sample containers be purged of all gases or vapors other than the natural gas. If the sample is to be accurate and useful to both the lease operator and the buyer, the technician taking the sample must be experienced and must follow standard procedures to the letter. Samples are shipped to a laboratory for analysis.

Although manual sampling methods are more common, automatic sampling devices connected to the pipeline are also used. These devices allow gas to flow into the sample container over a certain period of time at a constant rate. In some cases, automatic samplers are connected to analyzing equipment, but usually, field samples are sent to the laboratory for testing.

Gas Testing

Gas testing determines the liquid hydrocarbon content of the gas. The results of field and laboratory tests may determine how much a seller receives for his or her gas and whether the gas meets the buyer's contract specifications. Three of the most common methods for gas analysis are compression testing, charcoal testing, and fractional analysis. Compression testing is usually done in the field, while charcoal testing and fractional analysis are conducted in the laboratory.

Compression Testing

Compression testing is used extensively on casinghead gas (gas produced with oil). The sample is compressed in a portable compressor and then cooled in an ice-water bath or in a refrigerated condenser. This is the one form of gas testing that is done completely at the field location, and all equipment used must be in top operating condition so that the test is accurately repeatable. Those in the oil industry generally consider compression testing to be reliable.

Charcoal Testing

In a *charcoal test*, gas is drawn from the stream and passed over activated charcoal. The charcoal adsorbs the liquids, in the same way an adsorption dehydrator removes water from gas. After adsorption is complete, the charcoal is taken to a laboratory, which evaporates the adsorbed liquid with heat, condenses it, and measures it. This test is used when other methods of testing either give unsatisfactory results or are too expensive.

Fractional Analysis

Fractional analysis is used when the exact gas composition is important. In this procedure, a technician takes a sample of the gas stream in a metal container and ships it to a laboratory. Fractional analysis usually reveals not only the composition in percentage of each hydrocarbon present, but also the amount of each component in gallons per thousand cubic feet (litres per cubic metre) and the heating value of the gas. *Heating value* is the amount of heat the gas will produce when burned.

Gas Metering

Gas metering is the process of measuring the volume of natural gas flowing past a particular point. Since the temperature, specific gravity, and pressure of the gas affect the volume measurement, standards have been established for these variables to take differing pressures, for example, into account. These standards convert the individual properties of natural gas flowing from a particular well to what is called *standard condition*. This standardization of properties is necessary for metering.

Every gas purchase or sales contract made between major suppliers and consumers is concerned with metering methods, calibration, and equipment. Therefore, gas must be metered whenever there is a change of custody, or ownership. Measurements of natural gas are also used in regulatory reporting and reservoir studies.

Metering the Gas

A number of different types of meters can measure gas, but the most common is the *orifice meter* (fig. 5.59). Basically, an orifice meter has a finely machined plate with an orifice, or hole, in its center mounted vertically in a gas line. Because the orifice is smaller than the upstream tubing, the pressure of the gas flowing downstream through it drops. This pressure drop, called *differential pressure,* is measured and recorded on a chart. Also measured and recorded on the chart is the *static pressure*—the pressure in the line upstream from the orifice plate. By using these two measurements and other known factors, computer operators can calculate the rate of flow, the volume flow rate, and other data.

Other devices and instruments measure and record density, dew point, temperature, and pressure. *Recording gravitometers* measure and record the specific gravity of natural gas, which is needed to calculate gas flow.

Recording the Data

Electronic data recorders and processors gather information from the meters, which is then sent to an office for analysis. The recording devices are installed in a location so that the presence of liquids does not cause measuring problems. Other problems in gas measurement are gas pressure and flow

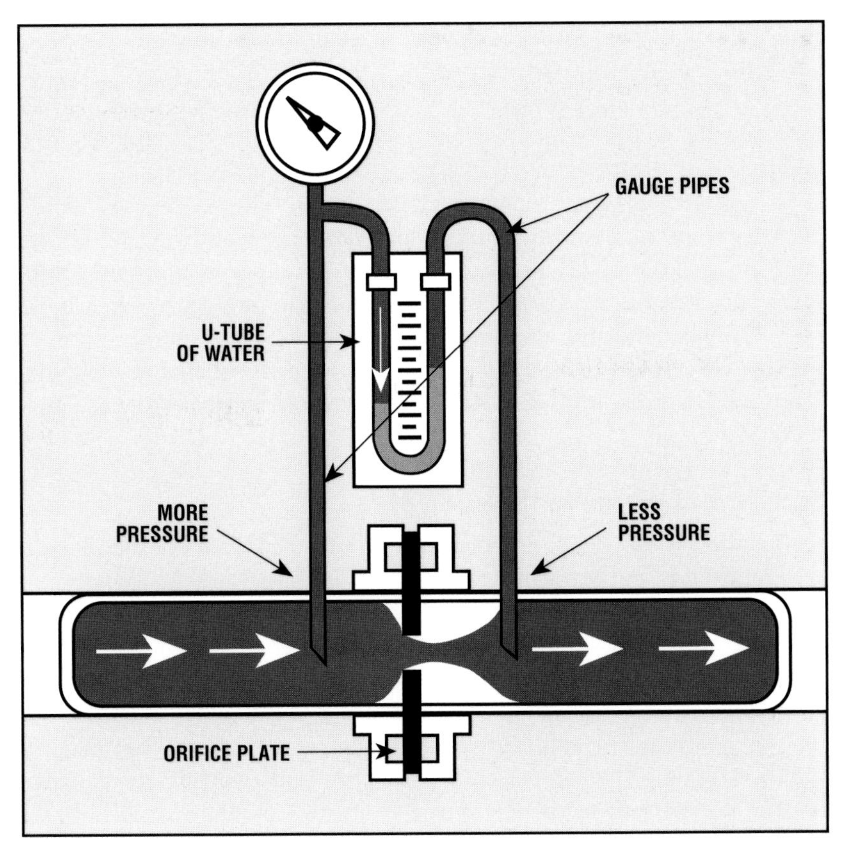

Figure 5.59 An orifice meter measures differential and static pressures of a stream of flowing gas.

pulsations. Pulsations are cyclical pressure surges caused by reciprocating pumps or compressors. Irregular pulsations also occur when valves are opened or closed on the line. Meter installations need to be as far away from pulsation sources as possible in order to prevent inaccurate readings. Some gas lines require *pulsation dampeners*—devices that minimize pressure surges. Specific gravity and temperature recording devices are placed where they will not interfere with meter measurements.

Standards

The American Gas Association's Gas Measurement Committee sets the standards for installing and operating field metering equipment. The committee was organized to combat the widespread variation in metering that existed among different localities and operators in the fledgling gas industry. The committee conducts research and reports its findings in AGA Committee Report No. 3, *Orifice Metering of Natural Gas,* which contains the most accurate information available on large-volume gas measurement. Operators accept these AGA committee recommendations as standards for gas measurement, ensuring uniform industry-wide practices.

WELL SERVICING AND WORKOVER

Wells that have already been producing for a long time and whose rate of flow has slowed or stopped altogether need to be repaired or worked on. The industry calls such repair or work well servicing or workover. Almost all producing wells eventually either have mechanical problems in the well equipment or need work to retrieve remaining oil or gas or both.

Well service is maintenance work. Service usually involves repairing equipment installed during drilling, completion, or workover. A servicing company—the company that does the repair—may also add new equipment, such as a pump, to restore the well's ability to produce hydrocarbons.

Workover includes any of several operations on a well to restore or increase production when a reservoir stops producing at the rate it should. Many workover jobs involve treating the reservoir rock rather than the equipment in the well.

Service and Workover Equipment

Until the 1950s and 1960s, drilling crews set up a permanent derrick at each well for drilling and maintenance of the well throughout its life. Now, however, since the whole drilling rig is moved to a new site when drilling is finished, the well is left with only a wellhead and sometimes a pump. Therefore, service and workover companies must bring the equipment they need to use.

The amount and type of equipment they need depends on the job. One job may require a light-duty rig and a couple of workers. The next well may need a somewhat larger rig with a tall mast and a crew of several workers. Another job may require extra crews to work around the clock and a rig capable of light drilling and heavy-duty hoisting.

Rigs

Service and workover rigs, like drilling rigs, are machines for hoisting pipe, wireline, and tools into and out of a well (fig. 5.60). They have a mast, a drawworks, and a power source. Unlike drilling rigs, not all of them have circulation or rotary systems.

They come in a variety of sizes. In general, servicing jobs require smaller rigs than workover jobs. The smallest rigs raise and lower a wireline or conductor line. Oilfield workers sometimes call these *wireline rigs*. Medium-duty rigs can pull sucker rods or a lightweight string of tubing. The heaviest rigs are workover rigs for drilling and deep well work.

Manufacturers build service and workover rigs on three different types of bases for transporting on land. Which base they use depends in part on the weight and size of the equipment it must carry. Deeper wells require masts that are stronger and heavier, larger drawworks drums, and more engine power, and therefore the rig needs a stronger base.

The smallest rigs usually rest on specially built trucks (*truck-mounted rigs*), larger rigs sit on trailers pulled by tractors (*trailer-mounted rigs*), and many are basically rigs on wheels with a steering mechanism, called *carrier rigs*. Portable rigs are relatively easy to move and quick to assemble in the oilfield.

Wireline Units

Wireline operations, which are an integral part of both completion and workover, are procedures performed with tools suspended on a wireline. A *wireline*, in essence, is a strong, thin length of wire mounted on a powered reel at the surface of a well. Wirelines vary in size from 0.072 to 0.1875 inch (1.8 to 4.8 millimetres) in diameter. The larger line is a braided electric line, while the lines from 0.072 to 0.092 inch (1.8 to 2.3 millimetres) are solid steel, or *slick*,

Figure 5.60 A single-size rig may function as a drilling rig, servicing rig, or workover rig for wells of different depths, depending on the amount of weight it must hoist.

lines. Among the more common jobs performed with wireline are depth measurement, logging, perforating, sand and paraffin control, fishing and junk retrieval, and the manipulation of subsurface well pressure and flow controls.

The wireline, along with the equipment required to perform the various wireline operations, are usually housed in a truck or a small portable house, and are called *wireline units*. On land these units are usually truck-mounted (fig. 5.61). Marine wireline units are mounted on skids and transferred to the

Figure 5.61 Reels of wireline, shown here mounted on the back of a truck, are used for many completion and workover jobs.

wellsite on boats or barges. Regardless of the type, a wireline unit contains a reel of wireline; a power system to let out and retrieve the line; instruments to indicate weight, line speed, and depth of the tool string; and the string of tools attached to the end of the line to perform the particular job. Equipment to contain well pressure is also common.

Coiled Tubing Units

A relatively new development in service and workover equipment is the *coiled tubing unit*. Coiled tubing is exactly what it sounds like: flexible tubing that can be coiled onto a drum and uncoiled for placement into a well inside the production tubing. Coiled tubing usually has a small diameter, ¾ to 2⅜ inches (19 to 60 millimetres). Service and workover crews use it in the same way they use wireline and regular tubing—to lower tools into the well and circulate fluids.

A coiled tubing unit is usually mounted on a trailer or *skid*, a platform that can be set on a trailer or barge (fig. 5.62). The components of the unit include a reel for the coiled tubing, an *injector head* to push the tubing down the well, a wellhead blowout preventer stack, a power source (usually a diesel engine and hydraulic pumps), and a control console. The reel can carry up to 20,000 feet (6,100 metres) of coiled tubing, enough to reach the bottom of most wells.

Figure 5.62 A coiled tubing unit is usually mounted on a trailer or skid.

Using a coiled tubing unit instead of a workover rig is faster and less expensive on many jobs. Rig-up time is usually about 1 hour, and coiled tubing can generally be pulled about three times faster than jointed tubing. It also allows the crew to work on the well without killing it. Coiled tubing is valuable in horizontal wells where it is impossible to use a wireline (which would be like pushing a noodle sideways). Although coiled tubing technology has been around since the 1960s, the bugs are only now being worked out of it so that it is becoming more popular.

Specialty Rigs

Service and workover rigs come in many variations to allow work on wells in any possible environment and using different combinations or types of equipment. All-terrain rigs are carrier rigs that have options designed for a particular environment. For example, they may have a steering system that allows tight turns, a heavy-duty cooling system and air cleaners to filter sand, oversized tires or tracks like a military tank for off-road transport, or an enclosure around the working area to protect the crew from extreme cold.

For wells that have been drilled on a slant rather than straight down, service and workover companies use a special rig called a *slant-hole rig* (fig. 5.63). It has a mast that can move forward and back as well as side to side so that the crew can adjust it to the angle of the wellbore.

Figure 5.63 The mast on this slant-hole rig is tilted to 45°, its maximum tilt.

A *spudder* is a cable drilling rig mounted on a truck or trailer. Instead of drilling by rotating a bit, it drills by repeatedly dropping a sharp, pointed bit onto the rock. It is no longer common as a drilling rig, but service and workover companies may still use a spudder to work on shallow wells. The advantage of a spudder is that a crew can move it in, rig it up, and be ready to work in a few hours.

On an offshore platform well with a permanent rig, no special service or workover rig is necessary—the crew uses the drilling rig. On offshore wells drilled by mobile rigs that have since moved on, ships or barges carry service and workover rigs to the well.

Auxiliary Equipment

Like rigs, auxiliary tools and equipment are mobile, often either on trucks or on skids. Some workover operations are 24-hour jobs. In this case, house trailers and provisions for the crews are standard auxiliary equipment.

Auxiliary equipment that a well servicing and workover rig may have includes a blowout preventer, makeup and breakout tongs, racks for storing pipe, a top drive or a rotary to rotate tubing, circulation equipment, and snubbing units.

Snubbing units are devices that force tubing (or drill pipe during drilling) into the well when the well pressure is high. Usually classified according to their power source, the two general types of snubbing units are the hydraulic unit and the rig-assisted unit. The hydraulic unit (fig. 5.64) operates with a self-contained hydraulic system and uses single or multiple hydraulic cylinders to move the pipe into or out of the well. The rig-assisted unit, on the other hand, uses the rig drawworks in a block-and-pulley arrangement to move the pipe.

Figure 5.64 This snubbing unit operates with a self-contained hydraulic system.

Servicing and Workover Fluids

The circulating fluid used during servicing and workovers is usually either salt water or a specially mixed drilling mud called *workover fluid*.

Workover crews very often use the salt water in the formation as the workover fluid because it is available and does not damage the formation. Or, like drilling mud, it can be a water-based or oil-based mud with additives that raise viscosity, add weight, or form a wall cake to prevent fluid loss.

Well Servicing and Repair

Wells require maintenance and repairs from time to time due to normal wear, age, and the hazards of the environment where the equipment is exposed. Downhole pumps, sucker rods, and gas-lift equipment all have moving parts, which wear out due to erosion from the fast-moving reservoir fluid that may contain sand or particles of metal left from perforating, for example. Production tubing also erodes. Both moving and stationary equipment can fail due to corrosion, scale, and paraffin deposits.

Well testers, operators, and production foremen are usually the first to notice abnormal conditions in the well that suggest the need for repair work. Routine tests and well reports on daily production, wellhead pressure, and percentage of water in the oil provide evidence of the need for maintenance work. The most common maintenance jobs include swabbing, and repairing or replacing wellhead parts, beam pumping equipment, production tubing, and packers.

Maintaining and Repairing Beam Pumping Equipment

Modern beam units perform for a long time with proper care and when not overloaded. Proper care means regular lubrication of the moving parts and seasonal oil changes. Many operators choose not to change the oil unless it appears to be dirty; however, they will have a worker pull a drain plug on the gear mechanism of the beam unit to drain any condensation in the lubricating oil. Another maintenance check is to assure that the cranks (see fig. 5.23) are correctly counterbalanced. Improper weight can cause damage to the gear teeth.

Sucker rods, their couplings, and downhole pumps can fail for any of several reasons or a combination of reasons. The couplings are the weakest points in a rod string, so they are the most likely points of failure. To service, repair, or replace the rods or downhole pump, the crew pulls the sucker rod string and the pump out of the well.

Corrosion due to oxygen (rust), hydrogen sulfide (sour corrosion), or carbon dioxide (sweet corrosion) in the well can cause metal parts to break (fig. 5.65). Metal also weakens and can break because of erosion from the moving fluids in the hole or because of wear from friction, when two metal surfaces rub together (fig. 5.66). When the completion or workover crew

Figure 5.66 Abrasion has damaged this sucker rod and its coupling.

Figure 5.65 Corroded sucker rods must be replaced to keep the pumping well in good condition.

makes up the sucker rod string, they lubricate each thread, but extreme pressure can squeeze the lubricant out. Friction then wears out the unlubricated thread and it breaks.

If the crew was not careful in handling the sucker rods, they may have damaged the metal by denting it. These dents become *stress risers*, which concentrate any destructive force that is present on that spot. For example, many sucker rods have a relatively soft coating of a special steel alloy that protects them from corrosion. Any break in this coating, caused by a hammer mark for instance, exposes the metal underneath and leaves a spot for corrosion to attack. Stress from the pump's repeated up-and-down movement is another cause of failure.

Repairing Production Tubing and Packers

A service crew can usually replace a packer fairly easily, but pulling tubing is a bigger job that requires a heavier rig. Packers and tubing fail for the same reasons the pump and sucker rods do: corrosion and erosion.

Paraffin deposits along the walls can reduce the diameter of the pipe so much that the production slows or tools cannot fit through it. The crew can clean out tubing with a *scraper* run in on a wireline (fig. 5.67) or by pumping hot oil or a chemical solvent down the hole to dissolve the paraffin or scale. When they use a solvent, the crew closes in the well for 24 to 72 hours to allow the solvent to dissolve the paraffin. Also, special chemicals that may inhibit paraffin formation are being developed. If none of these methods works, the crew may have to replace the tubing.

The tubing string must be intact and seal tightly over its entire length in order to contain the pressure of the formation fluids flowing up through it and to prevent them from leaking into the space between the tubing and casing. Leaks may develop in the middle of a tubing joint, but most often they occur in the couplings at each end. The reason for a leak may be a manufacturing defect or corrosion, or the rod string can wear a hole if it rubs against the tubing's inside wall. Some company representatives routinely test for tubing leaks in wells that use sucker rod pumps. They usually hire a company that specializes in pressure testing.

Swabbing

As mentioned in the section on well completion, a newly completed well is often swabbed to start the hydrocarbons flowing. When a service or workover crew pulls the tubing, they kill the well (stop production) by pumping brine into it. To restart the flow, they swab it. Swabbing in service jobs involves pulling a rubber-faced cylinder called a *swab* up the tubing, which reduces the pressure beneath the swab and sucks the fluids out. Another time the company representative may call in a swabbing rig (a light-duty service rig) for a producing well is when formation pressure drops and production slows. Sometimes swabbing will restart the flow.

Workover Operations

Workover jobs include cleaning sand out of the well and adding a means of preventing sand from entering the well, replacing liners, plugging the well, repairing casing, drilling deeper, and drilling around obstructions in the well. Some workover jobs require only a wireline to lower tools, but others need to rotate tubing or drill pipe, so the workover rig has equipment to rotate the pipe string. Operations that need to circulate workover fluid into the well require pumps and storage tanks.

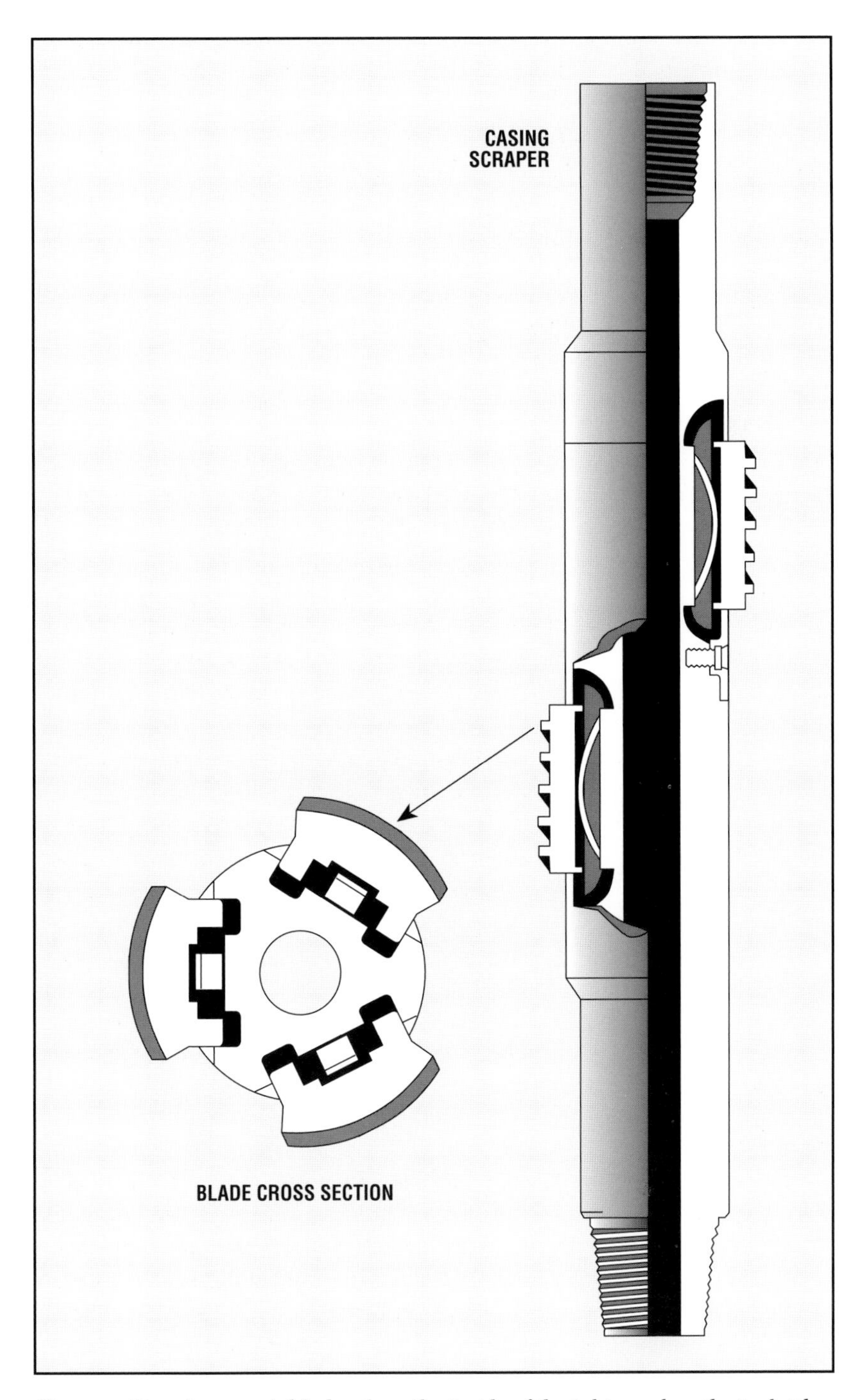

Figure 5.67 A scraper's blades clean the inside of the tubing when the tool either rotates or moves up or down.

Sand Cleanout

In a wire-wrapped screen completion, fine sand eventually infiltrates the gravel pack and the screen and fills up the inside of the slotted liner. Sometimes, however, in spite of every attempt to exclude it, sand enters the well and causes trouble. When this happens, a workover crew cleans out the well.

The method of cleaning out the sand depends on where the sand is and how tightly it is packed. All methods use circulation of a fluid, usually salt water, to flush the sand out.

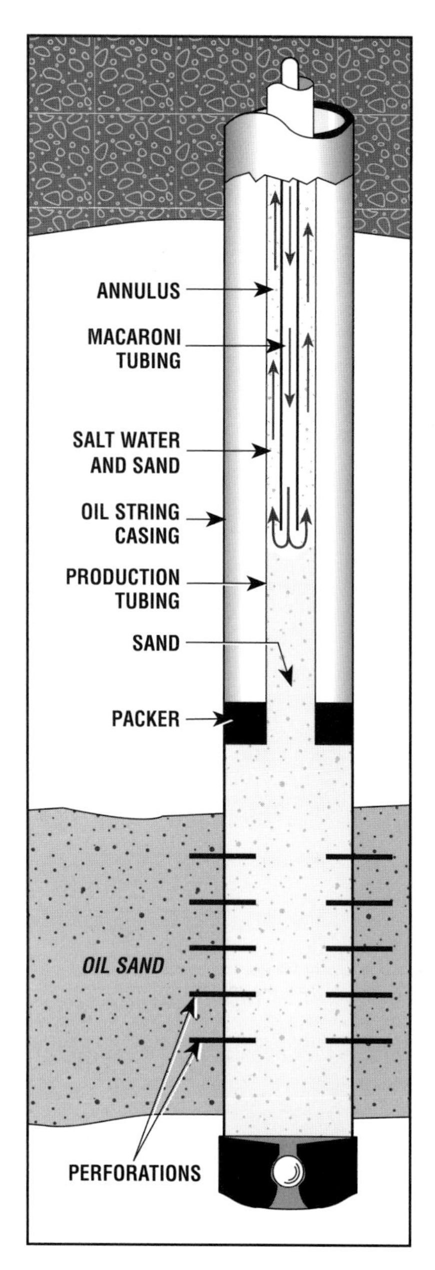

Figure 5.68 Macaroni tubing fits inside the production tubing to wash sand out.

The first method uses either a macaroni rig (fig. 5.68) or coiled tubing (fig. 5.69). A *macaroni rig* is a relatively small rig that handles special lightweight, small-diameter pipe called *macaroni*. The crew leaves the production tubing and packers in place and lowers the macaroni string or coiled tubing inside the production tubing until it is just to the top of the sand. Then they circulate salt water down the tubing at a high velocity, lowering the string as the sand washes out. This high-velocity salt water forces the sand to the surface through the annulus between the production tubing and the macaroni or coiled tubing. A macaroni rig is particularly useful for sand cleanout from a barge or offshore platform because the hoist and working string are lightweight.

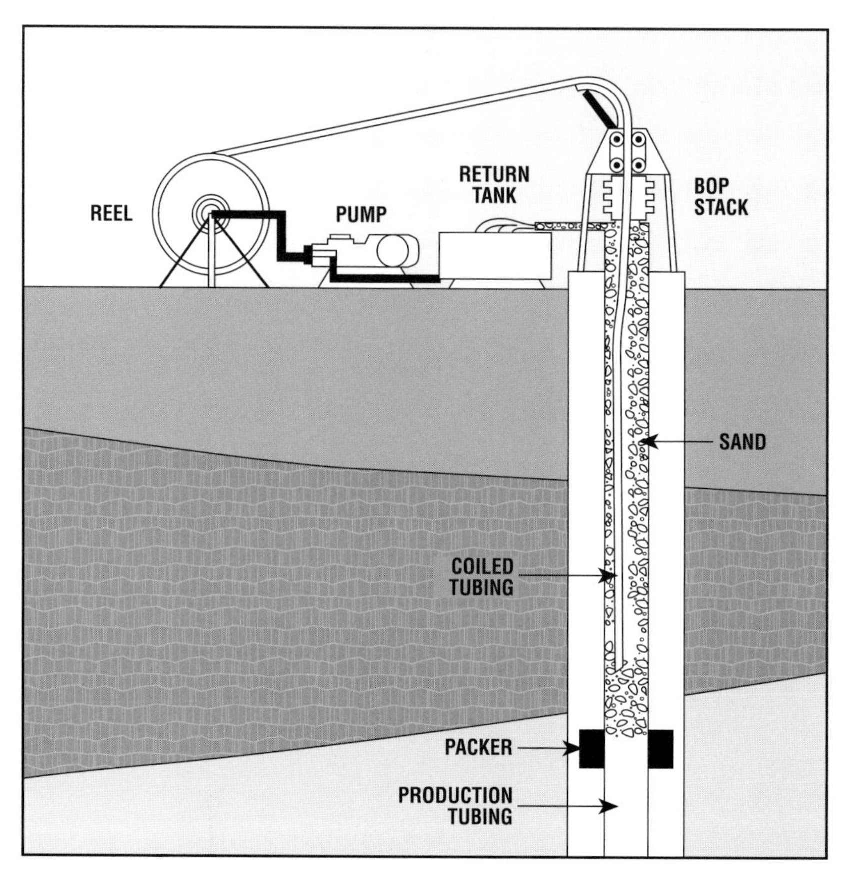

Figure 5.69 Coiled tubing also washes sand up the annulus between production tubing and coiled tubing.

In some cases, the sand gets packed solidly in the annulus between the tubing and casing. Sand can enter the annulus through a hole in the casing or through the perforations. The tubing gets stuck when this happens, preventing the crew from pulling it for service or workover jobs. To clean out the sand in this case, the workover crew first cuts the free portion of the tubing above the sand and pulls it out. Then they use a washover string that reaches from the surface all the way down to the packed sand (fig. 5.70). As the rotary shoe mills out the sand, circulation washes it up the annulus.

Replacing Screen Liner

A well that was completed with a gravel pack and screen liner may need repair after a time. As more and more sand passes into the slotted liner, the cutting action of the sand grains gradually enlarges the slots. Eventually, a workover crew must remove the damaged liner, either by pulling the tubing or cutting and fishing it if it is stuck. After removing the liner, they usually must clean out the well to the bottom of the open hole in order to condition the hole for the new liner.

Sand Control

Producers have tried for many years to devise a method of keeping sand out of a well. Wells in hard-rock country may never have sand trouble, but in some areas, such as California and the Gulf Coast, problems occur every day and are handled almost routinely. A company representative may call in a workover company to control sand in existing wells because the original completion was inadequate, because a crew has drilled the well deeper to a new formation that produces sand, or because the characteristics of the reservoir have changed.

Adding a *gravel pack* is the most common method of controlling sand in an existing well. Adding a gravel pack as a workover job is no different from installing it during completion. Another method is to bond the sand grains together using a plastic or a resin to "lock" them in place. This is known as *chemical consolidation*. A third method is placing a resin-coated sand pack. To form a resin-coated sand pack, the workover crew begins with a slurry (a solid suspended in a liquid) of sand and a resin. The crew pumps the slurry down the production tubing or coiled tubing into the casing and its perforations. When the resin hardens, it forms a plug that oil and gas, but not sand, can flow through.

Plug-Back Cementing

Plug-back cementing is placing a cement plug at one or more points in a well to shut off flow. The most common reason for plug-back cementing is to isolate a lower zone from the upper part of the well (fig. 5.71), usually because the lower zone is depleted of oil and gas. The cement plug shuts off the lower part of the formation so that only the higher area produces. The plug keeps produced fluids from migrating down into the lower zone. It also prevents salt water in the lower zone from migrating to the higher one.

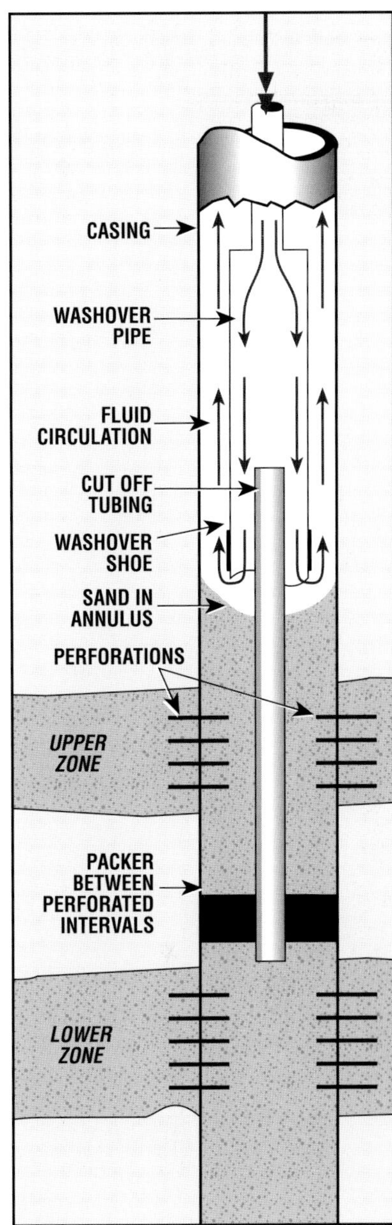

Figure 5.70 A washover pipe mills out packed sand around the tubing.

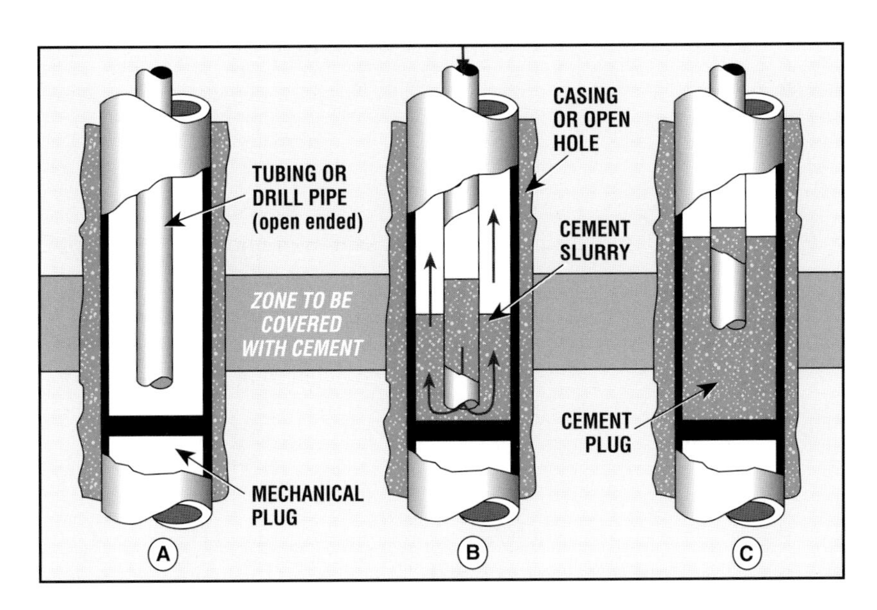

Figure 5.71 One reason to place a cement plug is to shut off a lower producing zone in order to produce a formation nearer the surface.

Other reasons for plugging back are to seal off a dry well before abandoning it and to plug the hole when preparing to sidetrack. *Sidetracking* is drilling a new hole to bypass something that is permanently stuck in the hole. The workover crew places the plug at the depth where they will start the sidetrack. Then they start drilling toward one side, and the plug seals off the old hole.

Casing and Production Liner Repair

Casing and production liners have the same problems and are repaired in the same ways. In this section, wherever the term *casing* is used, it generally means liners as well. Corrosion, abrasion, pressure, and other destructive forces can create holes or splits in the casing. One way that workover crew members know that the casing has a hole is when they find shale or sand inside the well after pulling the tubing.

Squeeze cementing is one way of repairing holes in casing. Other methods are patching it with a liner patch, replacing part of the string, or running a liner to cover the bad place in the casing string.

Squeeze Cementing

In general, in *squeeze cementing* the crew pumps cement down drill pipe or tubing to an opening in a well's casing and applies pressure. A tool on the end of the drill pipe or tubing called a *cement retainer* prevents the cement from flowing up the annulus. The pumping pressure forces the cement into the opening in the casing and against the formation behind it. The opening may be a hole that is the result of damage to the casing or an intentionally created hole, such as the perforations of a perforated completion (fig. 5.72). Workover crews can also squeeze cement against a part of the hole with no casing (as in an open hole completion). Squeeze cementing, like plug-back cementing, is useful for sealing off water in a lower zone from the oil in a higher zone as the water level rises. Other uses are to repair holes in the casing; to fill empty spots, or channels, behind casing where cement should be located but for some reason is not; and to seal off an upper depleted zone when the lease operator wishes to produce the well from a lower zone.

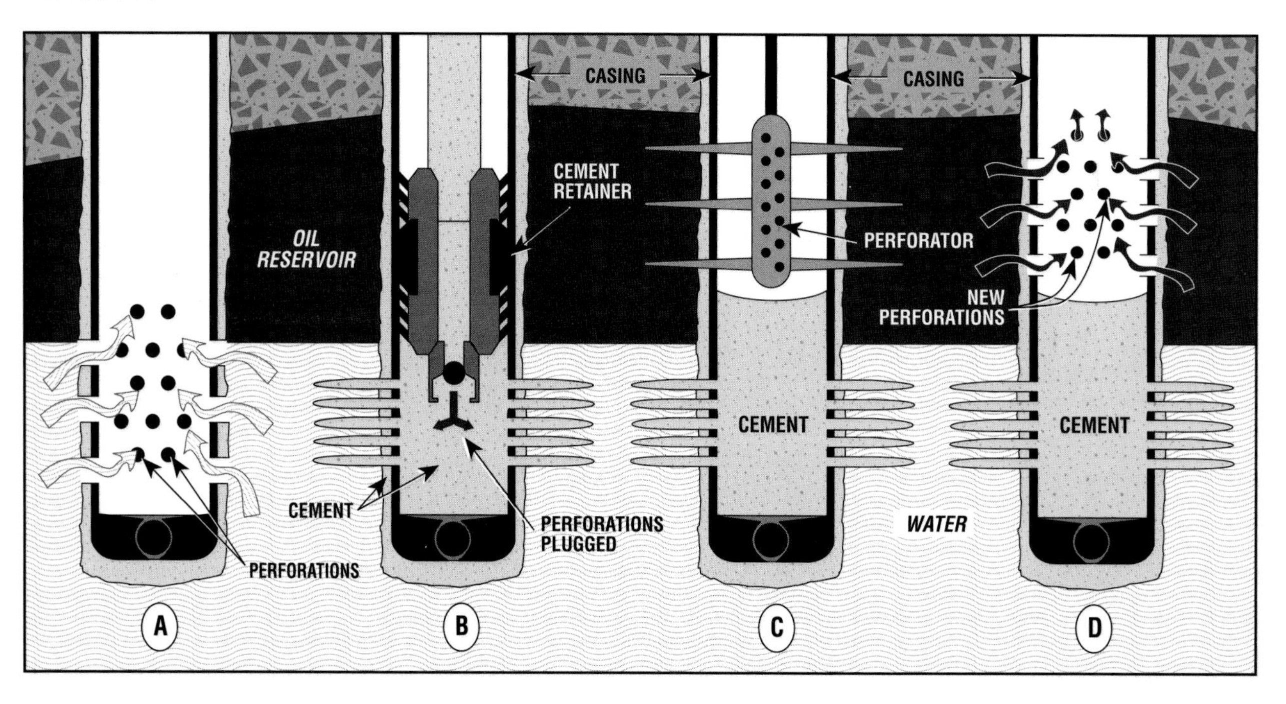

Figure 5.72 Squeezing cement through the lower perforations seals them (A, B). Then, the upper part of the zone is perforated (C, D).

OFFSHORE AND ARCTIC PRODUCTION

Hydrocarbons produced from offshore and arctic wells require the same types of separation and handling as those from land wells. The main differences result from the remoteness of the locations and the special challenges of the environments.

Modern Production Platforms

The drilling platform (see Chapter 4) becomes a production platform when one or more development wells have been drilled and production becomes the main activity of the platform. The operator may remove the drilling rig, but it usually remains on the platform for use in servicing the producing wells.

Besides the derrick, the modern platform houses all the usual production equipment. A platform generally has a helipad, one or more cranes, antipollution devices, and a myriad of safety devices such as firefighting equipment, life rafts, and escape capsules. Helicopters are used for emergency evacuation and to transport people and supplies to and from the platform (fig. 5.73).

Figure 5.73 Helicopters and cranes are common sights around an offshore operation. *(Courtesy of Petroleum Helicopters, Inc.)*

Figure 5.74 Work boats transport crew and cargo between shore and platform.

Figure 5.75 In surface completions, the wellhead is placed on the deck of the platform.

Crewboats, supply boats, and other types of work boats also move crew and cargo between the shore and the platform (fig. 5.74). Communications to and from the platform include remote monitoring of production equipment, closed-circuit television, radio transmissions, and recording devices of all kinds. Weather balloons and satellites give offshore platforms and auxiliary vessels early weather warnings. The use of interactive computer terminals in every phase of operation keeps the offshore drilling and production platforms in touch with the rest of the world.

Offshore Completions

A well completion offshore is much like one on land, except that it uses a different technique to control the fluids as they flow from the top of the well. The wellhead valves and fittings may be placed above the surface of the water for a *surface completion*, or they may be installed on the seafloor in a *subsea completion*. A well drilled from a platform is usually completed with the Christmas tree on the deck of the platform (fig. 5.75), while in a subsea completion, the wellhead is located on the seafloor. A subsea Christmas tree can be either *wet* or *dry*. A wet Christmas tree remains exposed to the surrounding water (fig. 5.76), while a dry one is installed in some type of housing that isolates it from the water (fig. 5.77).

Divers with special underwater clothing and breathing apparatus handle wet subsea repairs and maintenance. On a dry installation, workers may use a small submarine called a *submersible* or a *diving bell* to reach the

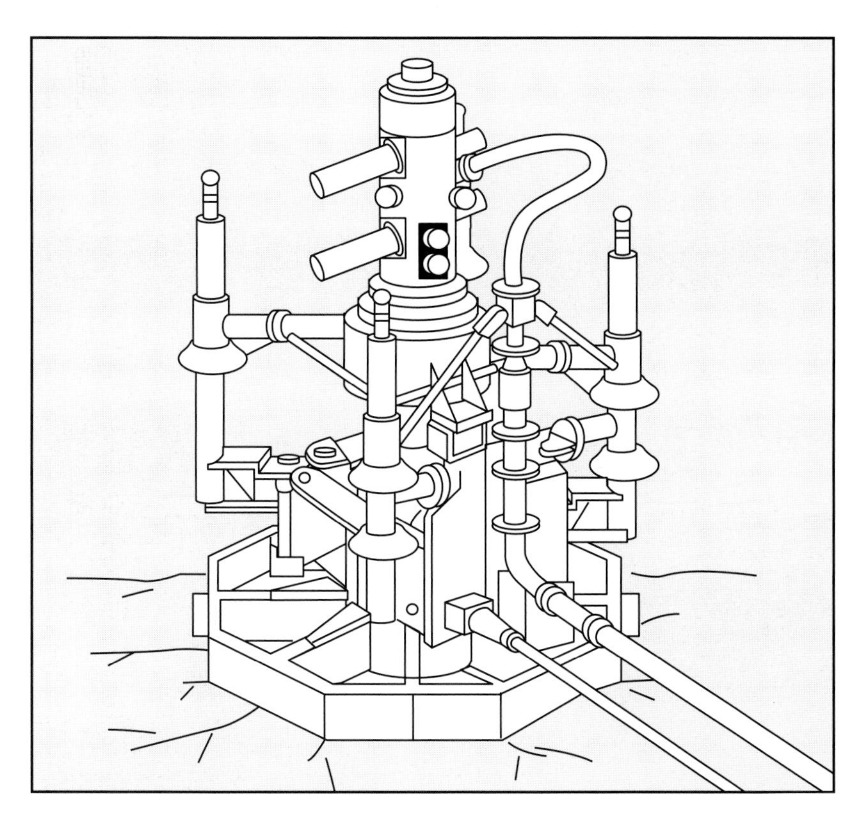

Figure 5.76 In a wet subsea completion, valves and equipment to control the flow of hydrocarbons are placed on the seafloor.

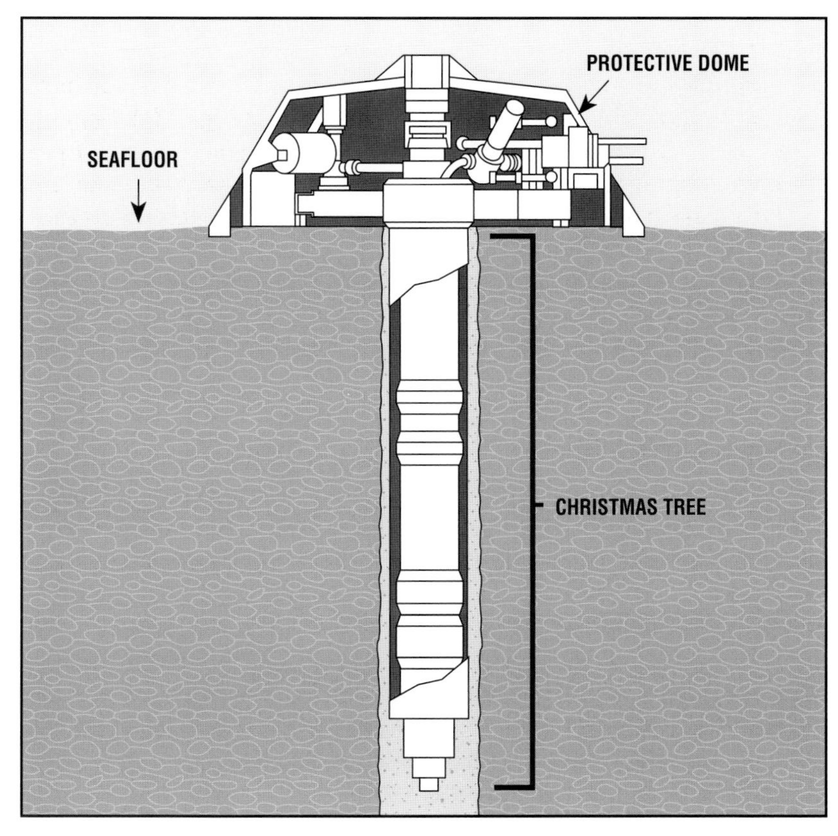

Figure 5.77 In a dry subsea completion, the Christmas tree is protected from the water.

Figure 5.78 Workers may use a diving bell to reach the seafloor for subsea repairs.

seafloor (fig. 5.78). There they can enter the housing and work without special underwater gear.

Handling Well Fluids Offshore

Generally, the offshore operator removes the water and other impurities from well fluids before the oil and gas are put into tankers or pipelines for transportation to shore. Therefore, the platform must contain separating and treating equipment, compressors, pumps, dehydrators, special storage tanks, and other fluid-handling equipment.

In some subsea completions, oil, gas, and water are routed into flow lines that run from the wells to a *production riser*, or pipe. Production flows up the riser to a floating buoy. An oil tanker can tie up to the buoy, take on the produced fluids, and transport them to shore for separation and treatment.

Sometimes, however, in deep waters where erecting a platform is not economically feasible, a processing and storage ship is moored to the buoy. Reservoir fluids are separated, treated, and stored on the ship. When the ship is full, it transports the hydrocarbons to a receiving terminal on shore.

Submerged Production System

A more elaborate subsea system consists of a grouping of satellite wells that supply natural gas or crude oil to a centralized production system (fig. 5.79). Flow lines attach these satellite wells, often widely scattered along the reservoir, to a central platform for treating and then to a buoy, where tankers take on the crude. Satellite wells allow the reservoir to be drained from widely separated points. This *submerged production system (SPS)* is more practical and economical than using only a single well attached to a conventional production platform.

The first successful SPS was tested in the Gulf of Mexico between 1974 and 1979. The SPS was created to handle all phases of production, from drilling to abandonment. The major components include using a seabed *template* (a steel pattern that lies on the seafloor with built-in accessories and guides for equipment) through which several satellite wells are drilled and completed. Pipelines connect the template to a conventional platform, and a *robot maintenance system (RMS)* services the underwater equipment.

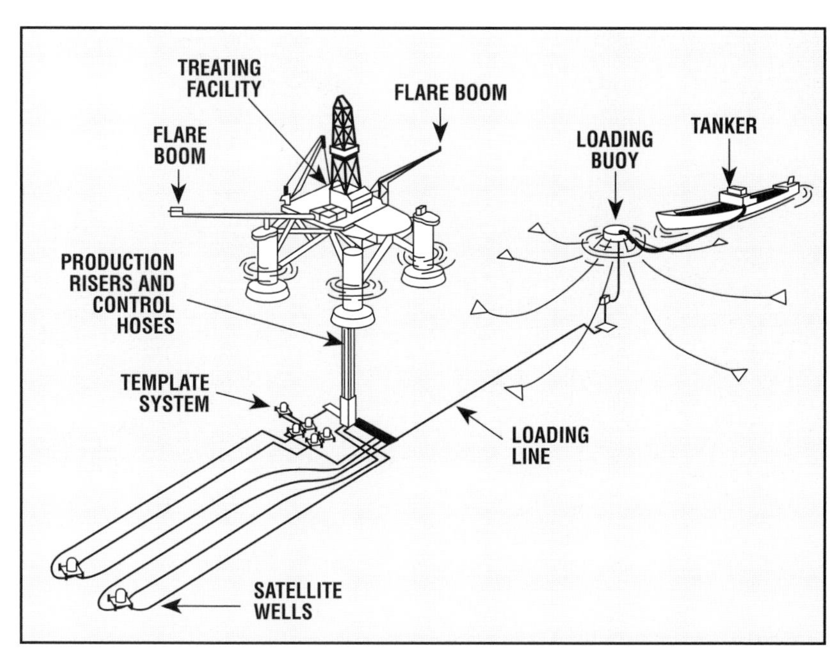

Figure 5.79 A submerged production system connects satellite wells to the production platform for treating and then to a mooring buoy, where tankers take on the oil.

Underwater Manifold Center

The successful use of the SPS in deepwater recovery inspired the technology for an *underwater manifold center (UMC)* for the North Sea. This revolutionary UMC system allows the exploitation of smaller offshore fields all over the world—once thought too unprofitable to produce in conventional ways. The UMC is connected by pipeline to a platform more than 4 miles (6.4 kilometres) away. The system has several functions, including acting as a template for wells and as a connection for satellite wells drilled some distance away. The UMC collects fluids from the wells and sends them to the platform. Other functions are to aid in enhanced recovery and pressure maintenance methods by the injection of seawater and to control and service individual wells with flow-line equipment.

Arctic Production

Production problems in the oil-rich Arctic ice fields are quite different from those posed by ordinary drilling environments. The frozen north, with its extremely low temperatures that can freeze human flesh in minutes, ice fogs, and overwhelming transportation problems, has challenged the technological skills of petroleum engineers and exploration companies worldwide.

One specific problem related to the completion of an Arctic well involves the presence of *permafrost*, permanently frozen soil or rock (fig. 5.80). Partial thawing and refreezing near the surface, which is common in permafrost regions, puts enormous strain on the casing and wellhead. Thawing causes the permafrost to deform and so puts pressure on the casing. The design of the lease site in the Arctic also must take permafrost into account. Thawing on a large scale underneath a well can lead to different frost heaving and settling pressures and can therefore adversely affect production. In the Arctic, Christmas trees are generally inside heated housings, as are other pieces of lease equipment, to make production possible in below-zero temperatures.

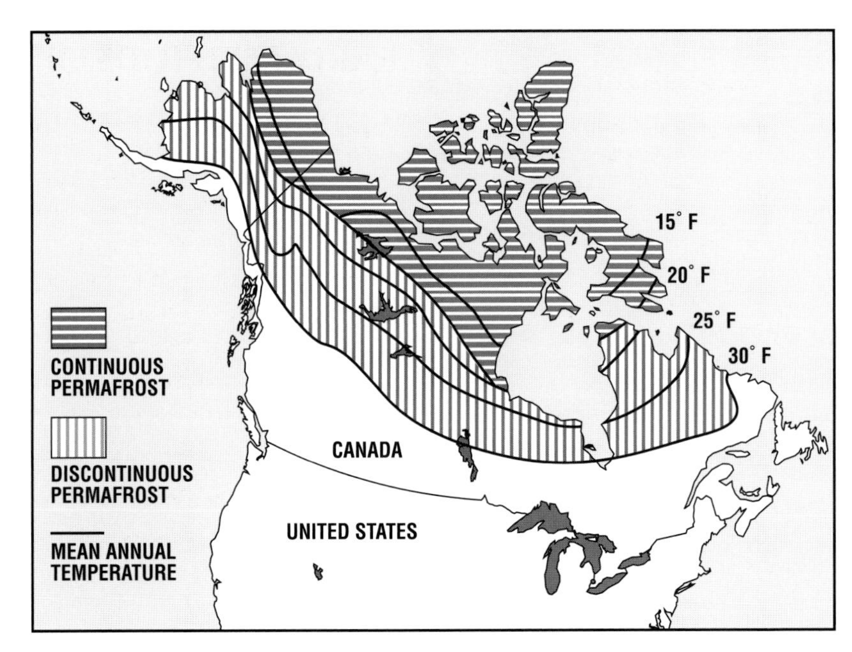

Figure 5.80 Arctic production must take into account areas of permafrost, shown here in North America.

6 Transportation

Moving oil and gas from a field to refining and processing plants and petroleum products from refineries to consumers requires a complex transportation system. Inland waterway barges, railway tank cars, transport trucks, oceangoing tankers, crude oil and products pipelines, and gas transmission pipelines all play an important part in the oil and gas transportation industry—an industry that started with horse-drawn wagons carrying wooden barrels of crude oil to nearby streams. As the industry grew, so did the methods of transportation, and today millions of barrels of crude oil, gasoline, fuel oils, and other petroleum products, as well as billions of cubic feet of natural gas, are moved from the wellhead to the plant, from one refinery to another, from offshore to onshore, and from continent to continent to reach the ultimate consumer.

EARLY METHODS OF TRANSPORTATION

Moving oil from well site to refinery became a necessity when the Drake well began producing 20 barrels (3 cubic metres) of crude oil per day and started an oil boom in the area of Oil Creek, a tributary of the Allegheny River. The first refinery was constructed at Oil City, some 11 miles (18 kilometres) from the field. Shortly thereafter, a refining complex developed on the Allegheny River at Pittsburgh, about 75 miles (121 kilometres) downstream, making Pittsburgh a major refining and oil transportation center. Demand for crude oil was pressing, and many competed for the job of hauling the oil to wherever it was needed—to the nearest waterway, railhead, pipeline dump tank, or refinery.

Wagons and Water

Because of the proximity of the rivers and streams, one solution to the transportation problem was a massive system of boats, barges, and wagons operated by teamsters. Horse-drawn wagons carried wooden barrels of oil to the streams, where they were loaded on barges and boats of all types and sizes. Local sawmill operators agreed to release the water normally kept for floating logs downstream, and the oil boats (and sometimes the barrels themselves) were floated down the Allegheny River to Pittsburgh.

Until the invention of the steamboat, barges laden with oil were pulled by horses walking on the river banks, or they were floated downstream, completely dependent on the river current for power (fig. 6.1). At the height of the operation, as many as a thousand boats were in use, including twenty to thirty steamers, passenger boats, and towboats pushing barges.

Rails and Tank Cars

After many barrels of oil were lost to the hazards of water travel, the oil industry turned to railroads as an alternative solution for moving petroleum and petroleum products. Railroad construction proceeded rapidly. The nearest rail outlet to Oil Creek prior to May 1861 was at Corry, Pennsylvania, some 20 miles (3 kilometres) north of Titusville. By the following October,

Figure 6.1 Horse-drawn wagons and barges were initially used to move oil to rail and water shipping points.

tracks had been laid to Titusville, and by 1864 the rail network provided crude oil to Cleveland and New York. Since the railheads were still located some distance from the wells, teamsters transported the crude in their wagons, each carrying from five to seven 360-pound (163-kilogram) barrels. As many as 2,000 teams a day lumbered through the mud into Titusville carrying oil to the railway for $1 to $5 per barrel. The wooden barrels of oil were then transferred to boxcars for shipment to refineries. But the barrels were expensive, were easily damaged or pilfered, tended to leak, and required too much handling.

These difficulties resulted in the development of the wooden tub tank car by Charles P. Hatch in 1865. In 1866, brothers Amos and James Densmore of Meadville, Pennsylvania, improved on the design and patented the first tank car. Their tank car consisted of two 1,700-gallon (6,435-litre) wooden tubs, glued and banded with iron hoops and mounted on a flatcar (fig. 6.2). The Densmore tank car was first used to ship oil from Pennsylvania to New York City. In 1869, a horizontal tank car with an expansion dome on top was developed and has remained the industry's basic design. The *expansion dome* is a space in which any gas that breaks out of the oil can collect. It prevents the buildup of pressure in the tank car that could cause damage.

Figure 6.2 The Densmore brothers patented this early railway tank car design.

The First Oil Pipelines

Pipelines are a major network of small and large arteries installed by a pipeline company (also called a *pipeline* for short) to give oil producers an outlet to market. The original stimulus for building pipelines was the high cost of moving oil from the field to shipping points.

After several unsuccessful attempts to pipe crude oil from the field to the nearest shipping points, in 1865 Samuel Van Syckel succeeded in building a pipeline and starting the first oil pipeline business. His Oil Transportation Association laid a 2-inch (5-centimetre) wrought-iron line from Pithole Creek (8 miles, or 13 kilometres, from Oil City) to the Miller Farm railroad station, 5 miles (8 kilometres) away. Sometime later he built a second line, and with his two lines he was able to move 2,000 barrels of oil per day at the low charge of $1.00 per barrel—much lower than the teamsters' charge of about $2.50 per barrel for the same distance. The teamsters reacted by tearing up portions of the line but were unsuccessful in deterring pipeliners as they developed crude oil gathering systems and trunklines.

Crude Oil Trunklines

The railroad companies were the first to construct and buy pipelines of their own or to form exclusive arrangements with pipeline transportation companies. In an effort to maintain monopolistic control over the oil transportation business, the railroads attempted to prevent other pipeline companies from crossing their rail lines (fig. 6.3). But public sentiment was against monopolies, and the Pennsylvania and Ohio legislatures passed laws in 1872 that granted common-carrier pipelines the privilege of eminent domain in obtaining their rights-of-way. Eminent domain is the right of the government to buy private property for public use. These acts were the prelude to the development of oil trunkline systems, because main trunk pipelines could connect the oil centers directly to the refineries and provide more economical transportation.

Until 1900, crude oil production was concentrated in Pennsylvania, West Virginia, Ohio, and other eastern states, and refineries were also centered around the eastern population centers. But between 1901 and 1905, oil was discovered in Texas, Kansas, Oklahoma, Louisiana, and California,

Figure 6.3 Denied the right to cross a railroad, one pipeline company used tank wagons to transfer the oil from the pipeline to holding tanks on the other side and then back into the pipeline.

Figure 6.4 An early pipeline pump station used steam power to build sufficient pressure to move the oil.

and the eastern refineries were forced to seek additional crude supplies from the western centers of production. By 1914 the mid-continent oilfields supplied over 60 percent of the crude oil to the eastern refineries, and by 1940 they were supplying over 85 percent. Pumping stations were installed along the pipelines to move the crude on its way (fig. 6.4).

A network of crude oil trunk pipelines was built to connect the source of supply with the refining centers, although a large part of transporting was done by ship as well. The combined system of ships and pipelines worked quite well until World War II, when enemy submarines sank many of the oil tankers destined for wartime industries of the East Coast. So the federal government joined forces with the petroleum industry and constructed the first large-diameter crude oil pipeline, called the *Big Inch* (fig. 6.5), and the refined products line, called the *Little Big Inch.* The Big Inch was 24 inches (61 centimetres) in diameter, while the Little Big Inch had a 20-inch (51-centimetre) diameter.

Gathering Systems

Early pipelines were not connected directly to the oilwells but received their oil from large dump tanks located at strategic shipping points. A *dump tank* is a calibrated metering tank that releases a certain amount of liquid automatically. Pioneer Henry Harley was the first to build a pipeline to carry oil from the wells to his dump tanks, a job generally handled by the teamsters. Soon after he built two lines from Benninghoff Run to tanks at the Oil Creek railroad loading depot, a longtime dispute between the teamsters and the pipeliners was brought to a head. A bitter fight resulted in much destruction, but the pipeliners persisted.

In 1866 A. W. Smiley and G. E. Coutant formed the Accommodation Pipeline Company and constructed the first *gathering lines,* a network of 2-inch (5-centimetre) pipe some 4 miles (6.4 kilometres) across the Pithole field to connect the tanks at the wells with the dump tanks of the pipelines. Later, gathering lines were connected directly into the trunklines, and elaborate accounting systems were developed to keep track of the owners.

Figure 6.5 As unit trains carrying oil sped east, pipeliners raced to finish the Big Inch crude oil pipeline.

Products Pipelines

The first *batching* of refined products (shipping one petroleum product after another through the same pipeline) was performed in 1901 in Pennsylvania, but construction of products pipelines did not begin in earnest until 1930 when the growth of large population centers in the Midwest had created new marketing areas for refined products. Before the use of pipelines became widespread, products were transported with horse-drawn carts and Model T trucks (fig. 6.6).

Figure 6.6 Petroleum products were transported with horse-drawn carts and Model T trucks before products pipelines became widespread. *(Courtesy of Humanities Research Center, The University of Texas at Austin)*

During the economic depression in the 1930s, the established oil marketers battled more fiercely than ever for a share of the refined products market. Competition demanded the least expensive method of transportation from the refinery to the market—pipelines that could transport many grades of refined products. As a result, 3,000 miles (4,830 kilometres) of products pipelines were put into operation during 1930 and 1931. In some instances, older crude lines were cleaned and the direction of flow reversed to move products from the East Coast to the Midwest.

Gas Transmission Pipelines

Although people knew about the uses of natural gas 3,000 years ago when the Chinese piped natural gas through hollow bamboo poles (fig. 6.7), widespread use of natural gas did not develop until it became easily available at competitive prices. Aaron Hart, a gunsmith, drilled the first American natural gas well in 1821. He had drilled only 17 feet (5 metres) when a hissing sound indicated that he had struck not water but something unexpected—gas. Hart promptly experimented with the gas to see what could be done with it. Before long he had strung together a few hollowed logs to serve as a gas pipeline. With this primitive system, he managed to light some buildings nearby, and this was the beginning of the natural gas industry—an industry that depended totally on an efficient method of transportation.

From then on, the problem that took longest to solve was finding a practical means of transmitting gas from the fields where it was produced to the cities and factories where it could be used. By the beginning of the twentieth century, natural gas was available as a fuel for homes and industries, but costs were high because the delivery system was still inefficient.

Figure 6.7 Using bamboo poles, the Chinese were among the first to pipe natural gas. *(Courtesy of American Gas Association)*

Figure 6.8 True horsepower was used in early-day construction of gas pipelines and compressor stations. *(Courtesy of El Paso Natural Gas Co.)*

Constructing the first natural gas pipeline was backbreaking work, especially since heavy machinery had not yet replaced horses and mules (fig. 6.8). Until the 1920s, construction was largely performed by hand labor, including clearing right-of-way, digging ditches, screwing ends of pipe together, lowering in, and backfilling (fig. 6.9).

The first large-scale natural gas transmission lines came about as a direct result of World War II. After the war, private industry bought the crude oil and products pipelines—the Big Inch and the Little Big Inch—and converted them to natural gas transmission lines, thereby solving the problem of supplying gas to the markets at rates competitive with other fuels.

Figure 6.9 Hand tongs were used by early pipeline construction workers.

Ships at Sea

During the 1860s a few sailing ships fitted with iron tanks had begun hauling crude across the Atlantic; however, bulk transportation did not replace barrels for some time, because the shifting of a load of oil in a big tank tended to throw a ship out of ballast, or balance. Also, loading and unloading large tanks was a problem. Petroleum, in barrels or in bulk, was a dangerous cargo with the threat of fire and explosion ever present. Barrels often leaked because the glue used to hold them together dissolved from exposure to crude or refined oil. In bulk containers, gases often accumulated—gases that were very susceptible to explosion.

When refining was promoted by European industrialists, oceangoing ships were designed to carry bulk crude from the producing countries (the United States and Russia) to Europe. In 1886 the *Gluckhauf,* hauling refined oil, made its maiden voyage from New York to Bremen, Germany. Designed by Wilhelm Riedemann, a German importer, the *Gluckhauf* was to be the first successful oceangoing tanker using the ship's hull as part of the storage compartments (fig. 6.10).

New designs that stored the oil in separate steel compartments with a pipe for loading and unloading and a pipe to vent gas accumulations, along with conversion from sail to steam power, revolutionized the tanker as a means of transporting petroleum and its products. By the turn of the century, companies were shipping products from their Gulf Coast plants to the East Coast and abroad. In 1901, at the time of the Spindletop strike, Sun Oil bought a cargo vessel, the *S. S. Paraguay,* and converted it into a tanker to ship oil from the Gulf Coast ports. Other companies followed suit and bought or leased tankers to ship their oil.

After World War I, steam-powered tankships were developed with cargo capacities of 9,000 deadweight tons (9,145 tonnes) and with double bottoms. (*Deadweight tonnage* is a ship's load carrying capacity, expressed in long tons. A long ton is 2,240 pounds, or 1,016 kilograms.) Diesel engines began to replace steam engines as the power source. In World War II, fleets of U.S. tankers, carrying some 34,000 deadweight tons (34,544 tonnes), played a key role in supplying petroleum for wartime industries on the East Coast, as well as products from the Western Hemisphere to the war zones. These tankers became prime targets for enemy submarines, forcing the United States to expand its pipeline network—a less vulnerable mode of transportation.

Figure 6.10 The *Gluckhauf* was the first successful oceangoing tanker that carried oil in storage tanks built into its hull.

Tank Trucks

When automotive vehicles came into use at the end of the nineteenth century, tank trucks, patterned after the tank wagons, were used to transport crude oil from the wellhead to the pipeline or other shipping outlet (fig. 6.11). With the requirements of World War I, better truck designs emerged, overcoming the handicaps of inadequate power, small size, and unsafe features. Welding replaced riveting as a means of fabricating the tanks, and soon after World War II, aluminum, stainless steel, and plastic tanks replaced many of the steel ones. The transport of chemicals and corrosive commodities and the development of new welding methods were largely responsible for this change in tank construction.

As gathering lines took over the job of transporting crude, tank trucks were used primarily for gasoline and fuel oil. During and after World War II, *liquefied petroleum gas (LPG)* became a major commodity for tank truck transportation. Tank trucks, specially designed to contain the pressures of liquefied gas, transported LPG not only to dealers throughout the country but from the dealers to their many rural customers.

Design improvements in tank trucks, as in all other methods of transporting petroleum and products, continued to accommodate the needs of the industry.

Figure 6.11 An early tank truck hauls crude oil from the well's wooden storage tank. *(Courtesy of Goldbeck Collection, Humanities Research Center, The University of Texas at Austin)*

BARGE TOWS

In the United States, petroleum and petroleum products are transported by barge wherever there is a waterway and a need for the products. Barges carrying gasoline, asphalt, crude oil, chemicals, and industrial fuel shuttle back and forth on a network of lakes, rivers, channels, and intracoastal canals. Levees, dikes, and other flood control devices and navigational aids have made the inland waterway system a safe and efficient method of moving huge amounts of crude oil and products to coastal tankers and refineries (fig. 6.12). A *barge tow* is a string of barges laced together with ratchets and steel cable.

Figure 6.12 This tide gate at Freeport, Texas, is one of the flood control devices that makes the Gulf Intracoastal Waterway safe for navigation. *(Photo by Greg White; Courtesy of* Texas Highways Magazine*)*

Figure 6.13 The draft marks on the side of this barge docked at a refinery show that it is virtually empty. Note the bow thruster, which aids the towboat in steering the barge. *(Courtesy of Valero Refining Co.)*

Barges

A *barge* is a flat-decked and usually flat-bottomed vessel. An empty barge rides high in the water, showing its draft marks, or waterline marks, on its side (fig. 6.13). As it is filled, the barge rides deeper and deeper in the water, and tonnage is estimated by reading the marks on the hull. Barges have no power to move on their own. They depend on tugboats, towboats, and water currents to move.

Oil barges were the first to travel the Gulf Intracoastal Waterway, which runs the length of the Gulf Coast from Brownsville, Texas, at the Mexican border more than 1,000 miles (1,610 kilometres) to Carrabelle, Florida. This waterway has contributed to making the Gulf Coast the nation's leading petrochemical region. Barges hauling huge spherical chemical tanks are a common site on the canal. Barges are considered one of the safest means of moving chemicals, since barrier islands, peninsulas, and inland routes in the area protect barge tows from the rough waters, squalls, and storms common in ocean waters.

Petroleum and fuel oil are still the largest cargoes on the Gulf waterway. Residual fuel is the most frequent cargo, followed by basic chemicals, gasoline, and distillate fuel oil.

Tugboats

Tugboats are strongly built boats, designed to push or pull other vessels in harbors, on inland waterways, and in coastal waters. Although comparatively small, a tugboat is powerful and can often be seen pushing a vessel many times its size. Tugboats help maneuver barges into line to make a barge tow or hold a barge in place at the dock for loading and unloading (fig. 6.14).

Figure 6.14 A tugboat holds a barge in place at a loading dock (left), while a towboat pushes a loaded barge out of the harbor. Note the keeled hull and bow of the tugboat and the square bow of the towboat. *(Courtesy of Saber Energy, Inc.)*

Towboats

Unlike the tugboat, the *towboat* is a relatively flat-bottomed boat with a square bow (front). Affixed to the bow are large upright *knees*, which are used for attaching and pushing barges (fig. 6.15). Sometimes a control barge called a *bow thruster* (see fig. 6.13) is used between the towboat and the barge to give more control in steering the barge down the waterway. Diesel towboats can push a string of barges carrying tons of petroleum products over great distances at low cost.

Figure 6.15 A towboat has large upright knees on its bow to facilitate pushing. *(Courtesy of Valero Refining Co.)*

A towboat may range in power from 600 to 2,400 horsepower (420 to 1,680 kilowatts). A large towboat, sometimes as tall as a four-story building, can push as many as forty barges in a tow, carrying some 18,000 tons (16,300 tonnes) of cargo. This string can "make tow" at about 6 miles per hour (10 kilometres per hour) or more, depending on the waterway—its current, crosscurrents, sandbars, wind, ice, and other traffic. On the river, the tow going with the current has the right-of-way, so those barges going against the river current must stop and wait for oncoming traffic to clear the waterway before they can proceed.

The towboat crew (typically, a captain, a ship's mate, a chief engineer, an assistant engineer, a tankerman, a pilot, a cook, and deckhands) build the tow as they stop along the way to pick up barges. Oil barges "may be strung out for more than 1,000 feet (305 metres) in barges 54 feet (16.5 metres) wide and almost 300 feet (91.5 metres) long, lashed tightly together so they can be steered as a single unit."[1] Such a barge tow may be loaded with as much as 60,000 barrels (9,540 cubic metres) of oil. A river tow may be as large as two side-by-side strings, each stretching some 800 feet (244 metres). With oil, chemicals, and other petroleum products on board, no one is inclined to take chances, especially when going through control gates, drawbridges, or swing bridges. The barges have no lights, and smoking is strictly prohibited on them. Visitors are not allowed on the towboat or the barges.

A towboat crew may be on a trip up the Mississippi River for as long as a month. A trip from Tulsa, Oklahoma, down through the rivers and canal to plants in Nederland, Texas, may take from 16 to 18 days. After picking up its string of barges, a towboat seldom stops. Fuel and supplies are furnished by service boats that come alongside the tow in midstream.

---

[1]Tommie Pinkard, "Gulf Intracoastal Waterway: Texas' Big Ditch," *Texas Highways*, August 1984, p. 22.

RAILWAY SYSTEMS

Compared to the amount of petroleum and products shipped by other means, the railroads handle only a small volume; however, rail is a vital link in the total petroleum transportation network. Railway systems in the United States transport petroleum feedstocks to refineries and refined products to consumers.

Petroleum Products Transported by Rail

The United States' chemical industry relies heavily on its fleet of tank cars to safely transport millions of gallons (thousands of cubic metres) of petroleum-based chemicals, or *petrochemicals*, used in virtually all manufacturing processes. Many of these chemicals are shipped from refineries along the Gulf Coast and in the Southwest to the Northeast and Midwest manufacturing centers for conversion into plastics, synthetic fibers, and paints. In addition to petrochemicals, hazardous materials are shipped by rail. Pressurized tank cars move petroleum-related hazardous materials across the nation and in Alaska and Mexico. Probably the hazardous material carried in greatest quantity by rail is liquefied petroleum gas (LPG), although most LPG now moves through pipelines.

Tank cars are also used to ship crude oil from waterways or ocean tankers to small inland refiners (fig. 6.16). Most refineries have rail lines running through their plants with loading and unloading depots. Crude oil produced in many states west of the refinery centers is shipped to the east coast by rail. Two factors contribute to this use of the rails: (1) most pipelines run from east to west, carrying *products* from refineries along the east coast to consumers in the west, and (2) local government regulations prevent the building of new refineries on the West Coast.

Figure 6.16 This general-purpose tank car is capable of carrying 10,000 gallons (38-cubic metres) of crude oil by rail. *(Courtesy of Union Tank Car Co.)*

Government Regulations

Since most petroleum derivatives are considered hazardous materials, the Federal Railroad Administration of the Department of Transportation (DOT) controls their transport within the United States. Diamond-shaped placards appearing on cars indicate that they contain hazardous materials. Tank car design is also regulated by the DOT.

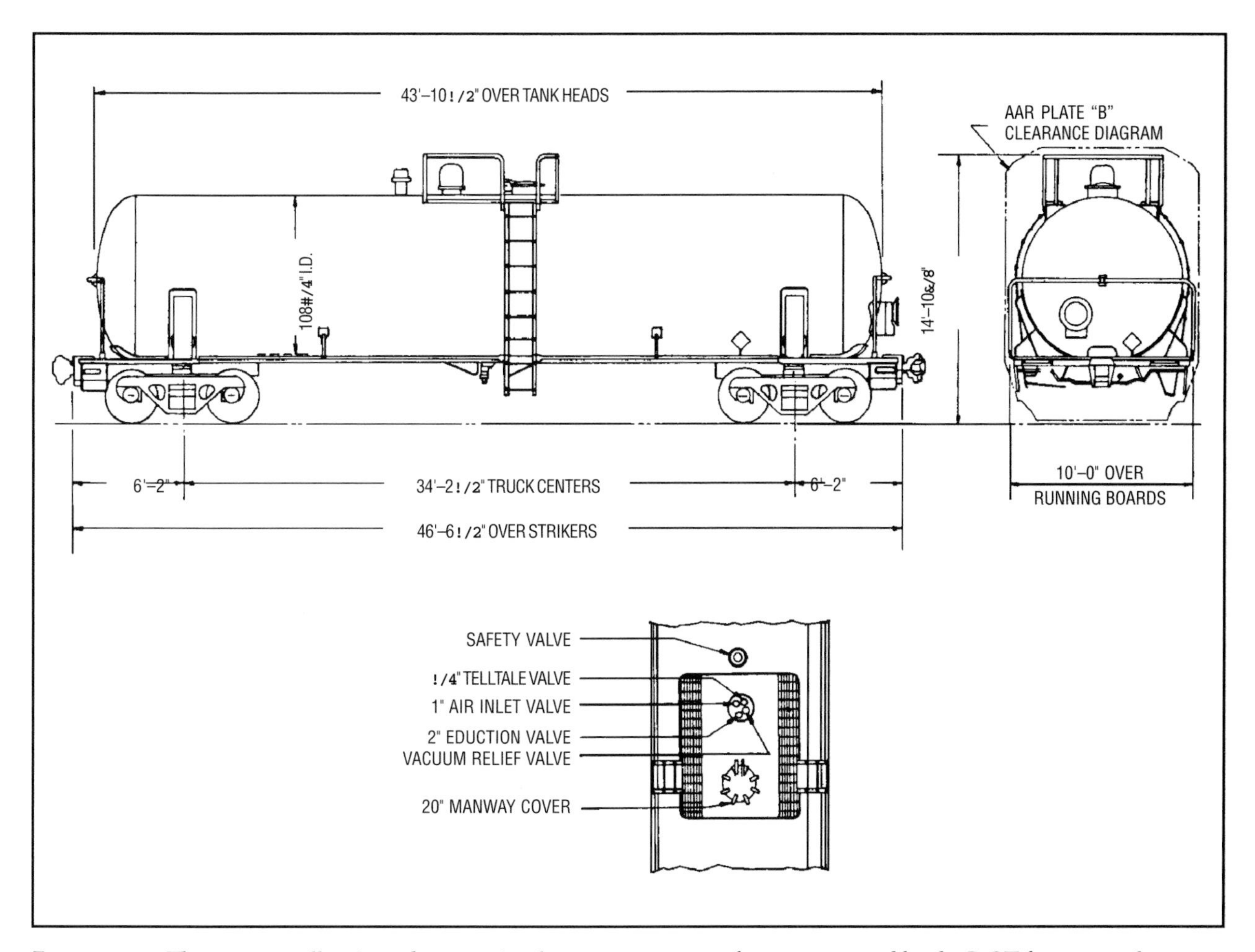

Figure 6.17 This 20,000-gallon (76-cubic metre) railway car meets specifications required by the DOT for a general-purpose tank car. *(Courtesy of Evans Products Co., Railcar Division)*

Tank Car Design and Manufacture

Although tank cars seem identical on the outside, they are tailored to carry specific products (fig. 6.17). Special linings and insulation, along with metals unaffected by corrosive liquids, are necessary for each type of product carried.

The huge cylindrical cars range in length from 24 to 64 feet (7.3 to 19.5 metres) and are up to 15½ feet (4.7 metres) high and 10½ feet (3.2 metres) wide. A general-purpose car carries 20,000 to 23,000 gallons (76 to 87 cubic metres) of liquid, compared with an LPG tank car, which carries about 33,500 gallons, or 127 cubic metres (fig. 6.18).

All tank cars are built in accordance with regulations of industry and governmental agencies that establish specifications for each tank car design before it is constructed. In addition to meeting DOT requirements, each tank car design must be reviewed and certified in writing by the Association of American Railroads (AAR) before it can be built and placed in service.

Other than the tank itself, most of the components used in building tank cars are common to all other railroad cars. For example, all tank car wheels, axles, brake systems, bearings, and couplers are built by specialty suppliers of all rail car builders. These specialty builders are also regulated by government and industry agencies.

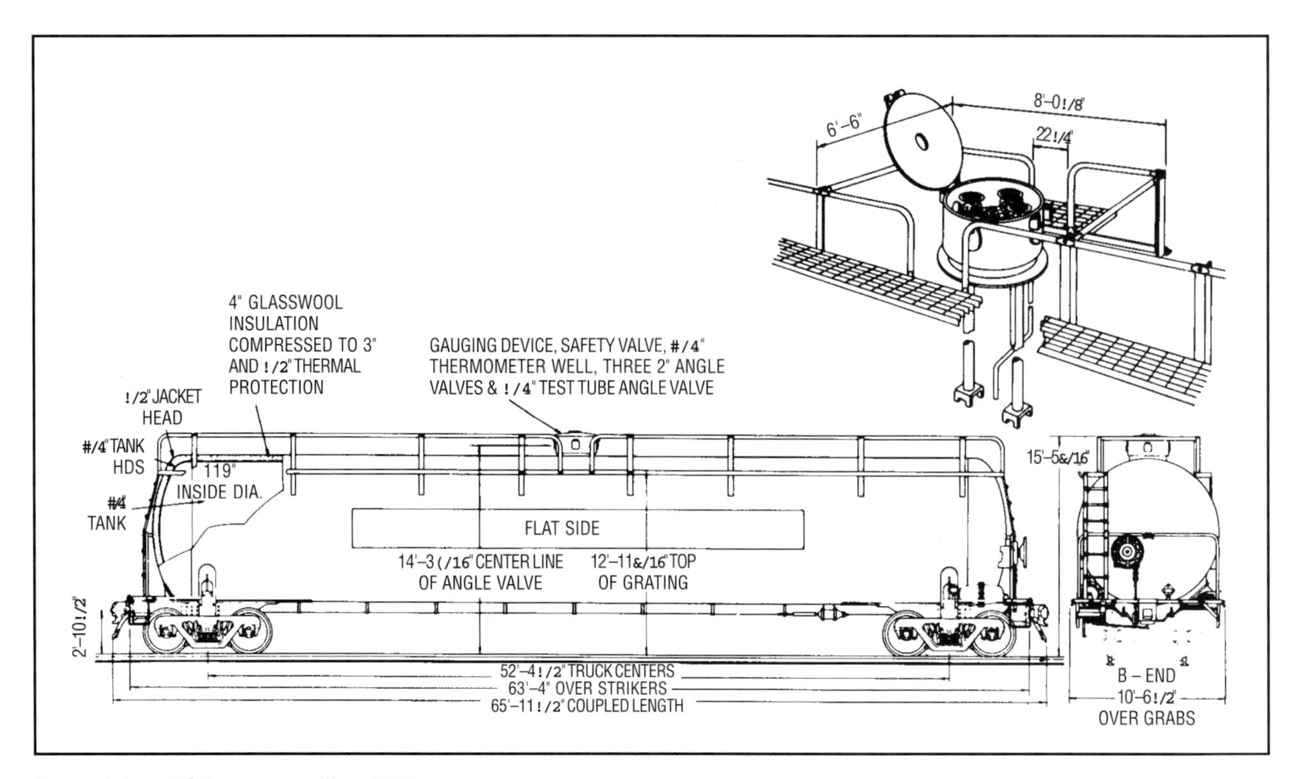

Figure 6.18 This 33,500-gallon (127-cubic metre) LPG tank car has special insulation and safety valves to meet DOT requirements.

Safety

A tank car is monitored throughout its operating life, primarily by the railroads that operate the cars. Safety inspections are required as well as regularly scheduled tests of pressure rating and checks of all valves for leaks. With the help of the Railroad Tank Car Safety Research and Test Project, tank car manufacturers have improved the safety of tank cars. Three design changes resulted from this project: (1) a new coupler that is less likely to disengage during derailments, (2) head shields mounted on both ends of the car to give added protection against punctures during derailments, and (3) thermal protection provided through either special steel jackets or thermal coating of the tank. By 1981, these modifications were in effect, and the industry spent some $200 million to make pressurized tank cars safer.

Tank car engineers and manufacturers continue to strive for the strongest metal alloy, the optimum tank capacity for each material, improved safety features, and the best valves and fittings.

Tank Car Strings

A unique design patented by General American Transportation Corporation (GATX) is the TankTrain®, a system of interconnected tank cars for transporting bulk liquids. A TankTrain® consists of a string or strings of interconnected cars, with each string being loaded or unloaded through a single connection, unlike the individual loading and unloading of conventional tank cars (fig. 6.19). Each TankTrain® car, with elbow pipes, protective housings, and tank isolation valves located at its top ends, is connected to the car in front of it by a large flexible hose. Petroleum is loaded into all of the cars through a connection to the first car. During loading, displaced vapors are collected in the last car and processed through a flare or scrubbing unit or piped to storage for recycling.

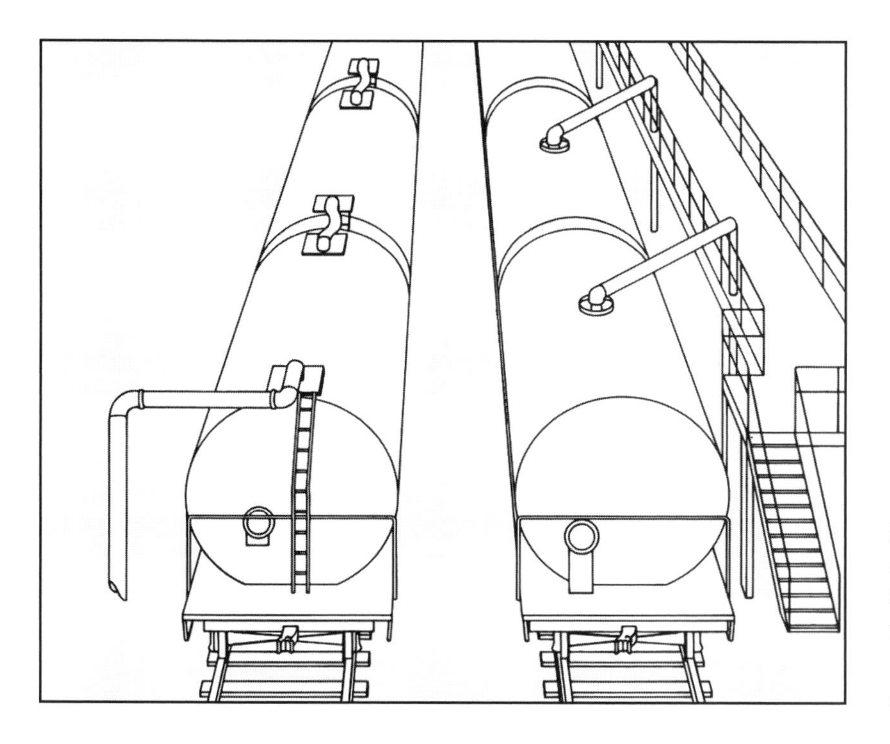

Figure 6.19 Conventional tank cars are loaded and unloaded individually (right), while interconnected tank cars may be handled from one loading rack. *(Courtesy of General American Transportation Corp.)*

At the unloading site, the entire car string is unloaded by injecting compressed air or inert gas, depending on the flammability of the liquid being transported, to push the liquid out (fig. 6.20). A pump is often used to increase the rate of flow. Two basic *loading and unloading systems—closed* and

Figure 6.20 The last car of this thirty-car string is attached to a compressed air supply (pipe in foreground), which is pumped into the last car at 60 psi, forcing the entire lading toward the front car and into a storage facility. *(Courtesy of General American Transportation Corp.)*

simplified—can handle petroleum products with a large range of viscosities and volatilities (fig. 6.21). The closed system is used when products have flammable or toxic vapors. The simplified system handles products with low volatility and no vapor hazard.

Before the advent of the TankTrain®, transporting crude oil from a production field to a refinery over a long distance by rail was too costly and time-consuming. But one petroleum company's system has proved successful in moving crude on a regular basis from its California oilfield to its refining complex some 250 miles (400 kilometres) away. The system consists of 72 insulated tank cars, each holding almost 550 barrels (87.5 cubic metres) of warm crude oil, making a single shipment of almost 40,000 barrels (6,360 cubic metres) of oil. Five 3,600-horsepower (2,520-kilowatt) diesel-electric locomotives lead the train, with six more locomotives interspersed throughout, providing 43,200 horsepower (30,240 kilowatts) to push and pull the fully loaded, 12,000-ton (10,900-tonne) train over its mountainous path.

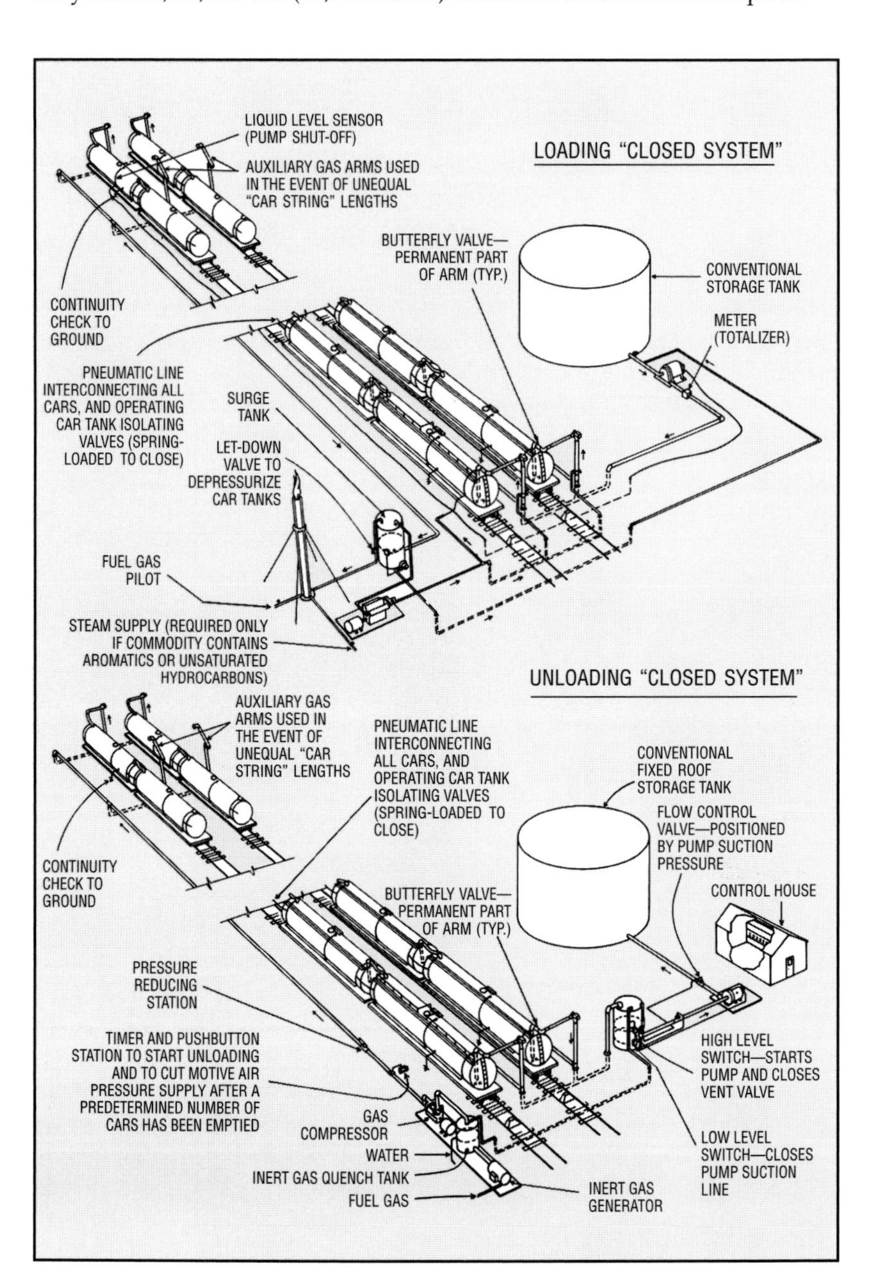

Figure 6.21 This closed system for loading and unloading maintains an average flow rate of 3,000 gallons (11.37 cubic metres) per minute for general petroleum products at a maximum pressure of 60 psi in the tank cars. *(Courtesy of General American Transportation Corp.)*

To handle such high-volume shipments efficiently, the company required some changes on the part of the railroad. Loading and unloading facilities were modified, heavy-duty roadbeds were laid, and new sections of track were installed to facilitate unloading six strings of twelve cars each simultaneously. A two-person crew at the refinery storage end of the train plus one operator can unload the complete train in about 5 hours (fig. 6.22). At the other end of the train, nitrogen—an inert gas—is used to pressure the heavy crude oil out of the tank cars.

Figure 6.22 Interconnected hoses and valves allow the entire train to be unloaded by a two-man crew in a short time. *(Courtesy of General American Transportation Corp.)*

MOTOR TRANSPORTATION

The design and engineering skill that has gone into the development of tank truck transports plus the expansion of the interstate highway system have made motor transportation important to the petroleum industry. Also contributing to the growth of the petroleum trucking industry are the large number of petroleum products being produced and the varied locations of production, refining, and distribution centers.

Types of Vehicles

Many types of motor vehicles transport petroleum, refined products, and gas. A *tank truck* is a self-propelled vehicle that carries a tank on its own chassis. A *truck-tractor* (simply called *tractor)* is usually diesel-powered and is designed to pull a *tank semitrailer* (often called *semi* or *trailer).* In combination, the *truck-tractor semitrailer rig* is generally known as a *tractor-trailer* but is also called a *semi* or an *eighteen-wheeler.* The semitrailer has part of its weight resting on the tractor and part on its own wheels. A *full trailer* is like a semitrailer except that its full weight is on the trailer wheels, and it is designed to be pulled by a truck or behind a tractor-trailer.

To carry petroleum products, all vehicles except the tractor have cylindrical, horizontal tanks. The type of tank used depends on the products it is designed to carry.

For purposes of this discussion, the term *truck* is used to mean any of the types of motor carriers just described.

Crude Oil Trucks

A lot of crude oil is moved by truck, especially from new fields where pipeline gathering lines have not been constructed. Large diesel-powered oil tank trucks can go almost anywhere to serve small lease tank batteries. From the central facility, the oil might go into a pipeline and on to a refinery.

Crude oil is carried in a general-purpose tank. This tank needs no special insulation, compartments, or multiple outlets since it carries only one product—crude oil.

Nonpressurized Refined Products Transport

Some refined products can be transported in general-purpose tanks, but others require tanks designed for special purposes. The general-purpose products tank is often lined with epoxy resin, divided into compartments, and fitted with a sump to ensure drainage. Each compartment has its own outlet, so the trucker can carry more than one product at a time. Gasoline, kerosene, and diesel are examples of products carried in general-purpose tank trucks (fig. 6.23). Constructed of steel, stainless steel, or aluminum, they are designed to operate at atmospheric (normal) pressure. Federal regulations require that trucks transporting hazardous materials display signs or placards stating that fact.

Special-purpose tanks must be used for highly corrosive products and products requiring special equipment. For example, heavy oils that have to be heated for loading and transporting require tanks that have flame tubes or steam coils, are usually insulated, and are made of nonconductive materials.

An asphalt tank also contains heating equipment and is insulated, since it may carry products of temperatures up to 400°F (204°C). Corrosives such as caustics and acids require tanks made of noncorrosive materials and with special linings and seals.

Figure 6.23 Tank tractor-trailers regularly line up at refineries to pick up refined products. *(Courtesy of Valero Refining Co.)*

Some refineries pipe their gasoline, kerosene, and diesel to remote terminals equipped with truck racks and meters that facilitate self-service loading by large tanker trucks (fig. 6.24). Authorized truck drivers can pull into the terminals and take on fuel at the computer-controlled truck loading rack anytime of the day or night. Billing and recordkeeping are automated, allowing truckers to load up to 10,000 gallons (38 cubic metres) of fuel in minutes and be on their way without even seeing a refinery or marketing employee.

Figure 6.24 At this terminal (A), a computerized meter (B) lets truckers load up with gasoline, day or night. *(Courtesy of* Phillips Pipe Line*)*

LPG Transport

Some products require refrigeration and compression tanks. Liquefied petroleum gases (butane and propane), ethylene, and similar products must be transported in tanks designed for pressures between 100 and 500 pounds per square inch (0.7 to 3.4 megapascals). These tanks are stronger, usually made of high-tensile steel, and not compartmentalized (fig. 6.25). Pressure and very low temperatures are used to liquefy the gases before they are loaded into the tanks, and they are kept refrigerated by insulation and controlled vapor release. Air compressors or pump units are used to discharge gas products from the tanks.

LPG is not only picked up by trucks at the refinery but also dispensed from remote storage systems. One company's propane flows through an 8-inch (20-centimetre) pipeline from storage tanks some 100 miles (161 kilometres) away directly into the propane trucks via a propane loading facility. This system eliminates area storage problems and makes tanking up easy for the drivers. The system includes computerization and special meters that measure product flow and volume, computer controls that automatically print bills of lading, and a security system that ensures safe operations.

Figure 6.25 A liquefied petroleum gas transport truck is used to carry propane to various LPG storage sites. *(Courtesy of Martin Gas Sales, Kilgore, Texas)*

U.S. Government Regulations

In addition to tank specifications, drivers and safety equipment must meet state and federal regulations. Regulations vary widely from state to state on the operation of motor vehicles carrying petroleum and petroleum products. Driver qualifications, driving rules, equipment and accessories, accident reporting, maintenance requirements, and weights and sizes are some of the things controlled by the government. The Interstate Commerce Commission governs those carriers engaged in interstate or foreign commerce. Many states have adopted ICC regulations to govern intrastate commerce, and the DOT sets the standards for safe operation.

OCEANGOING TANKERS

After World War II, oceangoing tankers became the major method of transporting oil around the world. Bigger and better crude carriers were built to meet the demand for oil by the industrialized nations. Although the global transportation of oil has slackened somewhat, modern tankers still transport over two-thirds of the millions of barrels (cubic metres) of petroleum produced in many areas of the world. Crude oil tankers range in size from the huge crude carriers with a carrying capacity of some 500,000 deadweight tons, or dwt, (508,000 tonnes) to the smaller tankers of 20,000 dwt (20,320 tonnes) or less. Tankers also carry refined products, natural gas liquids, and liquefied natural gas.

Supertankers

In 1973 the first half-million-ton (450,000-tonne) tanker was launched, and in 1979—the peak year of worldwide petroleum trade—some 190 million dwt (193 million tonnes) of petroleum products were transported by supertankers. Many companies were operating *very large crude carriers (VLCCs)* and *ultralarge crude carriers (ULCCs)*, although ports and waterways that could handle these mammoth vessels were limited.

A VLCC has a 100,000-dwt to 250,000-dwt (101,600-tonne to 254,000-tonne) range, while a ULCC ranges from 250,000 up to and over 500,000 dwt (254,000 to 508,000 tonnes). A typical supertanker is twice as wide as and hundreds of feet (metres) longer than the largest passenger ship and can hold some 2 million barrels (320,000 cubic metres) of oil. As one person explained it, a VLCC transports enough petroleum "to fill a 50-mile (80-kilometre) line of tank trucks or to heat the homes and power the cars in a city of 85,000 people for a year."[2] VLCCs and ULCCs are designed for the greatest capacity, safety, and fuel efficiency possible. Navigational aids, electronic control systems, automated power plants, satellite communications, and radar equipment allow for maximum security and a minimum crew, making this type of marine transportation an economical one for long-distance hauls.

During their prime, supertankers filled a specific need: they transported huge amounts of crude from new oilfields to the new refineries being built to supply an increasingly energy-hungry world. By the early 1980s, however, a worldwide slump in oil demand caused a surplus of supertankers, especially the very large ones. The oil shipping industry depression put about one-third of the world's supertankers out of the transportation business. Some companies have downsized their surplus supertankers, some have scrapped them rather than bear the cost of maintaining them, others have mothballed all or part of their fleets, and still others used them as offshore storage tanks. Of course, some supertankers are still in service.

Average-Size Tankers

Global supertanker fleets are gradually being replaced by smaller tankers ranging in size from 30,000 to 80,000 dwt (30,480 to 81,280 tonnes). The construction of new pipelines, the deepening and widening of canals and waterways that accommodate average-size tankers, the discovery of new fields that eliminated the necessity of many long hauls, and a use conservation program have all contributed to the popularity of smaller tankers (fig. 6.26).

---

[2] "How to Sail in an Oil Surplus," *Shield*, Vol. X, No. 2, 1985, p. 2.

Figure 6.26 The tanker *MT Garyville* takes on its first load of Brae Field crude oil at a terminal off Scotland. *(Courtesy of Marathon Oil Co.)*

U.S. tankers are constructed and operated under strict regulations. Using a communications network of telephone, telex, and satellite systems, a company can locate any of its tankers at any time. On board the ships, crews follow strict safety measures, minimizing loss of personnel and product. Automatic collision avoidance systems track approaching ships, alerting a tanker to an obstruction on its course.

Responsible shippers adhere to these environmental and safety regulations, although many countries have no building restrictions or rules for operation. In spite of the notoriety of oil spills, the percentage of crude lost to spills each year is small when compared with the total amount transported by tanker.

NGL and LNG Tankers

Tankers equipped with pressurized, refrigerated, and insulated tanks (*cryogenic* tankers) transport natural gas liquids (NGL) and liquefied natural gas (LNG). NGLs are hydrocarbon liquids that are gaseous in the reservoir but liquids in field facilities or in gas processing plants. Gas liquids that are not highly volatile (easily vaporized) are treated and shipped as natural gasoline or added to the crude oil. If separated before shipment, the other NGLs—butane, propane, and ethane—must be transported in tankers designed to

Figure 6.27 This cryogenic tanker carries 450,000 barrels (71,550 cubic metres) of liquefied natural gas per trip from Alaska's Kenai Peninsula to Tokyo. *(Courtesy of Marathon Oil Co.)*

handle the high pressures and low temperatures needed to keep the gases in a liquid state. NGL is usually transported as a mixture and separated into LPG, natural gasoline, and ethane after it reaches the refinery.

Natural gas is liquefied at the source before it is transported by special LNG tankers. LNG is then regasified and generally fed into a gas pipeline system when it reaches its destination. One company that furnishes natural gas to Tokyo liquefies the gas near its source by lowering its temperature to -259°F (-162°C). As the gas liquefies, it shrinks to 1⁄600 of its original volume. The LNG cargo is kept in its liquefied state during a 3,000-mile (4,830-kilometre) journey and is converted back to its gaseous state as it enters natural gas pipelines in Japan (fig. 6.27).

Icebreaking Tankers

In the Arctic Ocean, Baltic Sea, and other ice-laden seas, shipping oil and gas requires the use of the strongest tankers. The tankers must have hulls strong enough to prevent spills but avoid the high cost and excess weight of very thick steel hull plating. Some tankers used in these seas have double hulls, which manufacturers believed would be safer; however, the infamous Exxon Valdez was a double-hulled vessel.

Although U.S. companies are studying the feasibility of operating icebreaking tankers, they have not yet used them for petroleum transportation. Icebreakers designed by Wärtsilä Shipbuilding Division in Helsinki have been used since 1977 by Neste Oy, a government-owned oil company in Finland (fig. 6.28). Neste uses the *Lunni* and her sister ships to distribute oil products from the refineries in the south of Finland to the towns on the Finnish coast and to transport crude oil from the Baltic area to the refineries in Finland.

Figure 6.28 Hull strength is of vital importance to this icebreaking oil tanker used by Neste Oy of Finland. *(Courtesy of Wärtsilä Shipbuilding Division)*

Loading/Offloading Facilities

Because very few ports are large enough to accommodate VLCCs and ULCCs, oil entrepots and methods of discharging at sea have been developed. An *entrepot* is a center for the collection and distribution of goods, as a *depot* is a center for storing goods. Most ports capable of handling supertankers have offshore mooring facilities and underwater pipelines that transfer oil from ship to ship or from ship to barge. In the United States, these shuttle tankers and barges then take the oil or products to shore or inland through the 25,000 miles (40,250 kilometres) of major waterways maintained by the U.S. Army Corps of Engineers. Many of these waterways are deep-draft canals and channels, deep enough and wide enough to handle small tankers (fig. 6.29).

The only U.S. mainland deepwater port equipped to handle supertankers is the Louisiana Offshore Oil Port (LOOP), and its capability is limited to 150,000 dwt (152,400 tonnes). Oil discharged at LOOP is shipped by pipeline to company refineries, with computers controlling and supervising deliveries.

Figure 6.29 A tanker shuttles down the Sabine Ship Channel on the Gulf Intracoastal Waterway after unloading its cargo at an inland plant. *(Photo by Greg White; courtesy of* Texas Highways Magazine*)*

Figure 6.30 The *ARCO Fairbanks*, a 120,000-dwt (121,920-tonne) VLCC, docks at Alaska's Port Valdez terminal at the southern end of the Trans-Alaska Pipeline. *(Courtesy of Alyeska Pipeline Service Co.)*

Tankers of various sizes take on North Slope crude oil from the Trans-Alaska Pipeline at the Valdez terminal in southern Alaska (fig. 6.30).

Companies often use converted tankers as offshore, floating production, storage, and offloading (FPSO) units to facilitate the transfer of crude oil from production platforms to shuttle tankers. Well fluids are first transferred by subsea pipeline to a moored tanker outfitted with storage and treating equipment. Then the treated crude, which is sold by meter transfer, is picked up by shuttle tankers that dock at the FPSO unit for loading.

After its journey from the oilfield or from a supertanker to its destination, a shuttle tanker may offload its cargo at the refinery's loading/offloading docks (fig. 6.31). A products ship, which may carry more than one type of cargo, is often unloaded at different destinations; therefore, the ship must be loaded so that the trim (balance) is not upset when the first product is unloaded.

Figure 6.31 Crude oil and feedstocks are shipped in and products are shipped out from this refinery dock used for small- to average-size tankers. Note the loading/offloading arms, which swing out to reach a tanker's hold. *(Courtesy of Valero Refining Co.)*

CRUDE OIL PIPELINES

The construction of oil pipelines increased tremendously during and after World War II. From prewar deliveries of less than 50,000 barrels (7,950 cubic metres), daily pipeline shipments eastward soared to 754,000 barrels (120,000 cubic metres), with the companies east of the Rockies increasing their total daily movements about 60 percent between 1941 and 1945. Companies discovered that large-diameter trunklines had operating flexibility previously considered available only in smaller lines and that the larger lines reduced transportation costs per barrel (cubic metre) considerably. Pipelines devoted to the transportation of liquids increased from 124,000 miles (200,000 kilometres) in 1948 to more than 222,000 miles (357,000 kilometres) in 1976.

Since 1976, actual mileage has declined but total quantities shipped have steadily risen because of larger-diameter lines and the *looping of long lines,* that is, laying additional lines alongside existing pipelines (fig. 6.32). A major contributor to the increased quantities of oil transported is the Trans-Alaska Pipeline System (TAPS). This 48-inch-diameter (122-centimetre-diameter) pipeline, which runs from Prudhoe Bay to Port Valdez—a distance of some 801 miles (1,290 kilometres)—can deliver 2 million barrels (318,000 cubic metres) of oil per day to the Valdez terminal. Sophisticated handling facilities, including four tanker berths, permit tankers to load oil at rates up to almost 110,000 barrels (17,500 cubic metres) of oil per hour.

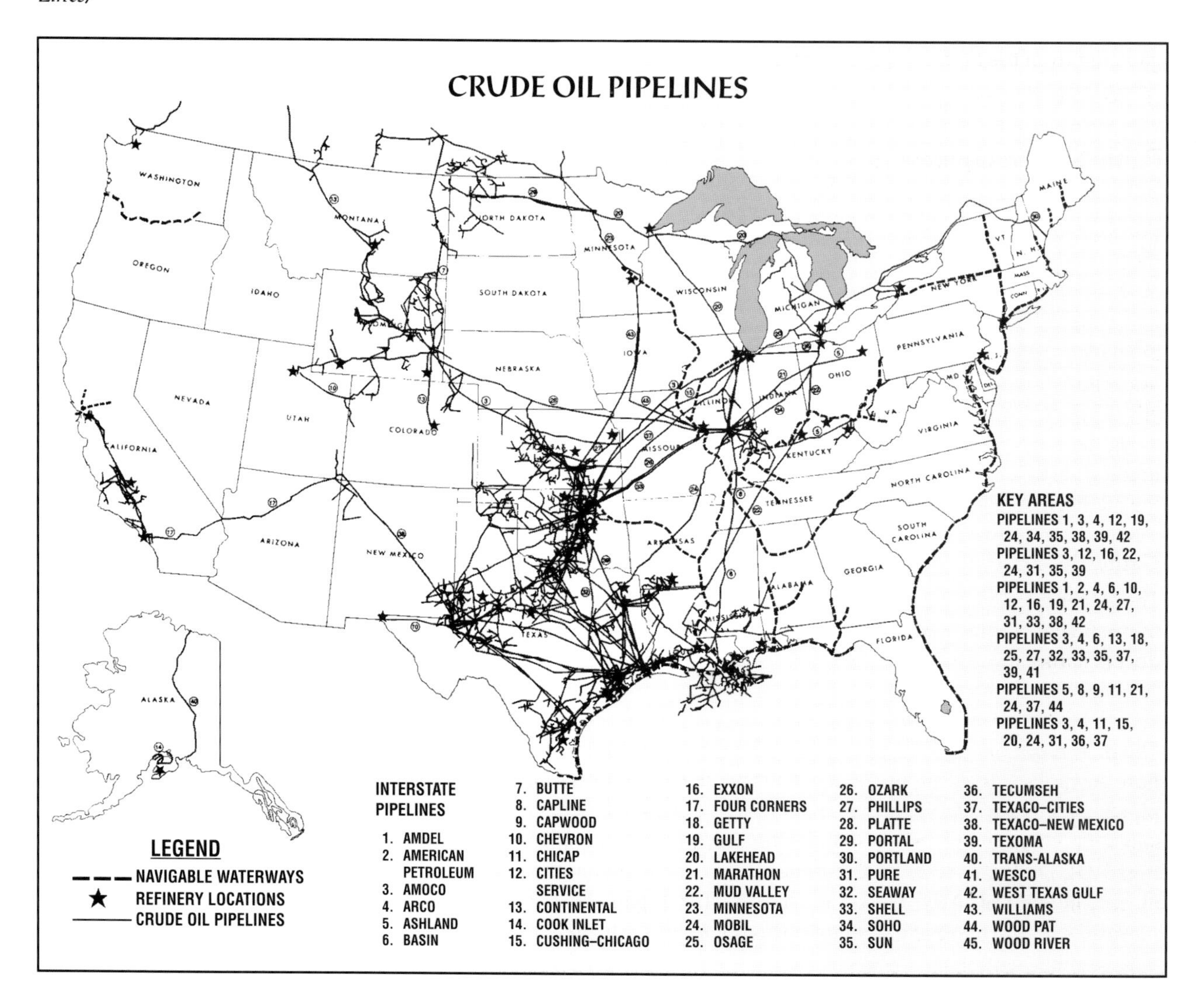

Figure 6.32 Shown as solid black lines, the main interstate pipelines of the United States connect key production areas with refineries (stars). Also shown on this map are the major inland waterways (dashed lines) used for transporting petroleum and petroleum products. *(Courtesy of Association of Oil Pipe Lines)*

State and Federal Regulations

Accurate recording of crude oil shipments sent and received can mean the difference not only between profit and loss, but also between legal and illegal operation. Some states in the United States require the pipeline companies to keep track of large oil company and independent shippers (a *shipper* is any company that transports oil or gas) to see that each company is complying with the established monthly schedules. Companies must file monthly forms with the appropriate state agency to show the quantity of oil taken from each lease.

The federal government also imposes stringent controls on various aspects of the pipeline industry. Oil pipelines are classified as common carriers and therefore operate under the U.S. Department of Transportation. At one time, they were under the jurisdiction of the Interstate Commerce Commission.

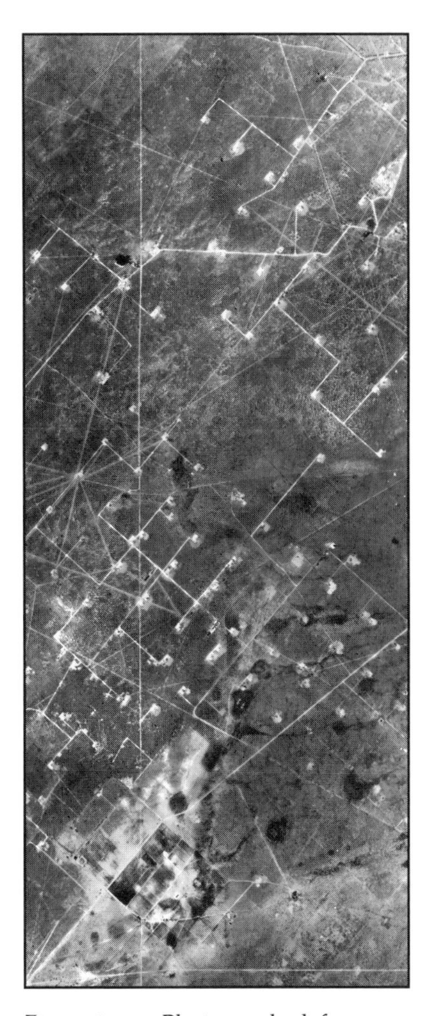

Figure 6.33 Photographed from an altitude of 27,000 feet (8,230 metres), this southwestern oilfield is a maze of wells (light spots) and flow lines (white lines).

Field Gathering Systems

An oilfield may have several hundred wells, with flow lines that carry crude oil from the wells to the lease tanks (fig. 6.33). Various tanks and treating vessels are often a part of the field gathering system, but lines can take crude oil away without having tanks or other ground equipment except for printing meters to register the number of barrels (cubic metres) taken. After the oil is gathered from the field, it is usually treated, measured, and tested before entering the pipeline.

Pump Station Operation

Pumps move the oil into and through the pipeline at a pump station. Functionally, pump stations are gathering stations, trunkline stations, or a combination of both. A *gathering station* is located in or near an oilfield and receives oil through a pipeline gathering system from the producers' tanks. From the gathering station, oil is relayed to a *trunkline station*, which is located on the main pipeline, or *trunkline*. Having a much greater capacity, the trunkline station relays the oil to refineries or shipping terminals. Since the pressure gradually drops as the oil moves through the line, booster pump stations are spaced along the trunkline as needed. *Tank farms* along the line serve as receiving and holding stations. A special set of lines and valves called a *manifold* regulates the oil that enters and leaves the tank farms and the pump stations.

Gathering Station

The gathering station may include one or more pumps and may move several thousand barrels (several hundred cubic metres) of oil daily from various producers' tanks and gathering station tanks. Pumping units usually consist of electric-drive reciprocating or centrifugal pumps. Internal-combustion engines have been almost totally replaced by electric motors as prime movers. The method of operating a gathering station depends on its size and function. At tanks where the oil changes hands, say from the producer to the pipeline company, the pumps are activated by a *gauger*, a person who measures the volume and draws samples for analysis before pumping begins (fig. 6.34). On leases with LACT units, pumps are started automatically or by remote control. After the oil is received at the gathering station, it is pumped either to another gathering station closer to the trunkline or directly to a trunkline station.

Figure 6.34 The gauger activates a battery of suction booster pumps that pull oil from the tanks and move it to the suction of the trunkline pumps.

Trunkline Station

A trunkline, or mainline, station is located on a main petroleum-carrying artery. It may serve as a gathering station if near the source, but usually a trunkline station relays or boosts the oil to the next pump station down the line. Pumps used in trunkline stations today are typically high-speed centrifugal units having a direct connection to electric motors (fig. 6.35).

Figure 6.35 A routine check of temperature and lubrication keeps this large centrifugal pump operating at a trunkline station.

Internal-combustion engines and gas turbines are used to a lesser degree because of increases in both fuel and maintenance costs. A large station usually has a series of pumps to multiply the pumping power available. Enough pumps are used to achieve the required pressure, and the oil is sent down the pipeline to the next station. Booster stations located along the trunkline assure that the work load for each station is approximately equal. This division of load is called *hydraulic balancing*. The work load is determined by the amount and speed of the fluid being pumped (barrels/hour/mile, or cubic metres/hour/kilometre), the density and viscosity of the fluid, and friction of the fluid on the inside surface of the pipe.

Station Tank Farm

A tank farm is to a pipeline what a railway yard is to a railroad. It is a place where oil in transit may be temporarily sidetracked for sorting, measuring, rerouting, or simply holding during repairs on a line or station. A tank farm may also be a receiving station where oil comes in from the producing fields or from other carriers to be injected into the pipeline transportation system (fig. 6.36). Trunklines bring crude oil belonging to many different shippers to the tank farm. By segregating shipments in the tanks, the pipeline company can make deliveries to various owners at refineries or terminals.

Figure 6.36 This pipeline junction has a tank farm, pump station, and converging pipelines. Note the positions of the floating roofs on the tanks.

Pump Station Manifold

Tank farm pipelines converge at a station manifold (fig. 6.37). A manifold is an accessory system of piping that divides the flow into several parts, combines several flows into one, or reroutes the flow to any one of several possible destinations. The manifold switches the oil stream to its proper destination. It can be very complex in a main station or simple in a booster station. A manifold might connect a mainline, a field gathering line, several tanks, and one or more pumps. This network of lines and valves, operated by the tank farm and station personnel, can (1) pump oil through the trunkline while the tanks are inactive, (2) receive production from the field into

Figure 6.37 The manifold at a pump station and tank farm permits routing of the oil stream to its destination.

any tank, (3) receive from a trunkline into any tank, (4) transfer from one tank to another, (5) pump from any tank into the trunkline, (6) isolate all pumps and tanks while the trunkline station upstream pumps through, or (7) inject oil from any tank into oil being pumped through the trunkline.

Control of Oil Movements

To keep costs to the shippers low, pipeline shipping of crude oil requires large shipments, high throughput, and efficient operation. For this reason, a single shipment, or *batch,* must consist of a minimum volume of several thousand barrels (several hundred cubic metres). Any single pipeline may handle several grades of oil for each of several shippers. The *pipeline scheduler* slates the shipment to leave a certain point of origin on a set date and to be delivered to a destination point on a set date. The shipper must have tank space to receive the grade of oil at the destination.

A typical shipment might accumulate over a period of several days. Each day a part of the shipment is delivered from the leases to pipeline company tanks and is held until the required volume is ready for delivery to the trunkline. The pipeline company takes custody at the lease where the oil is produced and holds it on account for the shipper who either produced it or bought it.

Suppose that the transit time takes ten days and ten different shippers regularly use a particular pipeline. At any given time each of these shippers might have several shipments in the line, and the pipeline company might have in the line and in its tanks more than a million barrels (159,000 cubic metres) of oil, all of which belongs to the shippers. As oil is pumped in at the upstream end, an equal volume is pumped out at the downstream end. Each shipper, therefore, is continually receiving deliveries and putting more oil in the line. Thus, the shipper always has a balance of oil in the system. While his or her oil is in transit, a shipper may trade it to another shipper for oil that was shipped earlier. Then the pipeline changes the route and delivers it to the new owner at a different destination. This buying, selling, trading, and rerouting of shipments is routine in pipeline operation.

Just as a railway freight shipment may travel on more than one railroad in transit, so a shipment of crude oil may travel on more than one pipeline system between its origin and destination. Requests for shipping space in one section of a line may be greater than those in other sections of the same line because of transfers to another pipeline system at some point downstream from the originating station. Because the oil cannot keep flowing if there are air gaps in the pipeline, the second line must be ready to receive the diverted shipment, and the first line must pump oil in from tanks at the point where the diverted shipment is being withdrawn from the line. Otherwise, oil movement downstream in the first line will come to a standstill.

Scheduling

To handle all of the operations involved in pipelining, a coordinator or scheduler works with the shippers to set up a schedule everyone agrees upon. On line with many pump stations and pumping units, computer programs determine which stations and exactly what pumping unit combinations will bring about the desired flow rate at minimum power cost. Coordinators schedule the movement of crude oil in the pipeline according to when they expect that produced oil will need to enter the system and when connecting pipelines will need to use it. Although tanks are common at production sites and at terminals, the fact that modern pipeline systems have few operating tanks along the line can be a major problem in pipeline scheduling. Determining how, when, and where a shipment must arrive at its final destination or intermediate delivery points for barge or tanker loading requires considerable skill on the part of the scheduler. Today, schedulers use computers to develop schedules, project the schedules to completion, and print out information about delivery times and dates (fig. 6.38).

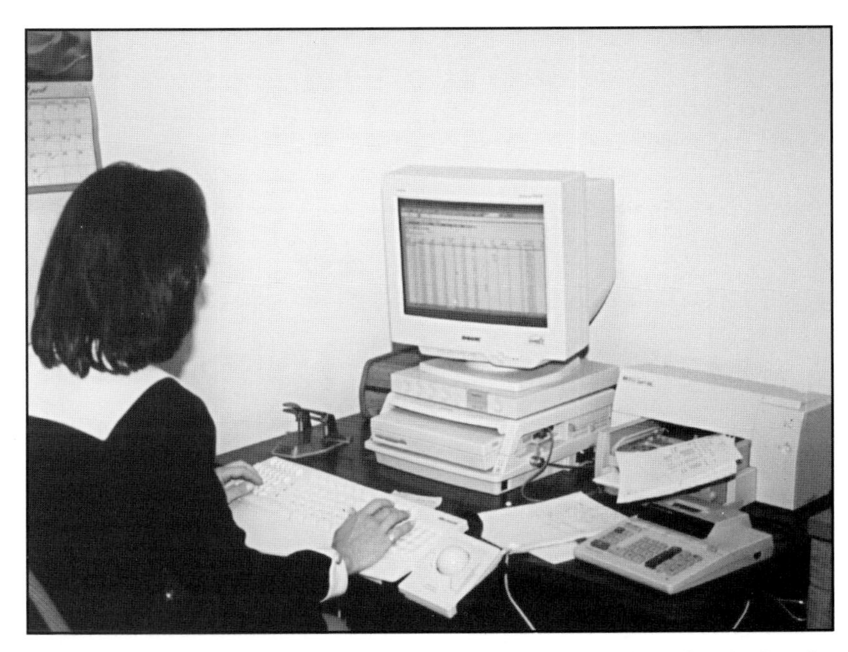

Figure 6.38 Schedules show point of origin, amount and grade of oil to be pumped, batch number, starting and ending time and place, and pumping rate of each batch. *(Courtesy of Shell Pipe Line Corp.)*

Dispatching

After the schedule has been determined, a *pipeline dispatcher*, sometimes called a *line operator* or *oil movement controller*, communicates with other points on the pipeline system. Using a computer console, the dispatcher controls the pumping stations hundreds of miles (kilometres) away and monitors line pressures, rates of flow into and out of the pipeline system, and gravities (densities) of the various crude oil streams. In executing the schedule, the dispatcher operates motor valves, starts and stops pumps, and performs other operations in the system through computer-assisted remote control. As the dispatcher executes the schedules, the computer records shipping information—the quantity and quality of oil moved, its origin and destination, and shipping and arrival dates. Dispatchers are also responsible for the security of the lines. They make sure that the liquid is not being lost because of a leak or theft.

Testing and Measuring

Along with the scheduler who devises the plan and the dispatcher who executes it, an *oil measurements group* has the job of preserving the quality and quantity of oil delivered to the pipeline by the shipper. This group checks the methods that field personnel use to measure the quantity of oil by meters and tank gauges. They constantly monitor testing procedures and sampling devices.

Accounting

The *run ticket* is the basic accounting document in the buying and selling of crude oil. Every time custody of oil changes hands, a run ticket is made and signed by the receiver and the deliverer (see fig. 5.54). This ticket serves as a record and a receipt for all oil that the pipeline handles. This control is essential to ensure the proper handling of oil movements. Records are maintained and used many months after the oil has been delivered to its destination.

PRODUCTS PIPELINES

Just as crude oil pipelines gather the oil, products pipelines distribute it, with refineries in the middle. A products pipeline transports refined products derived from crude oil, such as gasoline, diesel fuel, kerosene, jet fuel, heating oil, and other liquid hydrocarbons not produced by liquefying gas. Products pipelines usually begin at or near refineries and end at terminals in areas of high market demand for the products (fig. 6.39).

Most of the nation's products pipeline systems originate in and around the Gulf Coast area, since many of the country's refining companies and the petrochemical industry are concentrated there.

Figure 6.39 A products pipeline terminal at New York City's John F. Kennedy International Airport supplies an area of high market demand. Shown from left to right are a storage tank, prover loop, and filter vessels. *(Courtesy of Buckeye Pipe Line Co.)*

Control of Product Movement

A products pipeline operates much like a crude oil pipeline in reverse. A dispatcher generally directs the intake, movement, and delivery of products by remote control from the pipeline company's headquarters. Dispatchers work at computer terminals in the main control center to control all movements in the company's trunklines, or *mainlines* (fig. 6.40). Some products reach their destination through stublines. *Stublines* branch off from the mainlines and carry products to nearby areas. A backup control center, located away from the main control center, takes over operations in case of emergency.

Figure 6.40 A dispatcher controls product movement and monitors other operations from the main computer console. *(Courtesy of Shell Pipe Line Co.)*

Located on the mainline are booster stations (usually unmanned) and junction stations, generally operated by the main control center. Instruments at the control center keep the dispatcher informed of suction and discharge pump pressures and the destination and time of arrival of each shipment en route. A booster station usually consists of a central hydraulic system that operates valves and electric pumps that maintain the pressure needed to move products along.

Station operators at a pipeline's pumping station use computers to monitor all aspects of the station's operation (fig. 6.41). They initiate operations at input stations and handle local deliveries and transfers at delivery sites. Electricity supplies the power throughout the system.

Figure 6.41 Tricolor video displays enable an operator to view any aspect of the modern products pipeline pumping station. *(Courtesy of Phillips Pipe Line Co.)*

Breakout tanks, located at mainline junction stations, are tanks that hold products temporarily until they can be relayed to local shippers' tanks, shippers' tanks on stublines, or terminals farther up the mainline (fig. 6.42). Breakout tanks or the company's facilities may have manifolds installed.

Batching

A pipeline company takes care of many shippers and their various products, so batching is a very important aspect of products pipelining.

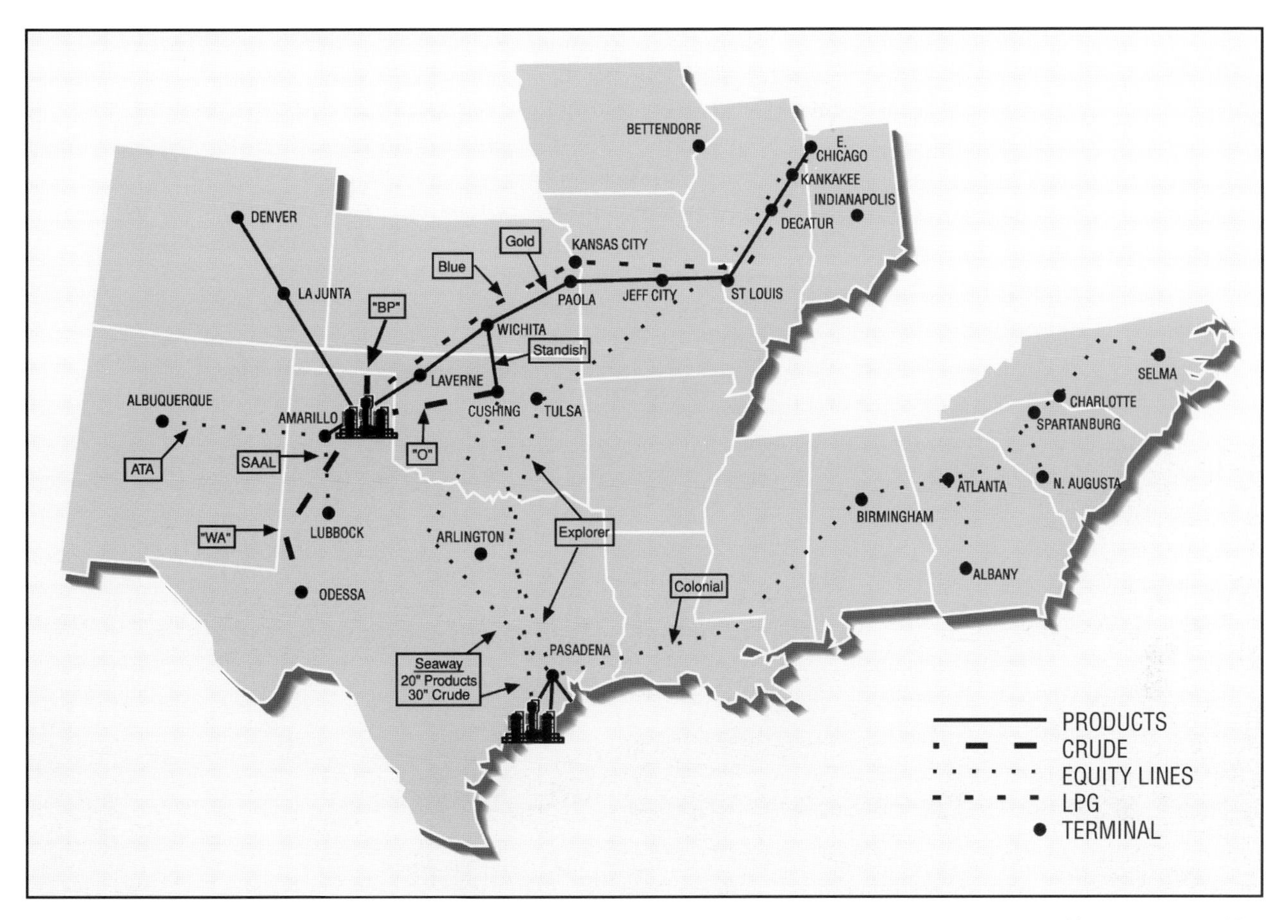

Figure 6.42 This pipeline system helps supply a sixeen-state area. (Equity lines are pipelines partially owned but not operated by a company—in this case, Phillips Pipe Line Company.) *(Courtesy of Phillips Pipe Line Co.)*

Batching is pumping a quantity of crude oil or product of one weight next to one of a different weight or type (fig. 6.43). Usually the two batches mix to some extent where they come into contact. One company, for instance, handles as many as thirty different grades of products. To operate such a process, the pipeline company operates on a ten-day cycle, meaning that every shipper is able to repeat shipments of any product grade at least every ten days. The sequence of products varies according to the types of products being shipped. The shipper's exact place in the cycle is determined by the kind of product being shipped and the products of other shippers.

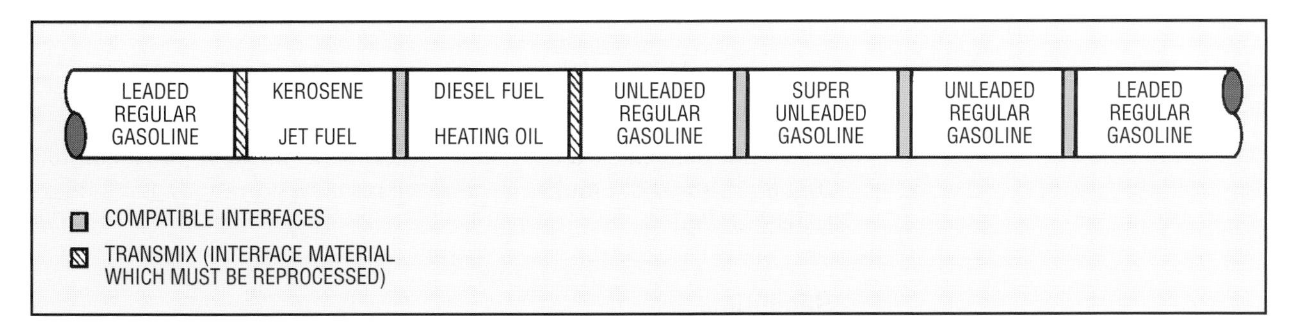

Figure 6.43 A products pipeline runs batches of different products in sequence similar to this typical cycle. *(Courtesy of Association of Oil Pipe Lines)*

Technicians sample each incoming batch of products to verify the arrival of a shipment. The gravity, color, and appearance of the product usually verifies the change from one shipment to another.

Sometimes batches of products are separated in the line by separators, such as batching spheres. A *batching sphere* is a ball made of neoprene, a synthetic rubber. Usually, however, mixing of products at interfaces between shipments is not a problem. Pipeline companies also use spheres and scrapers of various kinds to remove water and sediment from the pipeline, thereby preserving the quality of the products (fig. 6.44).

Figure 6.44 This worker is placing a sphere into a pipeline to clear water and sediment from it. *(Courtesy of Association of Oil Pipe Lines)*

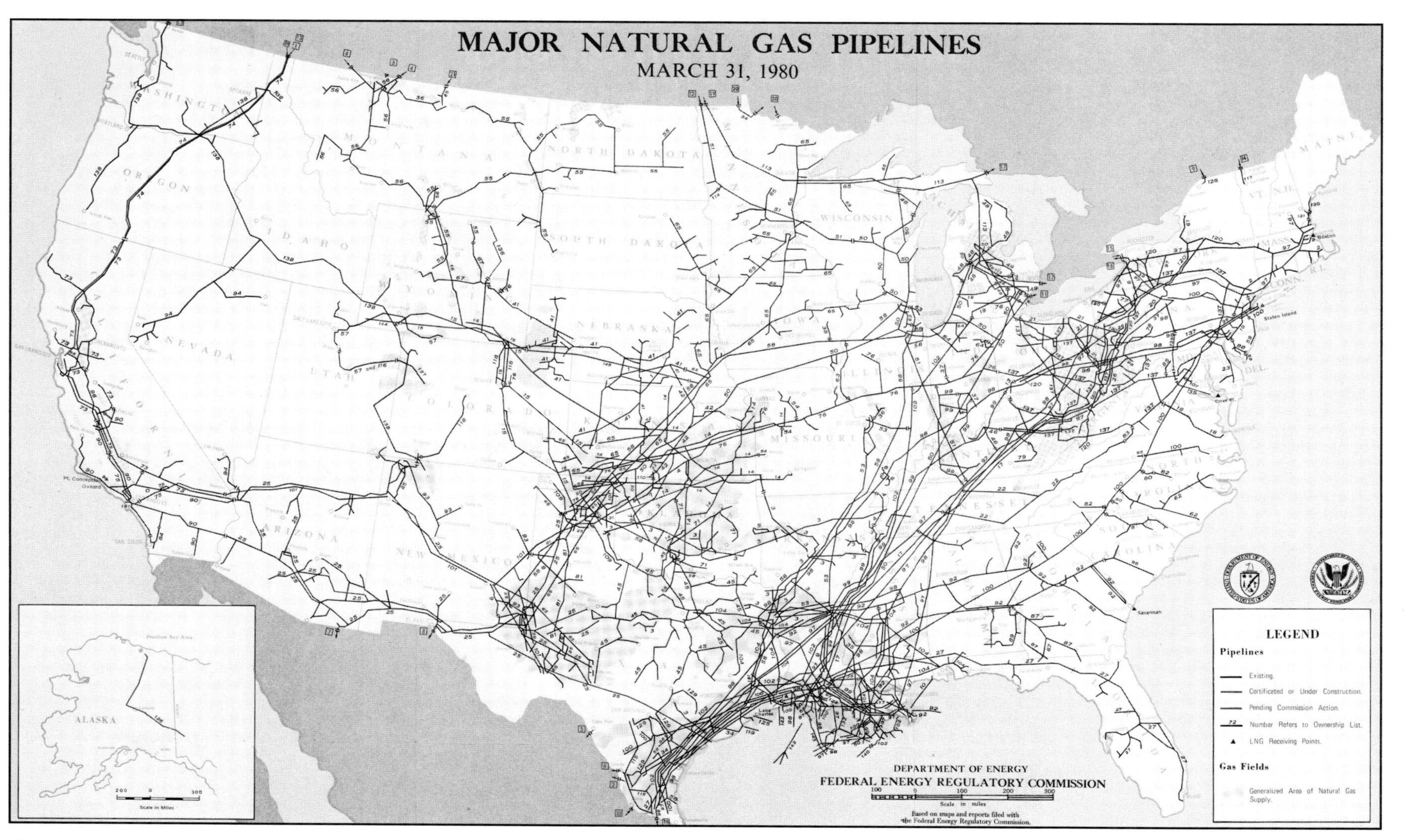

Figure 6.45 Major U.S. natural gas pipelines connect gas supply areas.

NATURAL GAS PIPELINES

Gas pipelines and oil pipelines historically have been separate industries, and they are different in several ways. A *gas pipeline* uses pressure from compressors as its driving force, while an oil pipeline uses pump pressure. Different agencies govern their operations. Before the establishment of the U.S. Department of Energy, gas pipelines were regarded as a utility and operated under the regulation of the *Federal Power Commission (FPC),* unlike oil pipelines that operated under ICC regulations. At present, the sale and transportation of gas are controlled by the *Federal Energy Regulatory Commission (FERC),* a branch of the U.S. Department of Energy, while oil and products pipelines are controlled by the Department of Transportation.

Modern Transmission Systems

Since 1950, the natural gas industry has grown tremendously as technological advances have cut the cost of installing pipe. From 1950 to 1975, the number of miles (kilometres) of gas pipeline tripled. Of the more than a million miles (1.6 million kilometres) of U.S. gas pipelines in use today, about 270,000 miles (434,700 kilometres) are transmission trunklines (fig. 6.45), with the remaining pipelines consisting largely of distribution, main, and gathering lines. Table 6.1 shows the number of miles of pipeline and the average number of gas customers by state.

TABLE 6.1

TRANSMISSION, DISTRIBUTION, AND FIELD GATHERING PIPELINES AND NATURAL GAS CUSTOMERS BY STATE

STATE	PIPELINES (MILES)	GAS CUSTOMERS (THOUSANDS)	STATE	PIPELINES (MILES)	GAS CUSTOMERS (THOUSANDS)
Alabama	19,644	674.1	Montana	8,538	184.9
Alaska	1,212	45.5	Nebraska	14,149	469.7
Arizona	16,817	585.2	Nevada	3,658	175.9
Arkansas	18,387	491.9	New Hampshire	1,162	52.2
California	80,837	7,410.7	New Jersey	23,106	1,850.1
Colorado	23,488	872.7	New Mexico	25,373	337.7
Connecticut	6,798	412.2	New York	41,681	3,943.3
Delaware	1,339	82.7	North Carolina	12,090	385.2
District of Columbia	1,162	147.8	North Dakota	2,950	88.6
Florida	12,651	441.9	Ohio	52,981	2,768.1
Georgia	25,147	1,053.6	Oklahoma	38,578	834.2
Hawaii	587	32.7	Oregon	10,070	283.9
Idaho	4,435	110.0	Pennsylvania	52,527	2,362.2
Illinois	53,501	3,317.8	Rhode Island	2,425	173.2
Indiana	27,762	1,266.8	South Carolina	9,973	286.2
Iowa	18,042	756.2	South Dakota	2,622	105.6
Kansas	37,318	735.5	Tennessee	16,502	515.0
Kentucky	20,110	630.1	Texas	120,880	3,260.6
Louisiana	42,256	1,006.7	Utah	8,504	386.5
Maine	389	15.3	Vermont	404	15.7
Maryland	8,764	783.0	Virginia	11,345	533.4
Massachusetts	16,915	1,093.5	Washington	11,883	380.6
Michigan	44,486	2,484.4	West Virginia	22,331	393.8
Minnesota	17,232	836.5	Wisconsin	22,197	1,060.2
Mississippi	18,101	400.3	Wyoming	8,457	79.9
Missouri	21,404	1,230.3	Total U.S.	1,064,190	47,844.1

Source: American Gas Association, *Gas Facts*, Arlington, Virginia.

Figure 6.46 This compressor plant speeds gas to consumers by pipeline.

The method used for collecting, conditioning, and transmitting gas to its primary destination depends on the individual situation, but any gas gathering system has specialized equipment for (1) conditioning the gas so it will flow safely and steadily and (2) controlling, measuring, and recording its flow through the pipeline.

Conditioning and Compressors

Conditioning equipment is another name for the processing equipment, such as separators, heaters, and dehydrators, located at the wellhead or at other locations in the field. In addition, large compressors compress the gas up to or in excess of a hundred times the normal atmospheric pressure. *Compressor stations* (fig. 6.46) usually use large *reciprocating compressors* driven by gas engines, but they may also use *centrifugal units* driven by gas turbines or electric motors (fig. 6.47). Large compressor stations along the pipeline are often fueled by natural gas from the pipeline. Regulators lower the high pressure of the gas used for fuel.

Figure 6.47 This 9,200-horsepower (6,440-kilowatt) gas turbine is typical of those used in gas pipeline compressor stations. *(Courtesy of American Petroleum Institute and Cities Service Co.)*

Figure 6.48 A computerized control center keeps tab on deliveries throughout the gas pipeline system. *(Courtesy of Shell Pipe Line Corp.)*

Automation

Many gas pipelines have become so automated that they are capable of operating under command of a computer system that coordinates the operation of valves, prime movers, and conditioning equipment. The computer receives input from each part of the system, including the conditioning and measuring equipment. In case of malfunction at any point in the process, the computer searches its programs for possible corrective actions and simultaneously sounds alarms at the appropriate control points (fig. 6.48).

Odorants

An interesting aspect of natural gas pipelines is the introduction of odorants into the gas system. Natural gas is almost odorless as it comes from the well or processing facility. If the gas is destined for use as a fuel in homes or industry, a chemical called a *mercaptan* is added to give the gas a distinctive odor so that people can easily smell it when its concentration in air reaches 1 percent. Gas and air mixed in this concentration are not hazardous, but a mixture containing 5 percent gas is explosive.

The odorant makes leaks or other unburned discharges of the gas quite evident long before a real hazard exists. Odorants injected into the gas burned in homes or industry do not create odors while burning, nor do they leave troublesome residue. Odorants are usually not introduced into gas sent to petrochemical plants where the gas is used as a feedstock for producing other commodities such as plastics, since the mercaptan will frequently interfere with the chemical process.

PIPELINE CONSTRUCTION ON LAND

Before construction begins on a modern pipeline, months and sometimes years of engineering studies and surveys of potential reservoirs and markets precede the final decision to build the line. Companies survey routes by aerial photography and surface mapping. During the surface mapping, the company must secure a right-of-way from each owner of the property that the pipeline will cross. Before a pipeline superintendent goes out to begin construction, he or she has several guidelines to determine all operations: maps, a line list, and contract specifications. Maps show in accurate detail all surface features affecting construction. The *line list* sets out in sequence the names of all landowners, the number of linear rods (metres) of line to be built on each piece of property, and any written limitations or restrictions. Included are copies of permits for crossing roads, rivers, and other public property. Specifications in the contract cover every planned operation and as many eventualities as possible.

Assembling the Spread

The pipeline contractor assembles the *spread*—the equipment and crew needed to build a pipeline. A spread may be composed of 250 to 300 workers in an average operation and up to 500 in a very large operation. The amount of construction equipment depends on the size of the pipeline to be built and the difficulty of the terrain. Stream crossings, marshes, bogs, heavily timbered forests, steep slopes, or rocky ground can require different pieces of machinery. The pipeline contractor might rent rather than buy the necessary machinery since outfitting a *big-inch* (large-diameter) pipeline spread can be a multimillion-dollar operation. A spread can move at a rate of 3 miles (4.8 kilometres) a day, with a distance of sometimes 10 or 15 miles (16 or 24 kilometres) separating the workers at the beginning of the crew from those at the end.

Clearing Right-of-Way

A *right-of-way* is the legal document granting a right of passage over another person's land. By common usage it has come to mean a cleared strip of land from 50 to 75 feet (15 to 23 metres) wide, depending on the size of the pipe and the type of terrain. The clearing crews open fences and build gates, cattle guards, and bridges as the first segment of the spread moves up on the job. Bulldozers are standard equipment for clearing timbered areas (fig. 6.49).

Figure 6.49 A bulldozer is used to clear a pipeline right-of-way.

Figure 6.50 In this aboveground section of the Trans-Alaska Pipeline, 15 miles (24 kilometres) south of Livengood, Alaska, the roadbed along the right-of-way supports working vehicles as well as the pipeline itself. *(Courtesy of Alyeska Pipeline Service Co.)*

Salable timber cut by the clearing crews is stacked; the rest is cut and burned. Grading and completion of a roadway capable of supporting all vehicles follow. The road must be large enough for the largest tractor and other necessary equipment. In rocky terrain, a machine equipped with a *ripper* that extends several feet into the ground is often used to loosen rocks for removal before digging the ditch for the pipe to rest in.

Construction of pipelines in arctic climates involves routes that cross permafrost (permanently frozen ground), forests, snow-covered mountains, tundra, rivers, and streams. The Trans-Alaska Pipeline, for example, runs through very diverse terrain. On the North Slope, the route required a special gravel and insulation workpad to prevent the subsurface permafrost from degrading or melting. The gravel road base acted as an insulator, protecting the permafrost as vehicles, equipment, and crew members of the spread built the aboveground portion of the pipeline (fig. 6.50). Parts of the line were laid in a zig-zag pattern to allow for expansion and contraction of the pipe during temperature changes.

Ditching

Digging a ditch in which to bury the pipe, or *ditching*, requires various kinds of equipment. For loose dirt or stable soil, the *wheel ditcher* is most

Figure 6.51 A wheel ditcher leaves a neat spoil bank (left) that aids in rapid backfilling.

common (fig. 6.51). Rotating toothed buckets on the wheel ditcher pick up the excavated dirt, or *spoil*, and pile it to the side of the ditch for later use as backfill. Rocky terrain requires the use of pneumatically drilled holes and dynamite, a *rock ditcher*, or a bulldozer with ripper. A ripper is a clawed, plowlike attachment. Sometimes the crew uses backhoes to dig through rock and to clear blasted rock out of the trench.

The ditch must be at least 12 inches (30.5 centimetres) wider than the pipe diameter and deep enough to ensure that it will not interfere with plowing and other normal land use. The Department of Transportation requires a minimum of 30 inches (76 centimetres) of cover across farmland in normal soil and 36 inches (91 centimetres) in municipal areas.

In arctic areas where permafrost conditions exist, pipeline contractors may elect not to use a ditch for the pipeline. The pipeline, or sections of it, may be built above the ground on specially designed horizontal crossbeams mounted between pairs of *vertical support members (VSMs)* (fig. 6.52A). Approximately 400 miles (644 kilometres) of the Trans-Alaska Pipeline is constructed in this manner. When the aboveground technique is used, the pipeline is surrounded by insulation panels made of steel-jacketed fiberglass with weatherproof expansion joints. Finned radiators atop the VSMs improve heat transfer between the atmosphere and the heat pipes to which they are attached (fig. 6.52B). The *heat pipes*, which are 2 inches (5 centimetres) in diameter, keep the soil stable in permafrost areas by cooling the ground. Drawing off heat is necessary because the friction of the flowing oil heats the ground enough that the pipe can sag. When a company decides to ditch in unstable permafrost soil, it may refrigerate the buried section of pipe to keep the subsurface soils in the area from melting. About 7 miles (11 kilometres) of the Trans-Alaska Pipeline is constructed in this special burial mode.

Figure 6.52 In aboveground construction, VSMs hold insulated pipe at various heights (A), while special finned radiators and heat pipes draw heat away from the ground (B). *(Courtesy of Alyeska Pipeline Service Co.)*

Figure 6.53 The pipeline contractor strings pipe joints along the right-of-way so that they are accessible but do not obstruct spread equipment and the work crews.

Stringing Pipe

Stringing is the process of delivering and distributing line pipe along the right-of-way (fig. 6.53). The pipeline contractor or its subcontractor purchases and strings the pipe. On some jobs the pipe is cleaned, primed, and coated before it is delivered to the right-of-way. This type of pipe requires special handling to prevent damage to the coating as well as to the pipe ends. To move it, a sling is formed from wire rope threaded through a boom, a long arm attached to a tractor. Special curved aluminum plates at the end of the sling fit inside the ends of the pipe, where contact with the coating and the beveled ends is kept to a minimum (fig. 6.54). As pipe is lifted off the trailer, the equipment operator must evenly distribute the weight of the load to prevent the pipe from buckling.

On jobs where the terrain is relatively flat and the road system permits it, the pipe may be double jointed before being strung along the right-of-way.

Figure 6.54 A stringing side boom unloads coated pipe and places it along the right-of-way. *(Courtesy of Sheehan Pipe Line Construction Co.)*

Figure 6.55 A pipe-bending machine shapes pipe joints to conform to the contours of the terrain.

Double jointing is the welding of two sections of pipe to form one piece approximately 80 feet (about 25 metres) long. The equipment operator generally uses a spreader bar between the two lifting lines to prevent these long joints from buckling.

Bending Pipe

When the ditching and stringing are finished, engineers measure the contours of the ditch to determine how many degrees of bend must be put in the pipe so that it will conform to the bottom of the ditch. Often as many as seven out of every ten pieces of pipe must be bent to fit the ditch. Special equipment allows the pipe to be bent on location (fig. 6.55). A *bending shoe*—an attachment that works off the side-boom tractor and winch—bends small-diameter pipe. Large-diameter pipe requires a *bending mandrel*, which keeps the pipe from buckling or wrinkling when it is bent.

Cleaning, Aligning, and Welding Pipe

Before the pipe is welded in place, the ends must be absolutely clean of all dirt, scale, and coating. *Line-up clamps* are used to align the ends of the pipe prior to welding. They ensure that the two ends match around the entire circumference and that proper spacing is maintained between the ends. The clamps fit on the outside of small-diameter pipe and inside large-diameter pipe. Side-boom tractors and winches lower the pipe into place and hold it while the crew puts skids under it and makes the initial welds (fig. 6.56).

Figure 6.56 Welders work on opposite sides of the pipe as they make the initial weld on two joints held together by an internal line-up clamp.

A technician then inspects the welds visually and by X-ray before the pipe is coated and wrapped.

Figure 6.57 When line travel tape is used, one machine performs the cleaning and priming and the coating and wrapping processes.

Coating and Wrapping Pipe

Steel pipe will corrode from the moisture in the ground, so it is coated and may be wrapped as well with a material that protects the metal. Three types of coatings used on pipe are fusion-bonded epoxy, enamel, and tape.

Fusion-bonded epoxy coatings are usually applied at the mill that manufactures the pipe because the pipe must be heated to around 400°F (204°C) or higher for application. Epoxy coatings do not require an outer wrap, are very resistant to handling damage, and can withstand high operating temperatures such as those found in gas transmission lines.

Enamel and tape coatings are applied to pipe while it is being laid in a process called *line travel applied coating*. When enamels are used, a self-propelled cleaning and priming machine moves along the top of the pipe, followed by a coating and wrapping machine. Coal tar enamels, tar-like substances, are particularly effective for line travel applied coating. The coating machine applies the heated liquid enamel, and then it wraps the coated pipe with fiberglass, felt, and kraft paper to protect it while it hardens and while the pipe is lowered in. Enamel coatings are most effective when the pipeline's operating temperature range is between 30° and 180°F (-1° and 82°C).

When line travel tape is used, one machine cleans, primes, coats, and wraps the pipe (fig. 6.57). The taping machine, with rolls of tape (polyethylene, polyvinyl, or other adhesive) mounted on spindles, wraps the tape in overlapping sections around the pipe as it moves along the line.

Lowering In and Backfilling

Two steps can be eliminated if the pipe is lowered into the ditch in conjunction with the coating and wrapping operation—lowering of the pipe to skids and picking it up later for final placement. When this special burial mode is used in arctic environments, the pipe is lowered into the ditch enclosed in a sheath of insulation. Refrigerant lines are laid along both sides of the pipe, and special padding is placed in the ditch to minimize melting in the permafrost layer. Side booms with nonmetallic slings place the pipe safely into the ditch (fig. 6.58).

Figure 6.58 Side-boom tractors with correctly spaced slings lower pipe into the ditch.

After pipe has been lowered into the ditch, the crew fills the soil back in both underneath and over the pipe. They often use an auger fitted to the front of a bulldozer to distribute the backfill evenly in the ditch. Backfilling requires breaking up rocky or frozen soil before returning it to the ditch to prevent it from damaging the pipe coating. Uneven distribution of frozen soil may leave sections of the pipe unsupported after the fill has thawed, causing cracks and leaks.

Specialty and Tie-In Crews

As the spread moves along, it bypasses portions of the line that require special construction techniques and equipment. A specialty crew, the *tie-in crew*, completes these unfinished sections of the line and ties them in to the main pipeline. The tie-in crew is a miniature spread that joins the segments into one continuous line.

Highway, railroad, and inland waterway crossings, swamps, and marshes are some of the environments that require tie-in crews. For example, a boring machine may be used to go under a highway, railroad, or pre-existing pipeline. In some areas, the crew must set casing to protect the pipe (fig. 6.59). Some operations require six or more side-boom tractors to maneuver a crossing. Creek crossings and steep terrain also require the skillful maneuvering of side-boom operators as they move long sections of pipe into position.

Figure 6.59 A boring machine pushes casing under a roadbed as its cutting head rotates just inside the front of the casing. Pipe will be installed inside the casing.

Figure 6.60 Flotation devices support sections of pipe during a river crossing.

One way to lay pipe across a river is to bury it under the riverbed. A directional drilling rig drills a controlled horizontal pilot hole in an inverted arc under the riverbed. Guided by computers, the crew pulls a pilot string of pipe through first, followed by a work string. Finally the rig bores the actual hole and the pipeline is pulled through. Another method of crossing water is to attach flotation devices to the pipe, then walk or pull the pipe across (fig. 6.60). The crew then lowers the pipe into an underwater ditch, which has been dredged out with a dredging unit.

The tie-in crew ties special sections in to the adjoining cross-country pipeline. Tie-in involves welding pipe ends together, testing them, and adding fabricated sections when necessary. Fabricated sections are special valve connections and fittings made up ahead of time and joined to the main pipe where needed.

As the pipeline is completed, the construction company takes special care to restore the right-of-way as nearly as possible to its original condition (fig. 6.61). Of course, they must keep the right-of-way clear of trees and brush to allow visual inspection for leaks.

Figure 6.61 After construction is completed, a neatly restored right-of-way allows this land to produce crops once again.

OFFSHORE PIPELINE CONSTRUCTION

Pipelining offshore in shallower depths has been well established for quite some time, and modern technology has made it possible to lay pipe in deep-sea areas. In all offshore pipeline construction, various types of barges play a key role.

The pipeline industry has also researched and developed techniques for building offshore pipelines in ice-laden arctic waters. Equipment that can detect the presence of ice features such as ice islands and pressure ridges that cause scouring is being used to reduce potential pipeline damage from contact with ice. To protect the pipeline against ice scouring, it must be buried so that the depth of the trench on the seafloor to the top of the pipe exceeds the deepest scour marks in the area.

Thawing can stress pipelines constructed in permafrost near the shore. To minimize thawing in the permafrost layer and to keep the stress on the pipeline at an acceptable level, the pipeline company can take one of these measures: (1) place pipelines on special beds of granular material to keep an insulating layer between the permafrost and the pipeline, (2) artificially reduce the temperature of the oil at the processing facility, (3) raise the pipeline above the surface of the water on piles or artificial berms, and (4) circulate refrigerant through cooling tubes surrounding pipeline enclosed in a jacket.

Conventional Lay Barges

Offshore lines are laid by pipe-laying barges, or *lay barges* (fig. 6.62). A lay barge is a complete seagoing plant that allows the pipeline to be assembled and laid continuously along the route either on top of the ocean floor or in dredged trenches on the seafloor. When used in conjunction with supporting tugs and an anchoring system, the lay barge can be self-sufficient for months at a time.

An eight-point mooring system of wire rope and anchors holds the lay barge on a precise heading to prevent the pipe from buckling as it is laid. The system also propels the barge as anchor lines are reeled in and out. As the barge progresses to the end of the lines, anchor-handling tugboats move the mooring system ahead.

Figure 6.62 The main deck of a lay barge, as seen on this center-slot vessel, may be thought of as an offshore pipeline spread. *(Courtesy of McDermott, Inc.)*

Figure 6.63 Welding, coating, and inspection are handled at stations on the lay barge.

A pipe barge brings the line pipe from yards on shore to the lay barge, which stores it until it is added to the lengthening pipeline. Welding, coating, and inspection are handled at stations, or work areas, along the length of the lay barge deck (fig. 6.63). The completed pipeline is lowered into the water by way of an inclined ramp and a stinger. The *stinger* is attached to the ramp and guides the pipeline to the seafloor at the proper angle (fig. 6.64).

The pipeline is coated with a high-density cement to overcome buoyancy so that it can be sunk into place. To prevent cracking of this concrete coating or buckling of the pipe when lowered, especially in deep water, the crew sometimes adds an attachment called a *pontoon* to the stinger. They flood the pontoon to lower the pipeline toward the seafloor at an angle that will not overstress it. Also, they sometimes use a *tensioner* in deepwater lay operations. The tensioner maintains an upward force in the pipe so that the pipe does not buckle under its own weight.

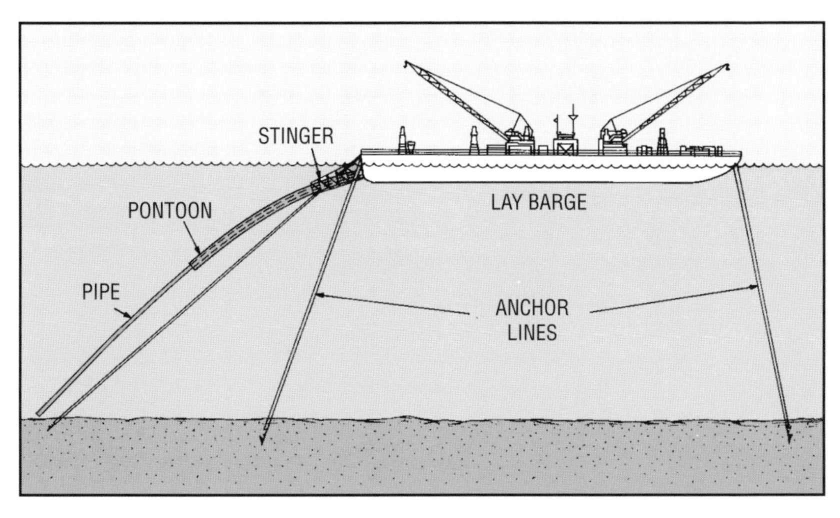

Figure 6.64 The stinger guides the pipeline safely to the seafloor while anchor lines move the vessel along.

The method used to keep the pipeline on the bottom and out of danger varies with the topography of the seafloor. On rocky bottoms, the line is fastened to the seafloor with pipeline anchors.

Bury Barges

Subsea pipelines that might be harmed by fishing or other marine operations are laid in a seafloor trench. On softer bottoms, a *bury barge*, or *pipe-trenching barge*, with a sled is used for this operation (fig. 6.65). The sled, which is attached to the submerged pipeline, moves along the line on the seafloor and uses a high-pressure jetting action to form a trench in which the pipeline will rest. Usually, the pipe is not covered at the time it is laid, but water movement eventually covers it with fill from the ocean bottom.

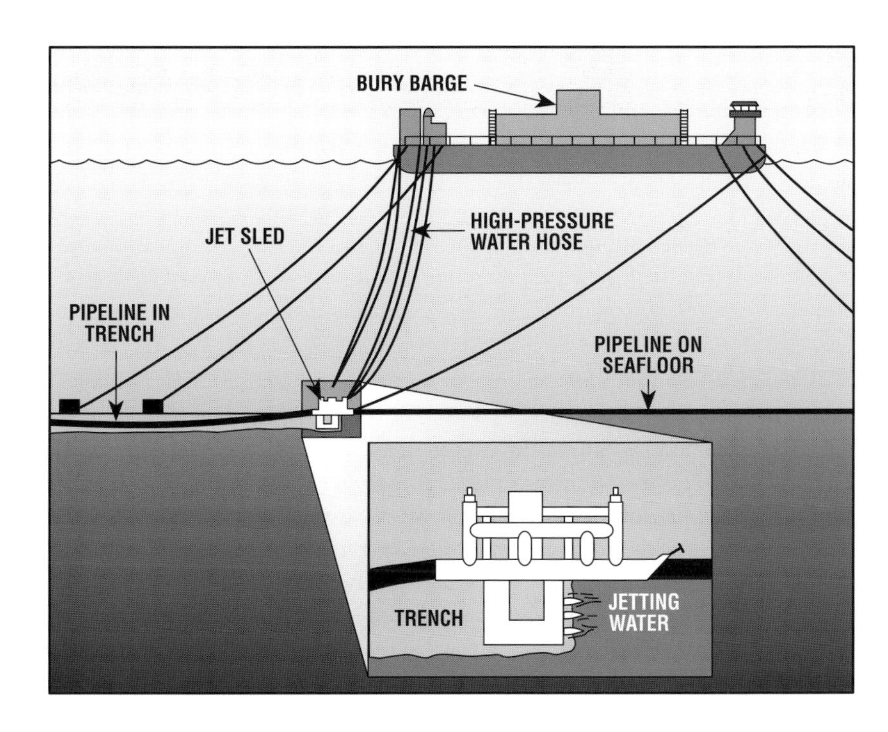

Figure 6.65 High-pressure jets of water from a jet sled dig a trench for a pipeline on the seafloor. The sled is operated from a bury barge as it moves forward by means of anchors.

Figure 6.66 Designed for use in rough, deep water, this superbarge houses 350 people and can store as much as 20,000 tons (18,144 tonnes) of pipe.

Superbarges

Designed for use in extremely rough and unusually deep water is a huge lay barge that might be called a *superbarge* (fig. 6.66). It is 650 feet long, 140 feet wide, and 50 feet deep (198 metres long, 43 metres wide, and 15 metres deep). This barge can lay 80-foot-long (24-metre-long) double-jointed pipe in deep water without the use of a pipe-laying pontoon.

The superbarge houses 350 people and can store as much as 20,000 tons (18,144 tonnes) of pipe. It has seven work stations for double-jointed pipe or nine work stations for 40-foot (12-metre) automatic or manual welding joints. It also includes a pipe-beveling station, two 125-ton (113-tonne) cranes, twelve 60,000-pound (27,216-kilogram) anchors, and a heliport.

Semisubmersible Barges

Coming to the forefront in offshore pipelining is the use of a semisubmersible lay barge, which operates in a manner similar to the semisubmersible drilling rig. The *semisubmersible lay barge* is designed to minimize the effects of wave and wind actions when operating in rough waters. For example, a semisubmersible barge crew can lay pipe in the North Sea during 35-foot (11-metre) waves and 70-mile-per-hour (113-kilometre-per-hour) winds—an impossible task with a conventional lay barge. The stability comes from the submerged portion of the vessel, which lowers the center of gravity and steadies the vessel against high winds and waves.

Reel Vessel

On a *reel vessel*, which may be a ship or a barge, the pipe is welded and spooled onto the giant reels at onshore facilities (fig. 6.67). This pipe must be flexible enough to withstand reeling and unreeling without buckling or cracking. In the early development of the reel vessel, pipe was limited to a small gathering line of 1½ to 2 inches (38 to 50 millimetres) in diameter. Now, however, reel vessels can lay lines with a diameter of 8 inches (20 centimetres) from the reels.

The spooled line, ranging from several hundred feet or metres to a couple of miles or kilometres in length (depending on the pipe diameter), is unwound by moving the barge forward. The line unspools cleanly and in a straight path. Giant unspooling devices located at the rear of the barge, where the line extends to the water, keep the line straightened and unwrinkled. When one spool of line ends, the end is held above the water surface while welders join a new spool of line to it. Then the submerged line is dropped, and unspooling continues with the new line.

Figure 6.67 A giant spool unreels continuous lengths of line pipe from a reel lay barge.

ECONOMICS AND SAFETY

Economics and safety generally determine which method of transportation is used to ship crude oil, petroleum products, and natural gas. Of the four methods the industry uses—pipelines, water carriers, motor carriers, and railroads—pipelines have been found to be the safest by the U.S. National Transportation Safety Board. Pipelines are also the most economical, provided that the line is available and can operate at full capacity. However, pipelines require a great deal of capital to build and operate, and once laid, they are immobile.

Since 1980, water carriers have had a slight edge over pipelines in crude oil transportation (table 6.2). This jump in water-carrier use is partially due to increased shipments of crude by tanker from Valdez, Alaska, to U.S. ports on the Gulf Coast, East Coast, and Caribbean. Most petroleum products, however, are shipped by pipeline, with water carriers running a close second.

TABLE 6.2

DOMESTIC TRANSPORTATION OF CRUDE PETROLEUM AND PETROLEUM PRODUCTS, 1973–1983 (IN BILLIONS OF TON-MILES[a])

MODE OF TRANSPORTATION	YEAR 1973	1974	1975	1976	1977	1978	1979	1980	1981	1982	1983
Pipelines											
Crude	302.0	303.0	288.0	303.0	326.6	359.5	372.2	362.3	333.1	335.1	332.4[b]
Products	205.0	203.0	219.0	212.0	219.4	226.3	236.1	225.6	230.6	230.6	223.7[b]
Total	507.0	506.0	507.0	515.0	546.0	585.8	608.0	588.2	563.7	565.7	556.1[b]
Water Carriers											
Crude	58.8	53.0	40.6	37.8	63.1	261.3	265.5	387.4	404.9	432.7	471.2
Products	238.0	244.0	257.4	269.1	270.2	269.3	257.4	230.4	212.3	184.2	159.3
Total	296.8	297.0	298.0	306.9	333.3	530.6	522.9	617.8	617.2	616.9	630.5
Motor Carriers[c]											
Crude	1.3	1.3	1.4	2.1	2.0	2.0	2.3	2.5	2.2	2.0	2.0
Products	23.7	27.7	26.2	30.4	27.6	28.6	27.8	24.3	22.7	20.7	22.2
Total	25.0	29.0	27.6	32.5	29.6	30.6	30.1	26.8	24.9	22.7	24.2
Railroads											
Crude	1.1	1.6	1.5	0.9	0.8	0.7	0.6	0.5	0.5	0.4	0.5
Products	13.7	14.1	12.6	12.4	13.7	12.5	12.9	12.0	12.1	12.5	11.3
Total	14.8	15.7	14.1	13.3	14.5	13.2	13.5	12.5	12.6	12.9	11.8
All Modes											
Crude	363.2	358.9	331.5	343.8	392.5	623.5	640.6	753.0	740.7	770.2	806.1
Products	480.4	488.8	515.2	523.9	530.9	536.7	534.2	492.3	477.7	448.0	416.5
Total	843.6	847.7	846.7	867.7	923.4	1160.2	1174.5	1245.3	1218.4	1218.2	1222.6

[a]A ton-mile is a standard measure for moving one ton of product one mile.
[b]Preliminary figures.
[c]Amounts carried by motor carriers are estimates.

Source: Association of Oil Pipe Lines, Washington, D.C.

The decrease in domestic petroleum products carried by water since 1981 is partly the result of an increase in imports of products from foreign refineries. Natural gas is almost totally shipped by pipeline, although LNG tankers and pressure transport trucks are used in special cases.

Regardless of the volume carried, each transportation method plays a key role in the functioning of a refining or processing plant—bringing feedstocks in and taking refined products out (fig. 6.68). Transportation is also vital to the marketing and distribution of consumer products made from petroleum.

Figure 6.68 Transportation of all kinds is evident at this refinery in Louisiana, located near a river and ship channel. A portion of the facility's crude oil feedstock is delivered by barge; products move out by railcar and tank trucks; and pipelines fill and empty the many storage tanks with feedstocks coming in and products going out. *(Courtesy of Cities Service Co.)*

7
Refining and Processing

Collected crude oil and natural gas are of little use in their raw state; their value lies in what is created from them—fuels, lubricating oils, waxes, asphalt, and petrochemicals. The refining world is a world of technology. To a traveler driving by, a *refinery* appears to be a strange conglomeration of illuminated towers and walls at night (fig. 7.1) and an equally strange maze of pipes and tanks during the day (fig. 7.2). In reality, a refinery is an organized and coordinated arrangement of equipment that separates the components in crude oil and gas and produces physical and chemical changes in them. These changes create salable products of the quality and quantity that consumers want. Refineries also include facilities to store crude oil and products and maintain equipment.

As *crude oil*, or *crude*, comes from the well, it contains hydrocarbon compounds and relatively small amounts of impurities such as oxygen, nitrogen, sulfur, salt, and water—plus traces of certain metals. The refinery

Figure 7.1 A refinery unit lighted against the backdrop of a Louisiana sky at dusk is an impressive sight. *(Courtesy of Cities Service Co.)*

Figure 7.2 A refinery is an organized and coordinated arrangement of processes linked together with miles of pipe carrying feedstocks in and products out. *(Courtesy of Coastal States Gas Corp.)*

removes any substance from the crude oil that is not a hydrocarbon and then breaks the oil down into various hydrocarbon components. Some of the components are further treated to give them more desirable properties. Finally, the resulting substances are sold as they are or blended into useful products.

Likewise, natural gas is a mixture of hydrocarbon gases, along with impurities such as water, nitrogen, and carbon dioxide. A refinery or a *gas processing plant* removes impurities and separates desired products.

Most refinery products are ready for consumers and industry to use. Some, however, go to petrochemical plants, where they are changed into chemicals used to manufacture an almost limitless number of products ranging from fertilizers to plastics.

STRUCTURE OF HYDROCARBONS IN OIL AND GAS

In other parts of the oil industry, the word petroleum can refer to both crude oil and natural gas. However, in the refining industry, *petroleum* refers only to crude oil, the heavier constituents that occur naturally in liquid form in the reservoir. The lighter constituents that occur naturally in gaseous form are simply called *gas*. A third, less common constituent is the natural gasoline. *Natural gasoline* is in a gaseous state in the reservoir but emerges from the oilwell as liquid petroleum.

Crude and natural gas, as they are piped in from the reservoir, are mixtures of many different hydrocarbons (see Chapter 1) ranging from light gases to heavy pitch, each with its unique chemical structure and properties. The crude oil varies in the type and amount of hydrocarbons as well as which impurities it contains, and the particular mixture determines such characteristics as color, viscosity, density, and boiling point. Some crude oils are light, flow easily, and make good gasoline. Others are heavy and thick and are excellent for producing asphalt. The refinery or gas processing plant does not separate each of these hydrocarbons; instead, it refines or processes groups of hydrocarbons with similar chemical structures and physical characteristics.

Refineries and gas processing plants separate hydrocarbons into four types—paraffins, olefins, naphthenes, and aromatics. All of these compounds are arrangements of carbon and hydrogen atoms, and sometimes they include other elements such as oxygen, sulfur, and trace metals.

Paraffins

Paraffin hydrocarbons make up natural gas and a substantial portion of crude oil (fig. 7.3A). The simplest paraffin is *methane*, the main ingredient of natural gas. A molecule (a group of atoms bonded together) of methane has one carbon atom attached to four hydrogen atoms. The other hydrocarbons in the paraffin group are chains of carbon atoms with attached hydrogen atoms. Examples are *ethane* and *propane*, also components of natural gas. Each carbon atom can bond, or attach, to four other atoms, and each hydrogen atom can bond with one other atom.

Paraffins always have names ending in *ane*. Chemically, they are very stable compounds. The paraffin group is large, varying in number of carbon atoms from one to ninety or more. Chemists express the number of carbon and hydrogen atoms in each hydrocarbon by a formula. For example, the formula for ethane is C_2H_6, meaning that each molecule is made up of two carbon atoms and six hydrogen atoms.

Hydrocarbons with very few carbon atoms (C_1 to C_4) are light in weight and are gases under normal atmospheric pressure. Hydrocarbons with more carbon atoms are heavier and are either liquid or solid under atmospheric pressure. As the number of carbon atoms increases, the boiling point also increases—more heat must be applied to change the hydrocarbon from a liquid or solid to a gaseous state.

Paraffins with four or more carbon atoms can have more than one arrangement of atoms. All the carbon atoms can form a single straight chain, or some of the carbon atoms can form branches attached to the main chain. *Butane,* for example, has two forms: *normal*, or *n-butane* (straight chain), and *iso*, or *i-butane* (branched chain). Isobutane is known as an *isomer* of normal butane because i-butane has exactly the same formula but a different arrangement of atoms. This difference is important because isobutane boils at a different temperature from normal butane and causes different chemical reactions in the refinery. The larger the paraffin molecule, the more isomers it can have.

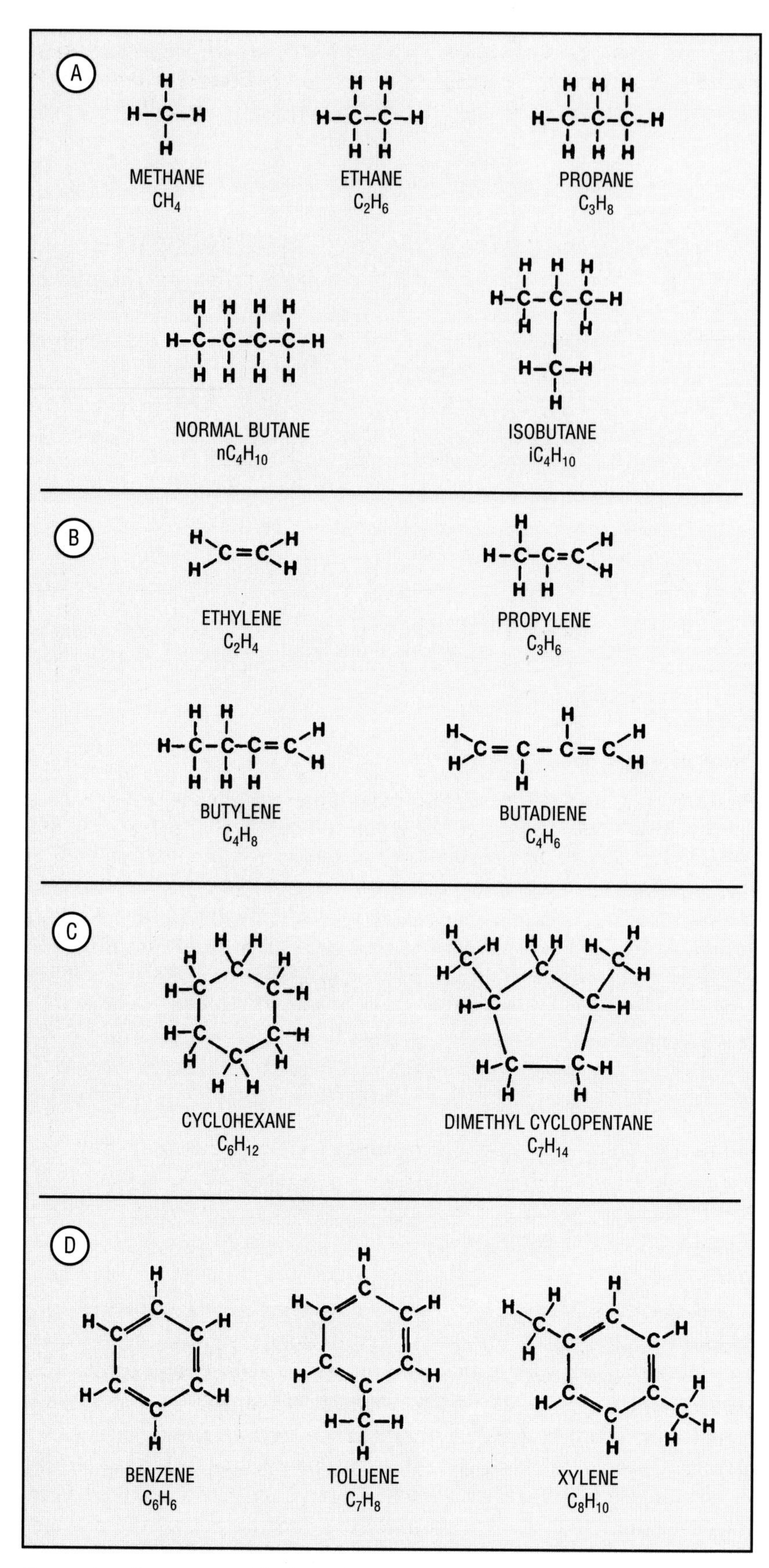

Figure 7.3 Four types of hydrocarbon structures are paraffins (A), olefins (B), naphthenes (C), and aromatics (D).

Other paraffins include *pentane* (C_5H_{12}), *hexane* (C_6H_{14}), *heptane* (C_7H_{16}), *octane* (C_8H_{18}), *nonane* (C_9H_{20}), and *decane* ($C_{10}H_{22}$). Because paraffins have the greatest possible number of hydrogen atoms, they are called *saturated hydrocarbons*.

Olefins

Like paraffins, *olefins* are chains of carbon atoms with attached hydrogen atoms (fig. 7.3B). However, olefin chains do not have the greatest possible number of hydrogen atoms. If two hydrogen atoms are missing, two carbon atoms in the chain will form a *double bond* to make up for the deficiency. An example is *ethylene,* an important petrochemical. If four hydrogen atoms are missing, the carbon chain will have two double bonds; an example is *butadiene,* another petrochemical. Olefins with one double bond have names ending in *ylene* or *ene,* and olefins with two double bonds have names ending in *adiene.* Other common olefins are *propylene* and *butylene.*

Olefins are known as *unsaturated hydrocarbons* because of their hydrogen deficiency. Because the carbon atoms of olefins are always seeking to attach their "empty" bonds to other atoms, olefins are unstable and are easily used to make new chemical compounds. They do not occur naturally in crude oil but are formed in the refinery by the breakdown of larger hydrocarbon molecules. Olefins are very useful in creating certain refinery products and petrochemicals.

Naphthenes

The carbon atoms of *naphthenes* form rings rather than chains, so naphthenes are called ring *compounds* or *cycloparaffins* (fig. 7.3C). Hydrocarbons in this group have names that begin with the prefix *cyclo* to indicate the ring structure. An example is *cyclohexane,* a hydrocarbon often occurring in natural gasoline. The carbon rings of naphthenes contain all the hydrogen atoms they can, so naphthenes are chemically stable. They occur naturally in crude oil and have properties similar to the paraffins.

Aromatics

Aromatic hydrocarbons, like naphthenes, contain a ring of six carbon atoms, and like olefins, have double and single bonds and so have fewer hydrogen atoms (fig. 7.3D). This type of structure is known as a benzene ring. The most important aromatics in refinery production are the BTXs—*benzene, toluene,* and *xylene.* Aromatics occur naturally in crude oil. Refinery and petrochemical plant processes also create them.

Other Elements

Anywhere from 2 to 50 percent of a crude oil may be compounds containing oxygen, nitrogen, sulfur, and metals. Oxygen content can be as high as 2 percent, and nitrogen content as high as 0.8 percent. Sulfur content ranges from traces to more than 5 percent. If a crude contains appreciable quantities of sulfur or sulfur compounds, it is called a *sour crude*; if it contains little or no sulfur, it is called a *sweet crude.* Trace metals contained in crude oil include sodium, magnesium, calcium, strontium, copper, silver, gold, aluminum, tin, lead, vanadium, chromium, manganese, iron, cobalt, nickel, platinum, uranium, boron, silicon, and phosphorus. Of these metals, nickel and vanadium occur in the greatest quantities.

REFINING CRUDE OIL

Today, many processes are available to refine crude oil (fig. 7.4). Which processes a refinery uses depend partly on the content and quality of the crude oil it receives, partly on consumer demand, and partly on existing plant facilities and the economics involved in changing them. Not all refineries produce all products or serve all markets.

However, every refinery begins the processing by separating crude oil into different components, called *fractions* or *cuts*—groups of hydrocarbons with the same boiling-point range and similar properties. Next it breaks down or rearranges the molecules of some of the separated fractions and separates them again. The final process is to blend the refined hydrocarbons into mixtures that have desirable qualities for certain purposes. These processes convert virtually all of the crude oil into salable products in amounts the market demands.

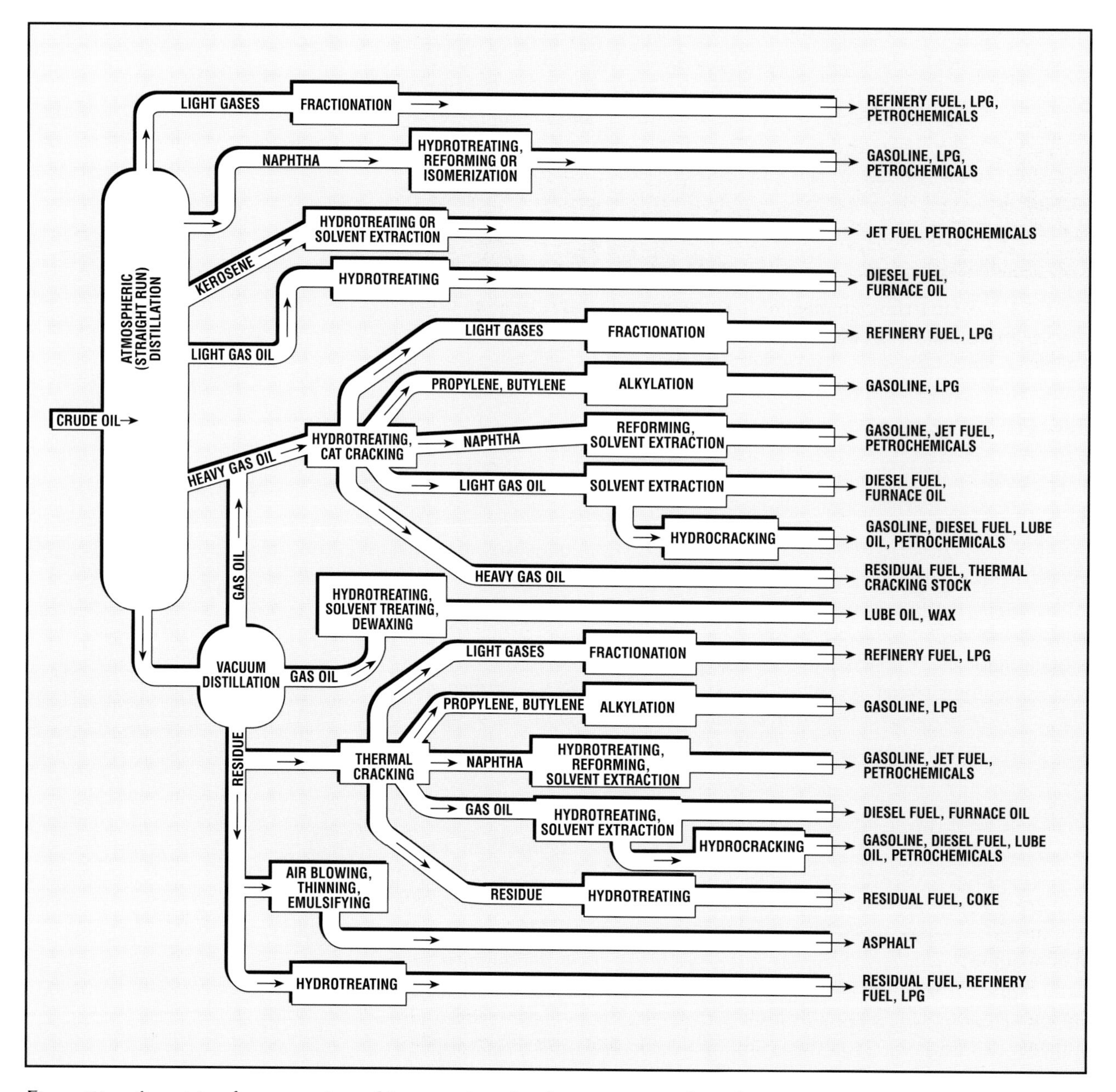

Figure 7.4 A variety of processes is used to convert crude oil into many products, beginning with distillation.

Figure 7.5 A lab technician monitors a crude assay—in this case, a simulation of vacuum distillation conducted in a refinery laboratory. *(Courtesy of Valero Refining Co.)*

Assays

Before a crude can be processed, it must be evaluated in a process called an *assay* (fig. 7.5). An assay determines exactly what hydrocarbons and impurities are present and in what amounts. The reasons for assaying include:

1. to plan the processing of the crude;
2. to determine whether a new crude oil will yield desirable amounts and qualities of products;
3. to detect changes over time in an oil whose properties are known; and
4. to compare the crude with the crude of competitors and assess its market value.

The completeness of the crude oil assay depends on the purpose. A few routine tests may be enough to monitor changes in well-known crude oils, but planning a refinery operation to process a large quantity of a new crude oil requires a full-scale evaluation.

A complete assay may evaluate the whole crude for properties such as API gravity (density), viscosity (resistance to flow), and sulfur content. Or just certain cuts may be evaluated for properties important to the products they will yield.

Some refineries purchase feedstock instead of whole crude oil. A *feedstock* is any material that is piped into the refinery for use in processing. In this case, the feedstock is partially distilled crude that comes from another refinery to be processed further. Regardless of their feedstock, most refineries conduct complete crude oil assays.

Classifications

An assay classifies a crude oil in one of three groups, based on the composition of its components: paraffin-base, asphalt-base, and mixed-base. *Paraffin-base crude* oil contains mostly paraffin hydrocarbons in its heavier fractions. It is a good source of paraffin wax, quality motor lubrication oils, and high-grade kerosene. *Asphalt-base crude* oil contains mostly asphaltic hydrocarbons in its heavier fractions. Asphalt is a dark solid or semisolid containing carbon, hydrogen, oxygen, sulfur, and sometimes nitrogen. This type of crude is particularly suitable for making high-quality gasoline and roofing and paving materials. The heavier fractions of some crude oils contain considerable amounts of both paraffins and asphalt. They are called *mixed-base crude* oils. Virtually all products can be obtained from them, although at lower yields than from the other two classes.

Refining Processes

Refinery processes (table 7.1) have developed also in response to changing market demands for certain products. Before the automobile, the primary product of refineries was kerosene for lamps and home heating. With the advent of the internal-combustion engine, the main job of refineries became the production of gasoline. At first a process called fractional distillation was the only method of producing gasoline. However, the amount of gasoline from distillation was soon insufficient to satisfy consumer demands (fig. 7.6).

At the same time, refineries had more heavy hydrocarbons than they needed for making lubricating oils and other types of fuel. So refineries began to look for ways of chemically converting a part of the heavier oils into synthetic gasoline. Two types of processes were developed to produce more and better quality gasoline: (1) cracking and (2) reshaping or rebuilding hydrocarbon molecules. The products of these two processes came to have many other uses as well.

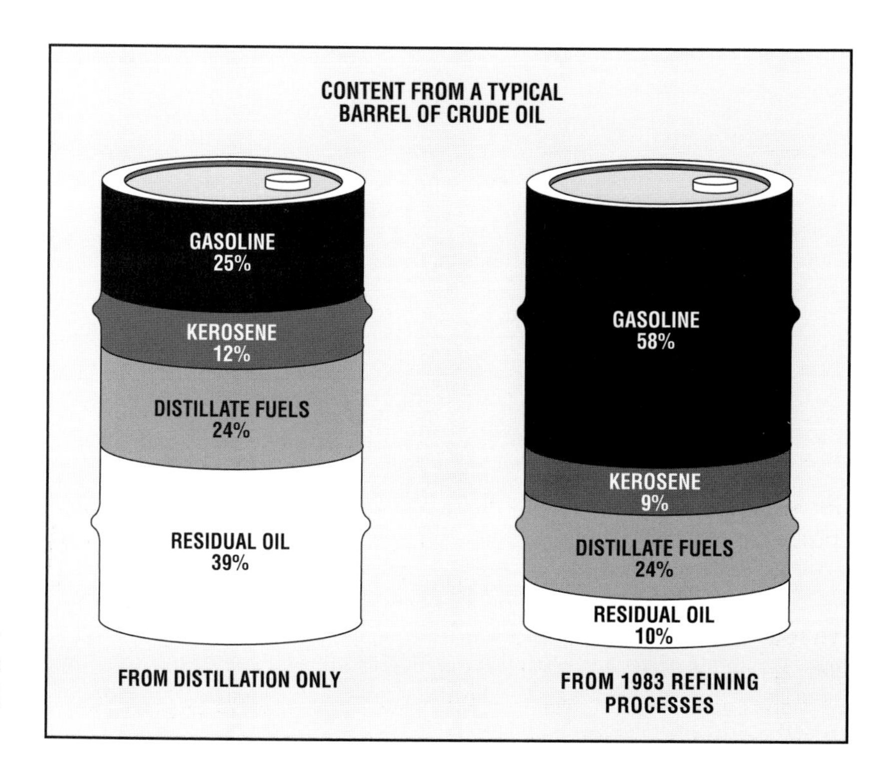

Figure 7.6 An important job of refineries today is the conversion of oil at the "bottom of the barrel" into gasoline and other more marketable products.

TABLE 7.1
REFINERY PROCESSES

Feed	Process	Primary Outputs
Crude oil	Atmospheric distillation	Straight-run fractions: light gases, naphtha, kerosene, gas oils, residue
Atmospheric residue	Vacuum distillation	Catalytic cracking and thermal cracking stocks Lube oil and wax Asphalt Residual fuel
Atmospheric and vacuum gas oils	Catalytic cracking	Alkylation, reforming, and hydrocracking stocks Residual fuel
Vacuum residue	Thermal cracking	Alkylation, reforming, and hydrocracking stocks Residual fuel Coke
Cracked gas oils	Hydrocracking	Gasoline Distillate fuels Lube oil
Cracked olefins and isobutane	Alkylation	Gasoline
C_5–C_6 gases	Isomerization	Gasoline Alkylation stock
Atmospheric and cracked naphthas	Reforming	Gasoline
Kerosenes, straight-run gas oils	Solvent extraction	Distillate fuels
Vacuum gas oils	Solvent extraction	Lube oil and wax
Reformed naphtha	Solvent extraction	Aromatic petrochemicals
Crude oil	Dehydration, desalting	Crude oil free of water and salt
Straight-run middle distillates, thermal-cracked fractions, residual oil	Hydrotreating	Oil free of sulfur, nitrogen, and metals Upgraded distillate fuels
Hydrogen sulfide and other sulfur compounds	Sulfur recovery	Elemental sulfur
Fuel oils, asphalts, lube oils and other products	Blending, additives	Products with improved performance qualities

Fractional Distillation

Every refinery begins the processing of crude oil by separating it into different cuts or fractions by distillation (see fig. 7.4). This is called *fractional distillation*, or *fractionation*, and the products may be called *distillates.*

Distillation is quite simple in principle—it is like turning water into steam in a teakettle. When the water is heated, it vaporizes into steam and escapes from the kettle. If the spout of the teakettle had a transparent cup over it, you could see the steam condensing on the inside of the cup and dripping down the sides as water. Because crude oil is a mixture of hydrocarbons that boils, or vaporizes, at different temperatures, heating it to a certain temperature allows one fraction to boil off while others remain liquid.

Refineries use two types of distillation: atmospheric and vacuum.

Atmospheric Distillation

In atmospheric distillation, the process takes place at atmospheric (normal) pressure in a piece of equipment known as a *distilling column*. The distilling column is a tall cylindrical steel tower (fig. 7.7).

Inside the tower, flat steel trays are welded to the sides every few feet. The trays have holes and slots in them so that liquid can flow down through them and the vapor can rise. Pipes called *downcomers* direct the liquids from higher trays to lower ones. To begin, crude oil is heated to a temperature at which some of the hydrocarbons vaporize, and the resulting mixture of vapor and liquid is piped into the distilling column. The liquid flows to the bottom of the column and is removed. The vapor rises, passing through perforated *bubble caps* on the trays (fig. 7.8). As the vapor rises, it cools and condenses, becoming a liquid that settles on the trays. Heavier hydrocarbons condense more quickly and settle on lower trays, and lighter hydrocarbons remain in the vapor state longer and condense on higher trays. The liquid fractions are

Figure 7.7 Lines carrying the distilled cuts from different heights surround this distilling column, or distillation tower.

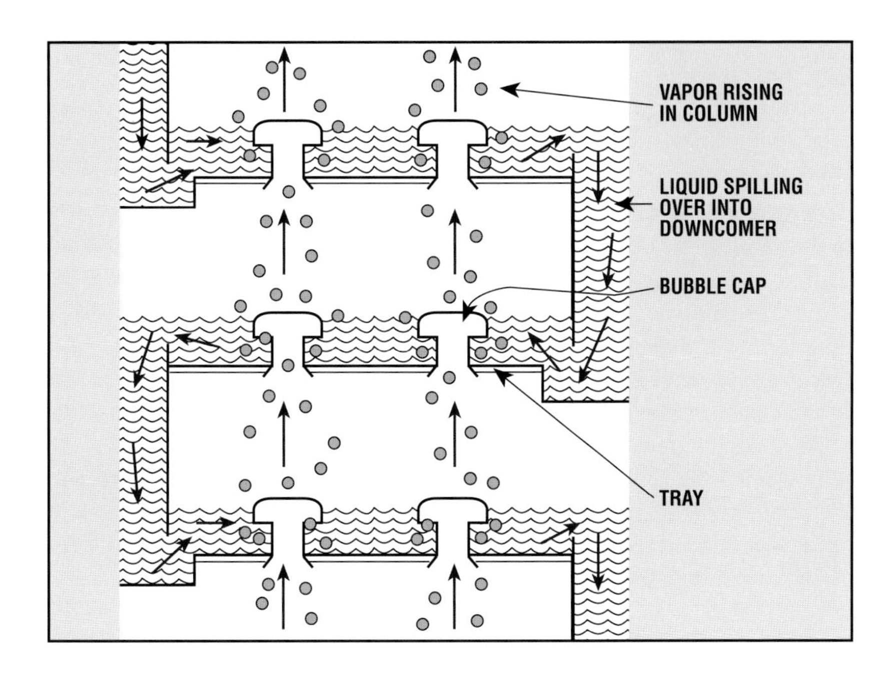

Figure 7.8 Vapor passes through bubble caps and settles on trays in the distilling column.

drawn from the trays and removed from the distilling column. The lightest hydrocarbons remain gaseous and pass out through the top of the column.

The distillates coming out from atmospheric distillation are useful in their new state, or they may be blended with other substances or processed further (table 7.2). For further processing, the gases go to a gas processing plant to be fractionated again. Light distillates include naphtha and kerosene. Naphtha is used to make gasoline and petrochemicals. (This naphtha is not the same as certain commercial solvents popularly known as naphthas.) Kerosene is used to make jet fuel. Middle distillates include light and heavy gas oils. Light gas oils are made into diesel fuels and furnace oils. Jet, diesel, and furnace fuels are sometimes called distillate fuels. Heavy gas oils are often cracked (see below) to produce naphtha and other products. The unvaporized residue left after atmospheric distillation, containing the heaviest hydrocarbons, usually goes to a vacuum distillation column for further fractionation.

Table 7.2

SOME FRACTIONS OBTAINED FROM ATMOSPHERIC DISTILLATION

Fraction	Carbon Atoms	Boiling Range, °F	State Under Atmospheric Pressure	Product Disposition
Gas	1–4	—258–31	Gas	Gas processing
Naphtha	4–12	31–400	Liquid	Cat reforming
Kerosene	10–14	356–500	Liquid, solid	Hydrotreating
Light gas oil	15–22	500–700	Liquid, solid	Distillate fuel blending
Heavy gas oil	10–35	640–875	Solid	Cat cracking
Residuum	36–90	875+	Solid	Flashing

Data Sources: "Petroleum Processing," *McGraw-Hill Encyclopedia of Science and Technology*, and Table 2.1, Cram and Hammond, *Organic Chemistry*.

Figure 7.9 A vacuum distillation unit turns residual oil into asphalt and other products. The heater preheats the feed before it goes to the vacuum distillation column.

Vacuum Distillation

Distillation at atmospheric pressure does not work for separating heavy hydrocarbons with boiling points of 900°F (482°C) and higher into fractions. The high temperatures required to vaporize them instead cause the molecules to break down into smaller molecules. To prevent this, the heated residue, or *residual oil*, from the atmospheric distillation column goes to a vacuum distillation tower (fig. 7.9). As the name implies, this tower creates a partial vacuum, which means that the pressure inside it is lowered. At a lower pressure, the residual oil's boiling point is lower, which allows the hydrocarbons to vaporize at lower temperatures and separate into fractions. Vacuum distillation separates light and heavy gas oil from the *bottoms*, the heaviest residue.

Light and heavy gas oils are further processed to produce products ranging from gasoline and distillate fuels to lubricating oils. Depending on its content, the bottoms will be processed to make fuels, asphalt, sealants, lubricating oils, wax, and coke.

Cracking

Heavier hydrocarbons from the "bottom of the barrel" are not very useful in their natural state. *Cracking* breaks down the heavy residues into lighter products such as gasoline and distillate fuels. Unlike distillation, which just separates hydrocarbons from each other, cracking is a chemical process that breaks long, heavy molecules into lighter, shorter ones. Cracking processes include catalytic cracking, thermal cracking, and hydrocracking.

Catalytic Cracking

Catalytic cracking, or *cat cracking*, uses a catalyst in a chemical reaction that converts gas oils obtained from atmospheric and vacuum distillation to gasoline (fig. 7.10). A *catalyst* is a substance that affects the speed of the reaction without itself being chemically changed or used up. The catalyst may be in the form of small pellets or a powder. Today, powders are more common.

Figure 7.10 Here, three cat crackers break down gas oils into lighter, more useful products.

Because the catalyst behaves like a fluid when mixed with the gas oil, the method is also known as *fluid catalytic cracking.*

Cat cracking produces gasoline, light gases, olefin compounds, cracked gas oils, a liquid residue called cycle oil, and a solid carbon residue known as coke. Cycle oil is usually recycled through the reaction chamber to break it down further. Coke is a solid that coats the catalyst and must be removed by burning so that the catalyst can be used again. One use for coke is as a fuel for in situ combustion (see Chapter 5). Other uses are in the manufacture of electrodes, graphite, and carbides. The other hydrocarbon products of the cracking reaction are sent to a fractionator to be separated, and most of them undergo further processing elsewhere in the refinery before they are ready for sale.

A few refineries now use a heavy oil cracker (HOC) to upgrade residues into unleaded gasoline and other products in demand (fig. 7.11). This large fluid catalytic cracking unit consists of a main fractionator, catalyst storage hoppers, vapor lines, heaters, and a bank of coolers. The fractionator has a regenerator at the bottom for removing water from the hydrocarbons and a reactor on top for catalytic cracking.

Figure 7.11 This heavy oil cracker, designed and engineered by the M.W. Kellogg Company, is used to produce unleaded gasoline and other light product yields from heavy crude oil. *(Courtesy of M. W. Kellogg Co. and Valero Refining Co.)*

Thermal Cracking

Thermal cracking uses heat to break down the bottoms from vacuum distillation and sometimes the heavy gas oils resulting from catalytic cracking. The lighter hydrocarbons produced by thermal cracking can be made into distillate fuels and gasoline. The heavy residue that remains is converted into a semiliquid residual oil or into coke. Two common types of thermal cracking are viscosity breaking, or visbreaking, and coking.

In *visbreaking*, a relatively mild form of cracking, a furnace heats the feedstock for a short period of time. Using high pressure and immediately quenching, or cooling, the cracked products prevent coke from forming. Then the pressure is lowered, allowing the lighter hydrocarbons to vaporize. They go to a fractionator to be separated. Part of the heavy residue is recycled back to the reaction vessel for further breakdown.

In *coking*, a more severe form of thermal cracking, the feedstock is heated and sent to insulated vessels called coke drums, where it remains under high heat for an additional period of time. The lighter cracked products vaporize and leave through the top of the coke drum, and the heavier hydrocarbons form solid coke. Periodically, high-pressure water jets into the coke to break it up so it can be removed from the drum.

The products of thermal cracking include both saturated and unsaturated light gases, naphtha, gas oil, and residual oil or coke. The lighter products are of low quality and are usually upgraded by further processing before being sold. Cracked gases and naphtha are converted to gasoline. Gas oil can be used as distillate fuel or can be converted to gasoline and other products. Coke is heated to drive off water and is crushed.

Hydrocracking

Whenever gas oils produced by catalytic and thermal cracking are not needed to make distillate fuels, they can be converted to high-grade gasoline by hydrocracking. *Hydrocracking* is catalytic cracking in the presence of hydrogen. The extra hydrogen adds hydrogen atoms to, or *hydrogenates*, the cracked hydrocarbon molecules. Hydrocracking is also a treating process, because the hydrogen combines with contaminants such as sulfur and nitrogen, allowing them to be removed.

A gas-oil feed is mixed with hydrogen, heated, and sent to a reactor vessel containing a catalyst in a bed (a fixed-bed catalyst), where cracking and hydrogenation take place. The excess hydrogen is separated from the cracked products and recycled to the reactor. The products go to a fractionator to be separated. Residue from the first reaction is mixed with hydrogen, reheated, and sent to a second reactor for further cracking under higher temperatures and pressures.

In addition to a large quantity of *hydrocrackate*—cracked naphtha for making gasoline—hydrocracking yields light gases useful as fuel for the refinery or for alkylation and as components for high-quality fuel oils, lubrication oils, and petrochemical feedstocks.

Some of the hydrogen for the hydrocracking process is a byproduct left over from reforming, but usually *steam methane reforming*, a series of chemical reactions that extract hydrogen from steam and methane, must supply additional hydrogen.

Rearranging Hydrocarbon Molecules

A refinery can use several chemical processes to rearrange hydrocarbon molecules or to build new molecules to produce high-quality gasoline, jet fuel, and petrochemicals. Common processes are alkylation, isomerization, and reforming.

Alkylation

Olefins such as propylene and butylene are *byproducts* (a side product of a process to produce a main product) of catalytic and thermal cracking. In the refining industry, *alkylation* refers to the chemical combining of these light molecules with isobutane to form larger branched-chain molecules (isoparaffins) that make high-octane gasoline. This process occurs in an alkylation unit (fig. 7.12).

Figure 7.12 This alkylation unit consists of a depropanizer, a main fractionator, and a reboiler heater. Hydrofluoric acid used in the unit is stored in the tank next to the heater. *(Courtesy of Phillips Petroleum Co. and Valero Refining Co.)*

Figure 7.13 These tanks store n-butane, isobutane, and propane until they are needed as a feedstock or are sold. Their spherical shape gives them strength to contain the high pressure necessary for storing liquefied petroleum gas. *(Courtesy of Valero Refining Co.)*

In the alkylation unit, olefins and isobutane are first mixed with an acid catalyst and cooled. They react to form liquid gasoline components known as alkylates. The reaction also produces some n-butane, isobutane, and propane, which are pressurized until they are liquid and then stored in dome-shaped tanks (fig. 7.13). After the reaction, the acid catalyst settles out of the liquid and is drawn off to be reused. The remaining liquid is treated to neutralize remaining traces of acid, and the products are separated in a series of distillation columns. The isobutane is recycled as a feedstock, and the butane and propane can be sold as liquid petroleum gas.

Isomerization

In the refining industry, *isomerization* is a chemical process that rearranges straight-chain hydrocarbons (paraffins) into hydrocarbons that have branches attached to the main chain (isoparaffins).

One use for isomerization is to create extra isobutane for use in alkylation. Catalytic cracking and hydrocracking produce the isobutane needed as a feedstock for alkylation. But if a refinery does not have a hydrocracker, it may use an isomerization unit to produce isobutane from normal butane by isomerization. Normal butane is mixed with a little hydrogen and a chloride, and the resulting reaction, in the presence of a catalyst, forms isobutane. Other products of this reaction include normal butane and a small amount of lighter gases, which are separated in a fractionator. The lighter gases are used as fuel for the refinery, and the normal butane is recycled as a feedstock.

Another use for isomerization is to improve the performance of straight-run pentanes and hexanes as a feedstock for blending gasoline. *Straight-run* products are those produced by the initial distillation process. Pentanes and hexanes are the lighter components of gasoline. Isomerization improves gasoline quality by converting these hydrocarbons to their higher-octane isomers. *Octane* in this instance is a classification of gasoline—a higher-octane gasoline has better anti-knock properties. The process is the same as for butane. The hydrocarbons react with hydrogen and chloride in the presence of a catalyst, and the reaction products are separated. Again, the light gases are used as refinery fuel, and the normal pentane and hexane are recycled as feedstocks.

Catalytic Reforming

Catalytic reforming is a process for upgrading naphthas into high-octane gasoline and petrochemical feedstocks. The feedstock can be naphthas from straight-run distillation, thermal cracking, or hydrocracking. Naphthas are hydrocarbon mixtures containing many paraffins (chain molecules) and naphthenes (ring molecules). Reforming converts a portion of these compounds to isoparaffins and aromatics, which make a higher-octane gasoline. Aromatic compounds are also an important feedstock for the petrochemical industry.

Reforming uses heat, pressure, and a catalyst to bring about the desired chemical reactions (fig. 7.14). Of the several ways of exposing the naphtha to the catalyst, the most common method is to send the naphtha through a series of reactors with fixed catalyst beds. There it is mixed with hydrogen, which helps to prevent coke from forming on the catalyst bed. Hydrogen does this by combining with free carbon to form carbon monoxide and carbon dioxide.

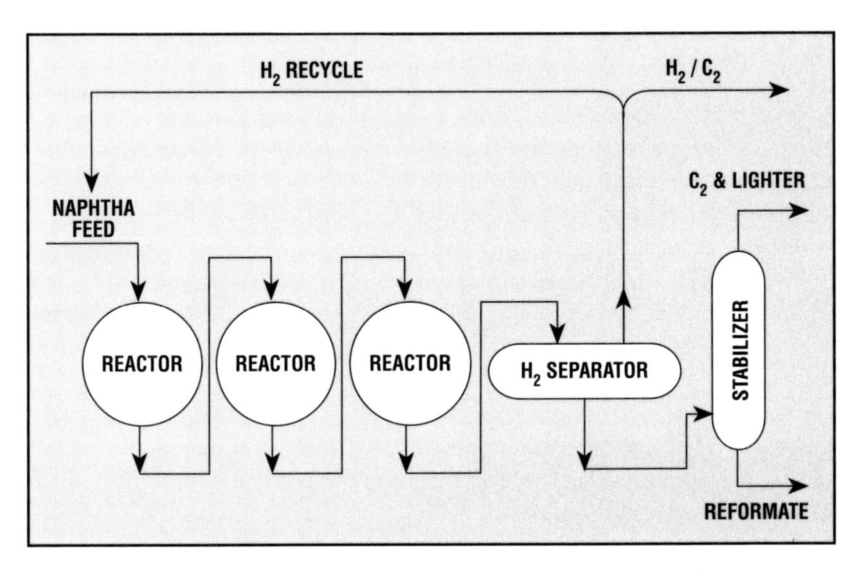

Figure 7.14 Catalytic reforming boosts the octane rating of naphtha.

Each reactor in the reforming series has a different temperature and pressure in order to promote a specific chemical reaction. Reaction products include hydrogen, light gases, and *reformate*—a liquid mixture of hydrocarbons. Hydrogen is separated from the other products by cooling, part is recycled as feed, and the remainder is used as a feedstock for hydrocracking. The gases are separated from the reformate in a fractionator and used as refinery fuel or as a feedstock for other plant operations.

Solvent Extraction

Solvent extraction is the use of a solvent to selectively dissolve a particular compound and remove it from a mixture of hydrocarbons. The method is especially useful when all the hydrocarbons in the mixture have the same boiling point and so cannot be separated by distillation. Solvent extraction has many applications in the refining industry. It is used most often to improve the burning qualities of fuels, to refine lubricating oils and waxes, and to recover aromatic hydrocarbons for use as gasoline components or petrochemicals.

Although there are many types of solvent extraction, the basic steps for all of them are the same. A solvent is selected that will dissolve only the *extract*, which is the substance that is to be separated from the *raffinate* (the remaining mixture). In a contacting tower, the solvent is thoroughly mixed with the feedstock, so that the solvent contacts and dissolves all of the extract. The extract solution is separated from the raffinate by settling, precipitation, cooling, and filtering, or by other means. Next, distillation or evaporation separates the solvent from the extract, and the solvent is recycled to the contacting tower. The raffinate contains traces of solvent, which must also be removed by distillation or other means.

Improving Fuels

Kerosenes and gas oils used to make jet, diesel, and burner fuels often contain aromatic hydrocarbons, sulfur, nitrogen, metallic compounds, and other substances that smoke when they are burned. Solvent extraction removes these hydrocarbons, which improves the burning quality of the fuels. Solvent extraction also improves the *cetane number* of diesel fuel, a measure of its ability to self-ignite.

Making Lubricating Oils and Waxes

Hydrocarbons used to make lubricating oils and waxes often contain substances that interfere with desirable product qualities. Solvent extraction methods such as deasphalting, treating, dewaxing, and deoiling can remove these substances and make good lubricating oils and waxes from almost any crude oil. *Deasphalting* is the removal of asphaltic substances that tend to form carbon deposits when lubricating oils are heated. It is an alternative to removing asphalt by vacuum distillation. *Solvent treating* is the removal of aromatics, naphthenes, and olefins that decrease the stability of a lubricating oil's viscosity under temperature changes and cause gummy deposits to form. *Dewaxing* removes waxes so that a lubricating oil will flow better at low temperatures. Wax that has been separated from lubricating oil is *deoiled* to remove traces of oil and compounds with a low melting point.

BTX recovery

A third use of solvent extraction is the recovery of the aromatic compounds benzene, toluene, and the xylenes—commonly known as *BTX recovery*. These aromatics are separated from reformed naphtha and used as stock for gasoline blending or sold as petrochemical feedstocks.

Treating

Treating either removes contaminants from crude oil or converts them into harmless compounds. Treating is often necessary in order to improve the qualities of finished refinery products, to prevent the destruction of expensive catalysts, to minimize corrosion of refining equipment, or to prevent environmental pollution. Sometimes treating may occur as a side benefit of a refinery process meant for other purposes. Examples of this are hydrocracking and solvent extraction. Other treating processes include dehydration and desalting, hydrotreating, and sulfur recovery.

Dehydration and Desalting

Salts, which are impurities in some crudes, break down during refinery processing and foul and corrode the equipment. They must be removed, along with any water in the crude, at the beginning of refinery processing, before distillation takes place. When the salts are suspended in the water, they can be removed by heating the oil and allowing the water-salt solution to settle out. However, if the oil and water are in the form of a stable emulsion that resists separation, chemicals or an electric current must be used to break the emulsion and allow the water-salt solution to separate from the oil.

Hydrotreating

Hydrotreating is a method of removing sulfur, nitrogen, and metals from crude oil fractions. It is used to treat straight-run naphthas, kerosenes, and gas oils; thermally cracked fractions; and residual fuel oils.

Hydrotreating is similar to hydrocracking. The feed is mixed with a stream of hydrogen, heated, and sent to a reactor where a series of chemical reactions takes place in the presence of a fixed-bed catalyst. The hydrogen combines with nitrogen and sulfur to form ammonia and hydrogen sulfide, and metals are deposited on the catalyst. The reaction conditions for hydrotreating are less severe (temperatures and pressures are lower) than for hydrocracking, because the purpose is to remove contaminants rather than to crack large hydrocarbon molecules. However, a certain amount of cracking and olefin saturation does take place. Heavier oils, in particular, have to undergo more chemical breakdown of the molecules in order to free contaminants from the molecules that contain them. In the case of kerosene hydrotreating, molecular changes are beneficial because they improve the burning qualities of diesel fuel. After hydrotreating, the reaction mixture is separated from the remaining hydrogen and sent to a fractionator to separate the oil from the contaminant and hydrocarbon gases.

Sulfur Recovery

Hydrogen sulfide, the highly poisonous gas sometimes contained in natural gas or created by hydrotreating, can be converted into sulfur. Sulfur is useful because it can be converted into ammonium thiosulfate, a fertilizer.

The most widely used and reliable process for sulfur recovery is the *Claus process.* In this process a furnace heats the hydrogen sulfide in the presence of a controlled amount of oxygen. Most of the hydrogen sulfide reacts chemically with the oxygen to form sulfur and water, but some of the hydrogen sulfide forms sulfur dioxide. The sulfur dioxide is sent to a series of two or three catalytic converters, which convert it into a sulfur vapor and water. The vaporous sulfur is cooled and sent through a condenser to change it into a liquid state. The liquid sulfur is then drained into a heated tank and stored until it is ready to be pumped into tank trucks or cars for delivery.

Other Methods

A number of other chemical and physical methods have been used for treating. They include washing oil with caustic (basic) or acidic solutions, exposing the oil to adsorbent clays, and oxidating (adding oxygen to) oil with catalysts. Many of these methods have been largely replaced by hydrotreating.

Blending and Using Additives

After crude oil has been refined into useful products, many of these products are further improved either by adding small amounts of other substances to modify their properties or by blending hydrocarbons with different characteristics. Blending is particularly important in refining gasoline, diesel fuel, furnace oils, and residual fuel oils.

Figure 7.15 In this gasoline blending department, the blender operator sets the control panel on the research method machine to obtain the octane reading. A similar machine is used to check the octane by the motor method. The average of the two readings constitutes the octane seen at the customer's pumps. *(Courtesy of Valero Refining Co.)*

Gasoline

The gasolines produced by straight-run distillation, catalytic cracking, hydrocracking, reforming, alkylation, and gas processing (see following) have different hydrocarbon contents and performance qualities. Refineries blend them into grades of products that will perform well under varying weather conditions, altitudes, and engine compressions to meet market demands. For example, the temperature at which it vaporizes is important. A gasoline has to vaporize enough to ignite in a cold engine but not enough to cause vapor-lock stalling in hot weather. It has to have an octane that prevents it from knocking or self-igniting (fig. 7.15). In modern refineries, computer operators make the complex calculations needed to plan production of the required amounts and types of gasoline components (fig. 7.16).

Figure 7.16 Computers are used to calculate the proper blend to meet the customer's specifications, to open and close valves of lines leading to equipment on the plant site, and to automatically gauge storage tanks on the site. *(Courtesy of Star Enterprises)*

Diesel Fuel

Diesel fuel is blended from straight-run light gas oils, cracked gas oils, and kerosene. It must have a cetane number that allows it to self-ignite readily, just the opposite of the antiknock quality needed in gasoline. Several grades are sold; the premium grades allow better cold starts and smoke less when burning.

Furnace and Residual Fuel Oils

Oils destined to be burned in furnaces for residential and commercial heating are blended from straight-run and cracked gas oils. Residual fuel oils, burned in industrial and ship boilers, are made from the residue of vacuum distillation and thermal cracking. These oils must be able to flow at the temperatures at which they are commonly used and must not vaporize enough to form mixtures with air that could start an accidental fire. Residual fuels are blended with heavy gas oils to make them pour more easily and to reduce sulfur content to acceptable levels (if they have not been hydrotreated).

Protecting the Environment

The U.S. and state governments have aimed much antipollution legislation at the petroleum industry, especially at refineries because the processes they use produce many toxic byproducts. For this reason, protecting the land, water, and air is a vital function in today's refineries, requiring environmental specialists on the job at every plant. Refineries in the U.S. spend billions of dollars to build, operate, and maintain antipollution facilities. See Chapter 8 for more detailed descriptions of environmental concerns.

GAS PROCESSING

As late as the 1930s, natural gas leaving the wellhead had to reach a market nearby or else be burned off, or *flared*. Huge amounts of natural gas have been flared in the United States. Flaring is still a common practice in remotely located oilfields when gas cannot be reinjected into the reservoir for gas lift or used locally as fuel. With the advent of gas pipelines (commonly called *transmission lines*), gas transport trucks, and field processing facilities for gas, gas production in the United States and elsewhere has become an industry in itself.

Natural gas straight from the well is processed in the field (see Chapter 5). The processing includes the removal of water, impurities, and excess hydrocarbon liquids as required by the sales contract. It also includes the control of delivery pressure. When it is economical to gather the gas from several wells to a central point, an operator may build a gas processing plant to do the same work as separate facilities next to each well would do. Often, these gas plants dehydrate the gas and remove hydrogen sulfide. In addition, they generally separate hydrocarbon mixtures or individual hydrocarbons from natural gas and recover sulfur and carbon dioxide.

In general, the larger the gas processing plant, the more economical it is to operate. However, large plants must be near fields that provide large volumes of natural gas. In recent years, manufacturers have developed portable skid-mounted plants to provide efficient, relatively inexpensive gas processing for smaller fields.

In addition, refineries have facilities to process the gases resulting from crude oil distillation, cracking, and reforming. Refinery gas processing provides fuel gas (methane, ethane, and ethylene) to power refinery operations. Refineries also separate individual natural gas liquids (NGLs), which may be used to make fuel products or may be sent to an alkylation unit for further processing.

Recovering NGL Mixtures

One function of a gas processing plant is to recover mixtures of natural gas liquids. NGLs include propane, butane, and natural gasoline. The four processes for doing this are straight refrigeration, cryogenic recovery, oil absorption, and dry-bed adsorption. In *absorption,* the molecules of one substance penetrate the inner structure of another, the way a sponge soaks up a liquid. In *adsorption,* one substance is attracted to and held onto the surface of another, as dust particles stick to flypaper.

Straight Refrigeration

The simplest way to produce NGLs is to cool natural gas until it becomes liquid. Smaller gas processing plants, particularly portable, skid-mounted plants that produce from 10 to 20 million cubic feet (2.8 to 5.6 million cubic metres) of product per day, often use straight refrigeration. Several designs are available; all of them use refrigerants such as Freon or propane.

Cryogenic Recovery

Cryogenic recovery processes are used in most of the newer gas processing plants. *Cryogenics* is a branch of physics that relates to producing and studying the effects of very low temperatures. Cryogenic processes use high pressures and extremely low temperatures to recover most of the ethane (90 to 95 percent) as well as all of the propane and the heavier hydrocarbons.

Because cryogenic processes use temperatures lower than -150°F (-101°C), they require special equipment and techniques. The construction materials

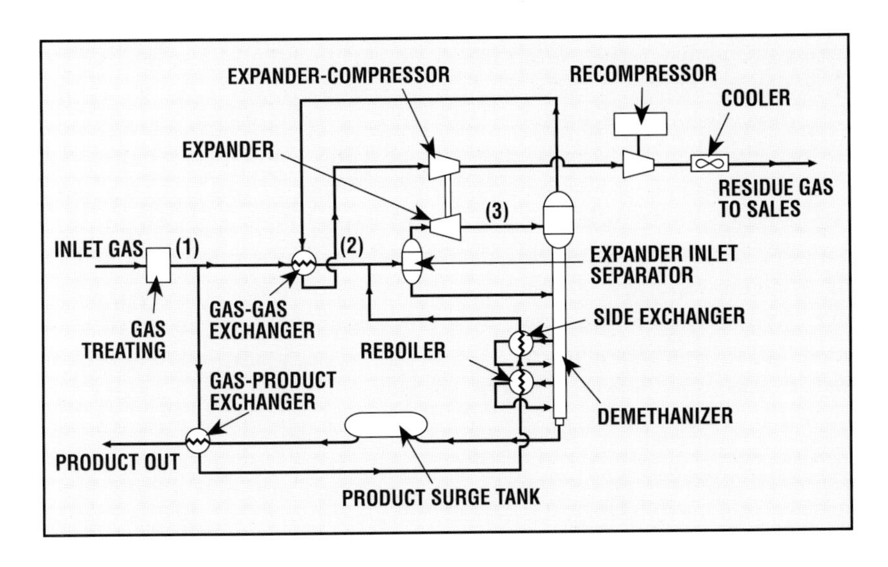

Figure 7.17 The expander process is one way to recover NGLs cryogenically.

must not become brittle in low temperatures, and cold boxes enclose and insulate the processing equipment. Control valves and pumps have extended shafts so that they can be located inside the cold box while motors on the outside operate them. The inlet gas requires special treating to remove contaminants before it can enter the cryogenic equipment. Salt water, wax, dirt, scale, iron sulfide, and oil carried over from wellhead separators and compressors will clog the heat exchangers and stop the flow of fluids. Water must be reduced to less than one part per million.

Two types of cryogenic recovery are used: the expander process and cascade refrigeration.

Expander Processing

The most commonly used cryogenic method is the *expander process* (fig. 7.17). After being treated, the dry, clean inlet gas is cooled. Under high pressure, the gas enters the separator chamber of an expander-compressor (fig. 7.18),

Figure 7.18 A medium-sized expander-compressor will generate up to 1,500 psi (10.3 megapascals) of pressure for use in a cryogenic natural gas plant. *(Courtesy of Rotoflow Corp.)*

where it expands and cools further. As the inlet gas cools, the NGLs condense and are separated from the methane vapor.

Because some methane remains dissolved in the NGLs, the liquid is then sent to a demethanizer to remove the methane. The demethanizer works similarly to a distillation column. It heats the liquid until some of it vaporizes and rises through the column. As the vapor contacts cold liquid descending from above, heavier hydrocarbons condense to liquid again, but the methane vapor continues to rise and leaves from the top. When separate, the methane is compressed to the proper pressure for sale, cooled, and delivered to its destination. The NGLs are piped to a fractionation facility to be separated into individual components.

Cascade Refrigeration

Cascade refrigeration uses two-stage refrigeration to separate methane from the NGLs. In the first stage, the treated inlet gas is cooled in heat exchangers by residue gas (methane that has already been separated) and then by a refrigeration system using propane as the refrigerant. In the second stage, residue gas and a refrigeration system using ethane or ethylene as the refrigerant cool the inlet gas further. The inlet gas, now at a temperature of -120°F (-84°C) or lower, enters a separator where the NGLs condense and separate from the methane vapor. As in expander processing, some of the methane remains dissolved in the liquid, so the liquid goes to a demethanizer for further separation.

Cascade refrigeration can also be used to liquefy methane. This requires three stages of refrigeration, with propane, ethylene, and methane as the refrigerants. Temperatures must be below -200°F (-129°C).

Oil Absorption

Oil absorption is a process used by older gas processing plants and by many gas plants in refineries (fig. 7.19). It may work alone or with refrigeration. Without refrigeration, oil absorption recovers about 70 percent of the propane and all of the butane and heavier hydrocarbons. With refrigeration, the process can recover about 50 to 75 percent of the ethane and all of the propane and heavier hydrocarbons.

The oil absorption process has three steps: NGL recovery, methane removal, and separation of the absorbent from the NGLs (fig. 7.20). Usually, a combination of refrigeration and oil absorption is used in the first step. The absorbent is a hydrocarbon liquid called *lean oil*. When the lean oil absorbs the heavier hydrocarbons, it is called *rich oil*. A refrigerant such as ammonia, Freon™, or propane cools both the inlet gas and the lean oil to a temperature between 0°F and -40°F (-18°C and -40°C). Since this step liquefies some of the methane along with the desired NGLs and methane is not a desirable component, the next step (the rejection step) removes methane from the NGL-rich oil mixture. Most plants use a device called a *rich-oil demethanizer (ROD)* to separate the methane. In the final step, the NGLs go to a fractionator to separate them from the rich oil. Exposed to steam or hot oil, the NGLs vaporize and flow out the top of the fractionator. The rich oil then becomes lean oil again and flows out the bottom to be recycled to the first step.

Dry-Bed Adsorption

Dry-bed adsorption can be used not only to dehydrate natural gas but also to remove some of the heavier hydrocarbons in order to prevent them from condensing in pipelines during transmission. The natural gas passes over a

Figure 7.19 This gas processing plant uses the oil absorption process.

Figure 7.20 The oil absorption process consists of NGL recovery, methane rejection, and separation of the absorbent.

bed of solid desiccant, which adsorbs some of the NGLs and separates them from the gas stream (see Chapter 5 for more detail). This process can recover 10 to 15 percent of the butane and 50 to 90 percent of the natural gasoline. Operators primarily use it to ensure that the NGL content of natural gas meets the specifications of the gas sales contract.

Fractionation of NGLs

Fractionation, or fractional distillation, separates a mixture of NGLs into salable individual products. Just as in fractionation of crude oil, NGLs can be separated because they have different boiling points. The process uses fractionation towers with reboilers and reflux equipment (fig. 7.21). A reboiler supplies heat to the system, while the reflux equipment returns the condensed stream to the column to cool it further. The hydrocarbon components usually separated in a fractionation plant are ethane, propane, isobutane, normal butane, pentane, and a remaining mixture of heavier hydrocarbons (pentane and higher).

The NGL products are separated in a series of fractionation towers, the lighter product being removed first. Fractionators are usually named for the main product being separated, known as the overhead or top product because it comes from the top of the fractionator. A deethanizer, for example, produces ethane as the top product, and a depropanizer produces propane.

Because individual NGL products have boiling points that are close together, they are difficult to separate. Fractionators for NGLs require a large number of trays in the tower, a substantial heat input by the reboiler, and heat removal by the reflux and condenser.

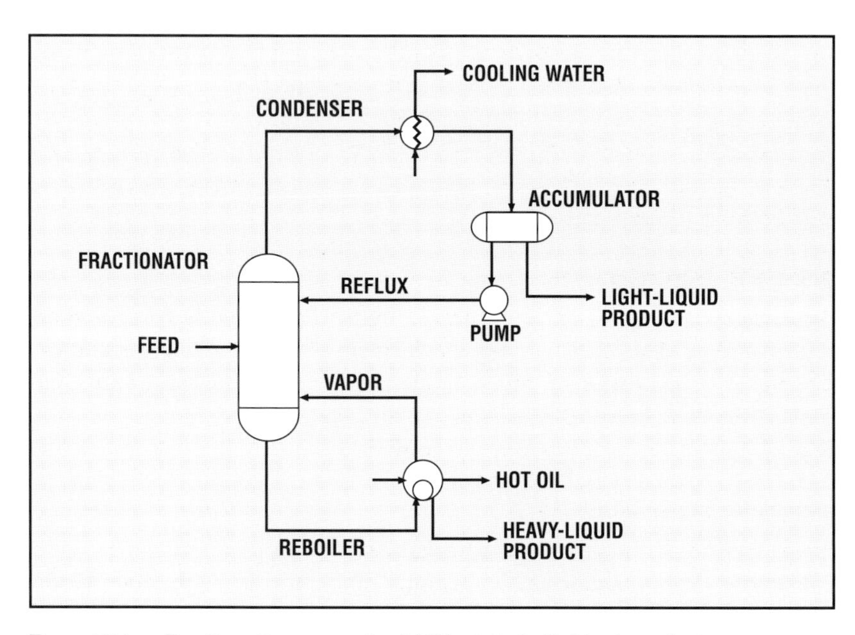

Figure 7.21 Fractionation separates NGLs into individual products.

PETROCHEMICALS

Although most petroleum products are used as energy sources, such as gasoline for cars, fuel oil for boats and ships, and natural gas for heating and fuel, petroleum components also provide the raw material for the manufacture of petrochemicals. A *petrochemical* is a chemical substance produced commercially from feedstocks derived from crude oil or natural gas. Consequently, many petrochemical plants are integral parts of large refining complexes and are often subsidiaries of major oil companies.

Types of Petrochemicals

The basic job of most petrochemical plants is to turn crude oil fractions or their cracked or processed derivatives into feedstocks that will ultimately be used in the manufacture of a host of other products, from liquid detergent to the plastic bottle containing it. These feedstocks fall into three groups, based on chemical composition and structure—aliphatic, aromatic, and inorganic.

Aliphatic Petrochemicals

Aliphatic petrochemicals are straight-chain hydrocarbons, either saturated (paraffins) or unsaturated (olefins). They make up over half the volume of all petrochemicals produced. Two major olefin feedstocks are ethylene and propylene. They are produced from hydrocarbons by thermal cracking in the presence of steam, a process known as *steam cracking*. The hydrocarbons used as raw materials range from natural gas to naphtha to gas oil. Ethylene in turn can be chemically converted to several other important feedstocks such as polyethylene, ethylene glycol, vinyl chloride, and styrene. Similarly, propylene can be used to make feedstocks such as polypropylene and isopropyl alcohol. These petrochemicals ultimately go to factories that make plastics, solvents, synthetic rubbers, and synthetic fibers. Other aliphatic petrochemicals include methanol, made from natural gas; and butadiene, derived from butanes and butylenes (C_4 hydrocarbons). Methanol is used to make resins, polyester fibers, and solvents. Butadiene is used to make rubber, resins, and plastics.

Aromatic Petrochemicals

Aromatic petrochemicals are unsaturated hydrocarbons with six carbon atoms in a ring. Major aromatic feedstocks include benzene, toluene, and the xylenes. A large portion of toluene is chemically converted to benzene, the most widely used aromatic feedstock. Aromatic petrochemicals are used to make plastics, resins, fibers, and elastomers.

Inorganic Petrochemicals

Inorganic petrochemicals include carbon black, sulfur, and ammonia. These substances do not contain carbon compounds and can be produced from nonhydrocarbon sources. Nevertheless, they are considered to be petrochemicals because the primary raw materials for their manufacture are substances derived from petroleum refining. Carbon black, made from natural gas, is used to manufacture synthetic rubbers, printing ink, and paint. Sulfur, produced from hydrogen sulfide and other sulfur compounds, is used to make sulfuric acid, which in turn is used in the manufacture of steel, fertilizer, paper, and other chemicals. Ammonia, synthesized from natural gas and naphtha, is an important feedstock for the manufacture of fertilizers, fibers, and plastics.

Figure 7.22 Polyethylene and polypropylene are used to make a host of products.

A Petrochemical Plant

Since each petrochemical plant is designed to convert particular petroleum fractions or other substances into specific petrochemicals, each plant uses different processes, procedures, facilities, and auxiliary operations. To get an insight into a petrochemical plant, consider the operation of Tex-Chem, a hypothetical plant that manufactures two basic plastics—polyethylene and polypropylene.

Situated in the heart of oil country where miles of pipelines crisscross underground and barges ply the waterways, Tex-Chem uses several basic raw materials derived from petroleum or natural gas. From these materials, the plant manufactures polyethylene and polypropylene, which are used to make thousands of products such as appliance and automotive parts, moldings, battery cases, luggage, bottles, plastic toys, transparent food wrappers, plastic containers, and a wide range of construction materials (fig. 7.22).

The plant operates twenty-four hours a day, seven days a week. Although the chemical processes themselves are complex, they involve the application of familiar principles: controlled temperature, pressure, and time, and sometimes the presence of a catalyst and a solvent. All operations take place in a labyrinth of pipes, valves, silos, and towers powered by heavy equipment (fig. 7.23), controlled by elaborate instruments, and monitored by trained technicians (fig. 7.24). The plant is organized in three systems: the olefin units, the polymer units, and the supporting facilities.

Figure 7.23 This 20,000-horsepower (14,000-kilowatt) centrifugal compressor, driven by a steam-powered turbine, is an example of the huge yet precise equipment in a petrochemical plant. *(Courtesy of ARCO Chemical Co.)*

Figure 7.24 A process technician can tell by checking a meter or chart what temperature or pressure is being applied in any section of a petrochemical unit, adjusting those conditions by touching a switch or turning a dial. Many plants today use computer consoles and keyboards to accomplish the same function.

Olefin Units

The primary job of the *olefin units* is to provide ethylene and propylene for the polymer units (polyethylene and polypropylene units), as well as for sale to other chemical or plastic producers. The olefin units, sometimes referred to as a petrochemical refinery because of their wide range of products, also produce butadiene, butylenes, hydrogen, and aromatics for use in the manufacture of other chemicals or gasoline products. These coproducts are sold to other chemical producers or used as feedstocks for other plants operated by the Tex-Chem Company.

The olefin units use liquefied ethane and propane and distillates such as natural gasoline and naphtha as raw materials. These hydrocarbons spend a brief period in a furnace where they are thermally cracked under high temperatures in the presence of steam. The resulting products are sent to a fractionator to be separated (fig. 7.25).

Other processes used in the olefin units include catalytic conversion of acetylenes, cryogenic separation of hydrogen, and adsorption. Catalytic reforming of straight-run naphtha produces aromatics. When the naphtha is passed over a catalyst such as platinum, it forms ring compounds with no double bonds. These compounds are changed into aromatic compounds by *dehydrogenation*—removing some of the hydrogen so the carbons will form double bonds.

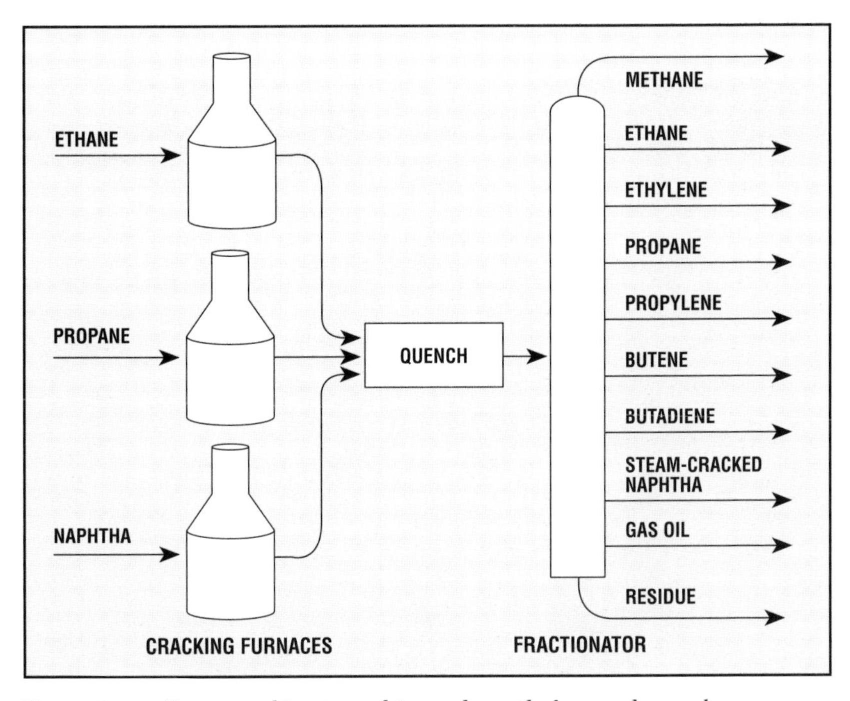

Figure 7.25 Steam cracking is used to produce ethylene and propylene.

Polymer Units

In the plant's *polymer units*, propylene and ethylene are combined into larger molecules (*polymerized*) to make polypropylene and polyethylene. The reaction takes place in a liquid solvent in the presence of a catalyst. After the polymers have been formed, they are removed from the reactor and sent to a purification facility to remove the catalyst and the solvent. The powdered product is blended with additives and sent through an extruder to form pellets (fig. 7.26).

Figure 7.26 Polypropylene is extruded into pellets about the size of BBs in its finished form.

Figure 7.27 Laboratory analysts run extensive tests on raw materials, process streams, and finished products to assure quality.

Supporting Facilities

Quality control technicians perform extensive laboratory tests to maintain quality during each phase of plant processes (fig. 7.27). Technicians check polyethylene and polypropylene pellets for proper content and consistency. If they meet the specifications, the pellets are prepared for shipping (fig. 7.28).

Figure 7.28 Polypropylene pellets are sacked and put in cardboard boxes for transporting to other plants.

Figure 7.29 In the aeration basin of the plant's water treatment facilities, millions of tiny bacteria (microorganisms) literally eat the hydrocarbon waste, purifying the water before it is reused in the plant or disposed of.

Plant operations are designed to conserve energy, air, and water. Onsite water-treatment facilities purify and recirculate water used to cool the process units (fig. 7.29). Only steam is allowed to rise from the plant's cooling towers and units. The steam is odorless, contains no contaminants, and rapidly disperses into the atmosphere. The intermittent discharge of waste hydrocarbon gas from the processing units—a potential source of air pollution—is burned in a flare system from tall, derricklike structures. Engineers design processing units to make full use of the energy consumed, and often the hydrocarbons created as byproducts are recovered and used as fuel.

Companion Plant

Linked closely with the operation of this Tex-Chem plant, the company runs another plant nearby. Using feedstocks from plant No. 1, Tex-Chem's plant No. 2 produces specialty chemicals and petrochemical intermediates for the manufacture of other products (table 7.3).

TABLE 7.3

SOME PETROCHEMICALS—THEIR FEEDSTOCKS AND FINAL PRODUCTS

Feedstock	Petrochemical	Typical Final Products
Styrene	Polystyrene	Cups and glasses, phonograph records, radio and TV cabinets, furniture, luggage, telephones, ice chests, lighting fixtures
Paraxylene	Terephthalic acid (TPA), dimethyl terephthalate (DMT)	Polyester fibers used to make wearing apparel, tire cords; polyester film used to make electronic recording tape, photographic film, cooking pouches, specialty packaging items
Metaxylene	Isophthalic acid (IPA)	Fiberglass-reinforced auto bodies, surfboards, snowmobile housings, outboard motor covers, cooling fans, vaulting poles, many paints and coatings
Propylene	Polypropylene	Appliance parts, bottles, safety helmets, disposable syringes, battery cases, construction materials, feed bags, carpet backing

8 Environmental and Health Concerns

Petroleum products are ubiquitous (see table 7.2), but their recovery, transport, processing, and use are all fraught with hazards to human health and the earth's ecology. Exploration, drilling, and production use poisonous chemicals that can pollute the air, water, and soil to yield a product that is poisonous to living things. Producing and transporting this poisonous product risks fire, explosions, and more pollution. Refining it produces still more noxious chemicals that must be changed into harmless compounds or disposed of in harmless ways. And some of the end products of all this labor pollute when they are used or thrown away.

For these reasons and because of the very size and importance of the petroleum industry, it is subject to much criticism. Various environmental groups monitor the industry and publicize any dangers and potential dangers they find or suspect, to the point of antagonizing many industry insiders into feeling "chased by the greens." Oil companies face a great challenge in terms of public relations in the environmental area. Although disasters have brought new laws, new technology, and innovative practices—necessity is the mother of invention—the public usually hears only about the damage done to people and the environment. The industry cannot prevent all accidents, but it can have an effect on consumers' opinion of it by publicizing its efforts to comply with environmental and safety laws and educating the public when it does things right.

FEDERAL LAWS AND REGULATIONS

In the United States, the oil industry, like all industries, is regulated by the federal and state governments. Many federal and state laws allow numerous agencies to oversee any aspect of the industry that could pollute air, water, or soil, or pose a risk to human health and safety. This regulation has done much to change unsafe practices in the past twenty-five years. Every petroleum-related company must be aware of these laws and how to comply with them. In addition, it needs to find ways to do this without going broke. The successful company must be ever flexible and open to ways that will help it produce petroleum more efficiently and cleanly at a profit.

A series of federal laws since the 1970s authorized the *Environmental Protection Agency (EPA)* and other agencies to draw up rules, called *guidelines* or *regulations*. Some of these regulations directly protect the quality of water, air, and land. Some protect the health and safety of employees, surrounding communities, and plant and animal life. Others regulate the use of hazardous chemicals and the disposal of hazardous waste. All of these types of regulations affect the oil industry. The list of laws described below is not comprehensive, but includes the most important ones to the industry as a whole.

Air Quality

In 1970, the *Clean Air Act* authorized the EPA to set standards for air quality and regulate emissions of pollutants into the air. Air pollutants include ozone, carbon monoxide, sulfur dioxide, lead, nitrogen dioxide, and particulates (fig. 8.1). Oil and gas exploration, drilling, and production produce several types of air emissions. Some examples are:

- dust from road construction, site clearing, and mud mixing;
- exhaust from heavy equipment, diesel engines, heater treaters, and other machinery;
- hydrogen sulfide (H_2S, or sour gas) from sour gas plants and treatment facilities; and
- H_2S from drilling into or producing from formations that contain it.

Some of the most visible EPA regulations pertain to emissions into the air from motor vehicles. Changing standards for the formulation of gasoline have affected the operations of refineries.

Figure 8.1 Smokestacks used to contribute to air pollution but U.S. regulations within the last 20 years have begun to change this.

Water Quality

Clean Water Act

The *Clean Water Act* charges the EPA with regulating pollutant discharges into surface waters, including oceans, rivers, streams, lakes, ponds, water holes, and wetlands (fig. 8.2). Some of the discharges that the petroleum industry must be concerned with are sewage, hot water, drilling fluids, drill cuttings, produced water, pollutants washed into water resources by storms, deck drainage from offshore units, blowout prevention fluid, and any fluids containing chemicals, suspended solids, or oil.

In order to prevent and clean up pollution, the Clean Water Act requires that all onshore facilities that could possibly discharge oil or other pollutants into navigable waters must abide by the Spill Prevention, Countermeasures and Control (SPCC) Plan. EPA guidelines say that this plan must include predictions of the type and quantities of petroleum that could potentially be released if equipment fails, and make available the following: descriptions of equipment that will prevent discharges from reaching a navigable watercourse, preventive training for employees, notification procedures, and descriptions of past spills, how they were handled, and plans for preventing a recurrence.

Figure 8.2 The operator of this offshore rig must follow EPA regulations concerning discharges into water.

Safe Drinking Water Act

Almost half of the people in the United States get their drinking water from underground aquifers (fig. 8.3). The *Safe Drinking Water Act (SDWA)* of 1974 protects underground waters. Under this law, the *Underground Injection Control (UIC)* program regulates injection wells for waste disposal. The petroleum industry uses injection wells to:

- dispose of produced water into depleted oil formations below drinking water sources;
- inject produced water produced from the reservoir back into the producing zone;
- inject fluids for enhanced recovery; and
- store hydrocarbons.

Petroleum operations can also contaminate drinking water if abandoned wells are not properly plugged, the casing is faulty, or fluids are accidentally injected into an aquifer. As a result, UIC has specific requirements for construction of wells, casing and cementing, plugging and abandonment, types of substances that can be injected, volume and pressure that can be injected, and the mechanical integrity of the well.

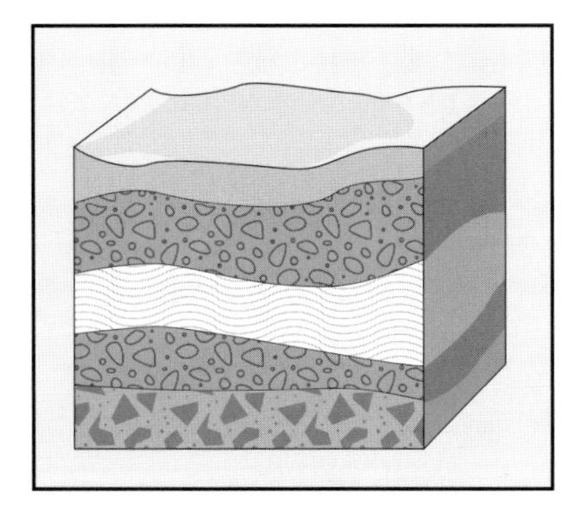

Figure 8.3 An aquifer is a stratum of rock, sand, or gravel that holds enough water to supply wells or springs.

Oil Pollution Act

The most recent water quality legislation is the *Oil Pollution Act (OPA)* of 1990. Congress enacted OPA in response to a number of large tanker spills. In 1988, 46 million gallons (174 million litres) were spilled in U.S. waters alone, in 16,000 reported incidents. The Exxon *Valdez* spill in 1989 was the final impetus for tougher regulations.

In the same way that EPA requires onshore facilities to have an SPCC plan for spills, OPA authorized the *Minerals Management Service (MMS)* to create guidelines for offshore spill plans. Under MMS regulations, owners and operators of onshore and offshore vessels and facilities must have contingency plans for containing and recovering a worst-case spill. A worst-case spill is, for example, the total release of a full cargo from the company's largest vessel in the worst weather conditions. OPA regulates owners and operators of drilling units, pipelines, storage units such as tanks, and any other facility that could potentially discharge oil or gas into federal waters.

Outer Continental Shelf Lands Act

The *Outer Continental Shelf Lands Act* of 1953, amended in 1978 and 1985, authorizes the Secretary of the Interior to grant mineral leases and regulate oil and gas activities on the Outer Continental Shelf. The *Outer Continental Shelf* consists of all lands under offshore waters within U.S. jurisdiction (fig. 8.4). Leases here often require the exploration and development company to plan measures to protect the environment, train its employees in environmentally safe operations, and dispose of its wastes in a safe manner.

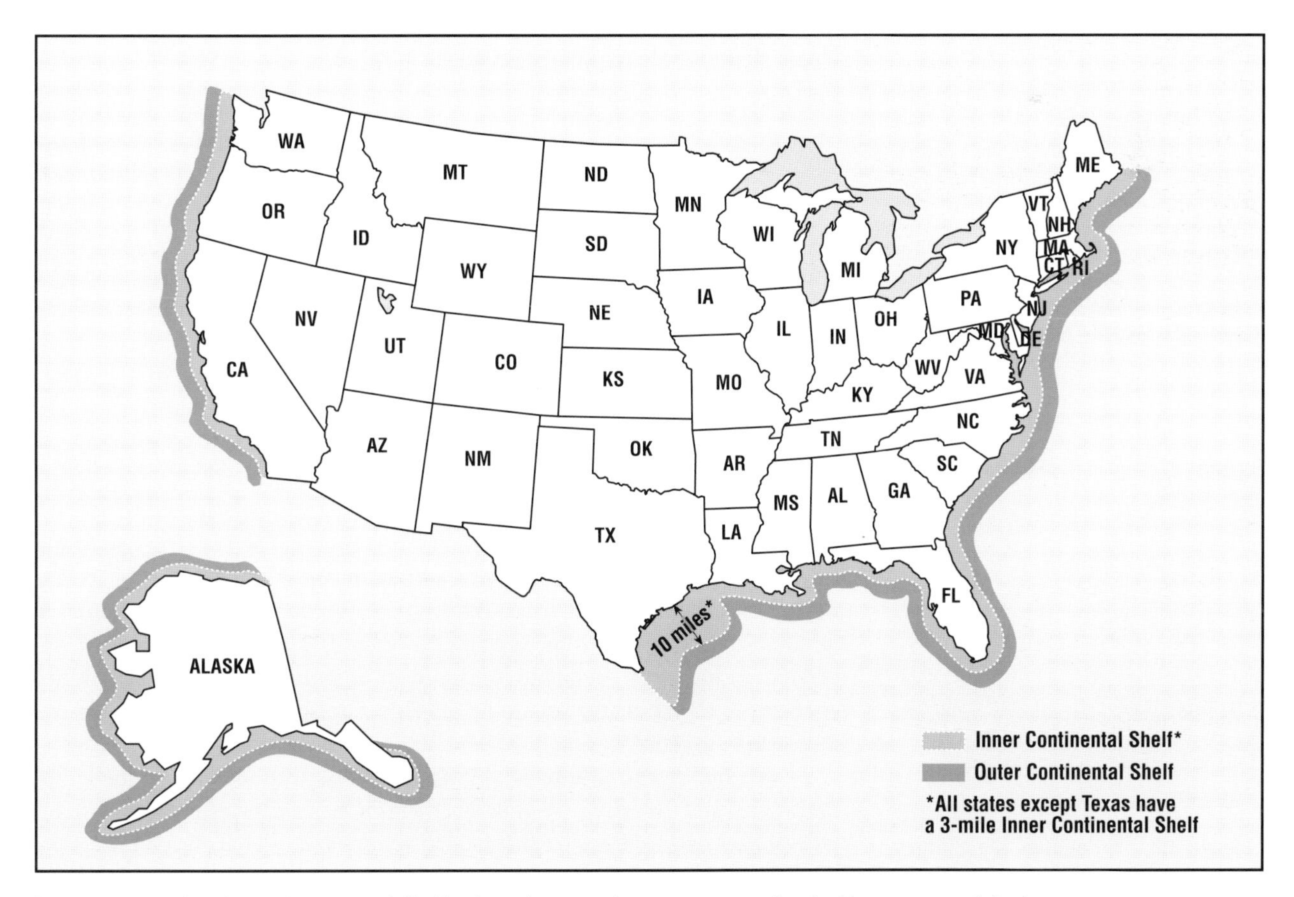

Figure 8.4 The Outer Continental Shelf is leased to petroleum companies by the Department of the Interior.

Figure 8.5 These waste disposal containers are clearly labeled according to OSHA regulations.

Protection of Human and Other Life

OSHA

The *Occupational Health and Safety Act* of 1970 established the *Occupational Safety and Health Administration (OSHA)* to regulate safety and health in the workplace. One of OSHA's guidelines that affects the petroleum industry is the *Hazard Communication, or "Employee Right-to-Know" Standard*. It guarantees employees the right to know about chemical hazards on the job and how to protect themselves from those hazards. Employers must list hazardous materials, label them, and train employees in handling them (fig. 8.5).

Another OSHA regulation, the *Hazardous Waste Operations and Emergency Response Standard (HAZWOPER)*, requires employers to protect workers who are responding to emergencies involving hazardous wastes.

Wildlife Protection

Two main acts protect wildlife in the United States. The Department of the Interior's Fish and Wildlife Service administer both. One is the *Migratory Bird Treaty Act (MBTA)*, a name that covers five treaties signed with other nations since 1918 to protect migrating birds. Provisions that concern the petroleum industry cover birds injured in pits or open-topped tanks that contain oil, oil products, caustic chemicals, or poisonous contaminants such as boron or arsenic.

The *Endangered Species Act (ESA)* of 1973 protects species on the *List of Endangered and Threatened Wildlife and Plants*. Once a species is on the list, all federal agencies must ensure that federal actions will not harm it or its habitat. Although oil companies are not federal agencies, they often drill on federal lands, so they are subject to this act.

Hazardous Materials and Wastes

Since we use petroleum products every day, we may not think of gas and oil as being hazardous materials. But they are toxic when breathed or ingested. The industry also uses many different chemicals in drilling, production, processing, and refining that are dangerous to people and other life.

The *Toxic Substances Control Act (TSCA)* regulates the manufacture, processing, distribution, use, and disposal of chemicals. EPA maintains a list of chemicals that are or once were used in commerce, and no facility can manufacture a chemical that is not on the list. Industries that use these chemicals must report where and how much they use or produce. Naturally occurring chemicals are exempted, so a pipeline company need not report oil transported from the well to the refinery. However, refineries must report the production of, for example, benzene, toluene, and xylene from the oil.

Another law, the *Resource Conservation and Recovery Act (RCRA)*, regulates the disposal of hazardous wastes. Most wastes from exploration, drilling, and production operations are exempt from this law, but the same materials produced or used during transportation or refining are not exempt. For example, unused fracturing fluids or acids, refinery wastes, caustic or acid cleaners, pipeline waste, laboratory wastes, incinerator ash, and used drums and insulation must all be disposed of in proper containers, labeled, and shipped to a special dump site.

International Laws and Treaties

Over the years, there have been 13 international treaties regarding safety and pollution prevention to which the United States has been a signatory, beginning with the *Safety of Life at Sea Treaty* in response to the Titanic disaster in 1912 and continuing to the recent *Oil Spill Preparedness and Response Treaty*. Many other countries that produce and use petroleum products have laws to protect the environment and human health. For example, Great Britain passed the Environment Act in 1995, giving an environment secretary authority to draft standards. The European Economic Community has also worked with industry to create standards for European producers and refiners. Norway's industry and energy minister has created a campaign to solve his country's offshore environmental problems. He has involved the fishing and petroleum industries and environmentalists to try to avoid a confrontational approach.

SAMPLE PROBLEMS AND THEIR SOLUTIONS

Since environmental and safety regulations affect many aspects of the petroleum industry, companies are continuously researching ways to comply with the laws and still make money. Changing from old tried-and-true technologies to new ones is a challenge that sometimes brings unexpected benefits in recycling and efficiency.

The following sections relate some examples of environmental hazards and methods that the drilling, production, transportation, and refining industries are using, or in some cases trying out, in order to reduce the harmful effects of petroleum operations, processes, and products.

Closed-Loop Drilling System

A *closed-loop drilling system* recirculates drilling mud, making a reserve pit unnecessary. States have various regulations regarding reserve pits. Some require the operator to reclaim the site after drilling or use an impermeable lining or dispose of liquids off the site. The advantages of a closed-loop system are that it lowers construction and reclamation costs and uses less water as well. The solids that remain after drilling can sometimes be reused to line irrigation ditches, feed lots, and landfills if the operator has not added too many chemicals to the mud.

Synthetic-Based Drilling Fluid

Spent drilling fluids and cuttings comprise the largest quantity of waste from exploration and drilling. Their impact is especially significant offshore, where muds that do not meet EPA standards must be barged away for disposal on land. EPA standards do not allow the discharge of oil or of mud with a high particulates, cadmium, or mercury content, for example. No drilling mud or cuttings at all can be discharged within 3 miles (4.8 kilometres) of shore.

Figure 8.6 Oil that has washed ashore from a spill is thick and sticky. *(Courtesy of U.S. Coast Guard)*

Drilling fluids are generally either water based or oil based. Onshore drilling operations usually use water-based muds, but offshore horizontal drilling often requires oil-based muds. Oil-based muds have greater impact if they are disposed of on site. Oil and cuttings cut off oxygen to sediment where they are dumped, and recovery is slow. At many drill sites in the North Sea, cuttings lie in huge piles 100-150 feet (30-45 metres) high and 200 feet (60 metres) across. Oil-coated sand and pebbles smother bottom-dwelling life (fig. 8.6), and oil leaches out as far as 2.5 miles (4 kilometres) from one site. If the mud and cuttings are not dumped on site but transported out, the transporting causes air pollution and uses energy, disposal threatens groundwater, and worker safety during loading and unloading is an issue.

Since 1990, mud suppliers have offered nontoxic, biodegradable, synthetic-based muds, which may be a solution to the high costs and problems of disposing of mud offshore. Synthetic-based mud is reusable and recyclable, so the quantity of spent mud and cuttings is lower. It also allows faster drilling, especially in the horizontal wells common in offshore drilling, which reduces air emissions from the prime movers (table 8.1). Although it costs more than oil-based mud, the cost may be offset if the EPA will allow disposal onsite. Tests are showing it is less toxic than oil-based mud, so EPA approval may be forthcoming for this new technology.

TABLE 8.1

COMPARISON OF WELLS USING WATER-BASED MUD AND SYNTHETIC-BASED MUD

WBM Wells	Footage Drilled	Footage Per Day	Mud Cost, Million $	Cost per Foot, $/ft	Total Well Cost, Million $	Total Days
1	17,981	138	1.329	73.93	11.568	163
2	16,928	63	2.538	149.97	18.268	326
3	17,540	82	—	—	9.563	214
14	17,142	101	1.550	90.44	12.671	197
15	17,381	215	—	—	10.092	77
SBM Wells						
5	16,842	301	0.806	48	4.976	50
6	18,122	275	1.707	94	7.848	75
7	17,250	431	0.776	45	3.703	33

Source: Marathon Oil Co.

Mud Additives from Waste

On the flip side, Russian companies have developed chemical additives for drilling mud from the waste of other industries. After the collapse of the Soviet Union, the usual sources for such chemicals were outside Russia and became expensive to import. In searching for a new source for these chemicals, researchers looked at industrial wastes and by-products that contain the raw components for drilling mud additives. Some of the wastes they are using are cake from sewage disposal plants, starches from expired biological materials, and by-products of the metallurgical, perfume, vitamin, and microbiology industries.

Detecting Contaminated Water and Soil

Soil can become contaminated with oil and hazardous chemicals because of pipeline breaks, leaking aboveground or underground tanks, truck spills, and leaking refinery and storage facilities. A method called the passive soil-gas technique is a new way to identify the presence of volatile organic compounds, chlorinated hydrocarbons, semivolatile organic compounds, and other contaminants in groundwater, surface water, and soils. It maps areas of suspected pollution—where the contaminants are going and whether or not they arise from multiple sources. It also identifies complex mixtures of contaminants. Because the way the information is collected offsets the influence of variations in temperature, barometric pressure, and moisture, the technique is useful in Arctic as well as tropical environments and in deserts as well as in bottom sediments. It is also inexpensive and provides quality data.

Cleaning Contaminated Soil

A number of ways exist to clean, or *remediate*, contaminated soil so that it can be safely left in place or reused as fill.

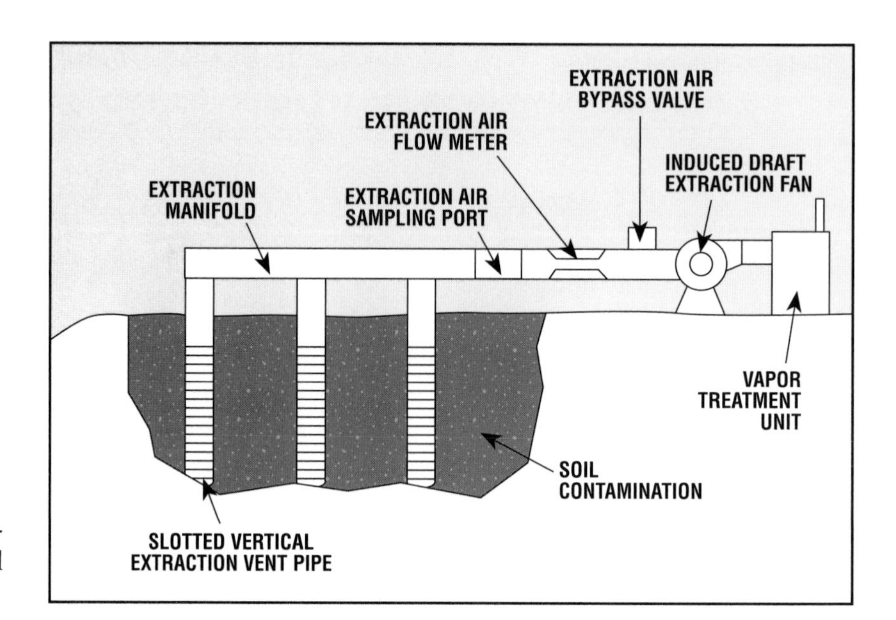

Figure 8.7 A blower extracts the volatile components from soil contaminated with oil.

Mechanical Methods

In a process called *volatilization*, or *soil venting*, workers place slotted pipes down into the soil (fig. 8.7). The slotted pipes meet a horizontal pipe at the surface attached to a fan that blows out, creating a partial vacuum in the soil. As a result, the oil *volatilizes*, or evaporates, passes into the slotted pipes, and flows out to a treatment unit on the surface. Volatilization is a limited form of remediation; only the lighter fractions of the oil volatilize, and the process does nothing to remove metals and nonvolatile compounds.

In *leaching*, a chemical is mixed with water and added to the soil (fig. 8.8). The chemical either collects or reacts with the contaminant as the mixture soaks down through the soil. Before it reaches the water table, the mixture is collected, usually using recovery wells. The recovered water mixture then must be purified by fractionation, centrifugation, solvent extraction, or some other means. None of these purification methods is totally effective, and research continues.

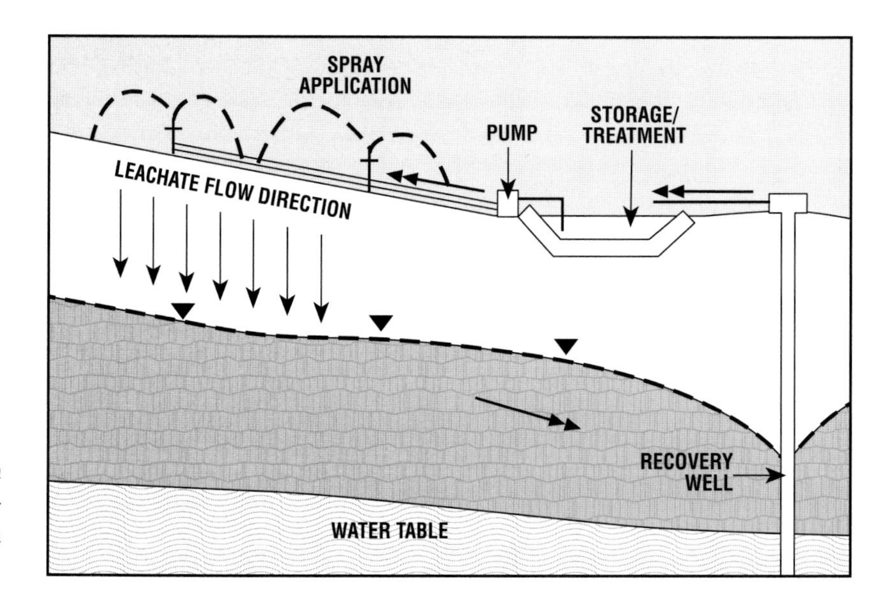

Figure 8.8 Chemicals mixed with water and sprayed into the soil leach contaminants from the soil and into a recovery well.

Contaminated soil can also be treated with heat or electricity. In a thermal treatment, the soil is excavated and incinerated. This method requires air pollution control systems and disposal of the ash or soil that remains. *Vitrification* is a treatment in which electricity is applied to the soil to melt it. Hydrocarbons are volatilized or encapsulated in hardened soil. The soil takes one to two weeks to cool, and the process generates gases that must be treated.

Biodegradation

A wide variety of bacteria break down petroleum hydrocarbons, essentially by eating them. In addition to hydrocarbons, these bacteria need oxygen and water. A reclamation company can promote the action of such bacteria by mixing fertilizer with water and spraying or irrigating the soil. An oxygen source, such as hydrogen peroxide, can be added to the water, or the soil can be tilled periodically to supply oxygen to the bacteria. This process of *biodegradation* is similar to composting.

Where the water table is close to the surface, making a water source readily available, biodegradation is economical without moving the soil (fig. 8.9). In other sites, the reclaimer may remove the soil and put it onto a lined pad. These biodegradation methods take from three to eighteen months.

To hasten biodegradation, a reclaimer can excavate the contaminated soil and pile it on top of a plastic liner with layers of crisscrossing slotted pipes every 1 to 2 feet (30 to 60 centimetres). Fertilizer may be added between rows of pipes. Black plastic, which absorbs heat, covers the pile. The heat volatilizes the lighter hydrocarbons and speeds biodegradation of the rest. The pipes carry the produced gases away and can also be used to add air to the pile. In a few months, the reclaimer can return the soil to the site as fill.

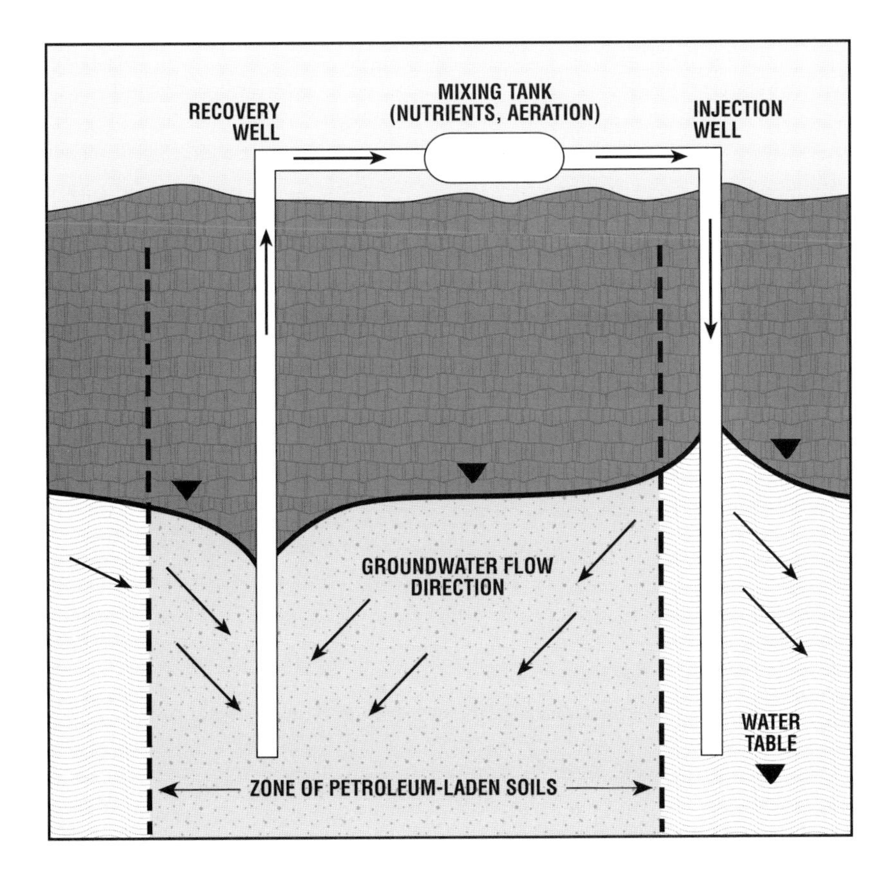

Figure 8.9 In biodegradation on site, groundwater is pumped to the surface, fertilizer is added, and then the groundwater is returned to the soil.

Figure 8.10 A blowout and subsequent fire can completely destroy a drilling rig. Although spectacular and dangerous, blowouts are relatively rare events.

Recycling

If an asphalt plant is nearby, the reclaimer may take contaminated soil there to be added to the asphalt. Because of the high heat involved in producing asphalt, the lighter fractions volatilize. The rest becomes part of the asphalt. There are concerns that some contaminants, leaded gasoline for example, may leach out of the asphalt during use.

Passive Methods

If environmental regulations allow, the company responsible for the contaminated soil may do nothing but test it from time to time to see if the site is cleaning up on its own by means of natural biodegradation, volatilization, photolysis (degradation by light), leaching, and adsorption. They may use liners or impermeable caps to contain the contaminated soil and prevent the contaminants from migrating due to the action of rain.

Blowouts

In addition to the initial explosion and the possibility of fire (fig. 8.10), blowouts can cause a toxic cloud of hydrogen sulfide gas (H_2S) to accumulate close to the ground, especially in low spots. In low concentrations, H_2S smells like rotten eggs but it quickly deadens the sense of smell. It is colorless, corrosive, and flammable, producing sulfur dioxide when it burns. It kills by paralyzing the nerve centers that control breathing (table 8.2).

Table 8.2
EFFECTS OF HYDROGEN SULFIDE

Concentration	Reaction
100 ppm	Coughing, eye irritation, loss of smell after 2–5 minutes
200 ppm	Marked eye and respiratory tract irritation after 1-hour exposure
500 ppm	Loss of consciousness and possibly death in 30 minutes to 1 hour
700 ppm	Rapid unconsciousness, cessation of breathing, and death
1,000 ppm	Unconsciousness with early cessation of breathing and death in a few minutes even if removed to fresh air at once

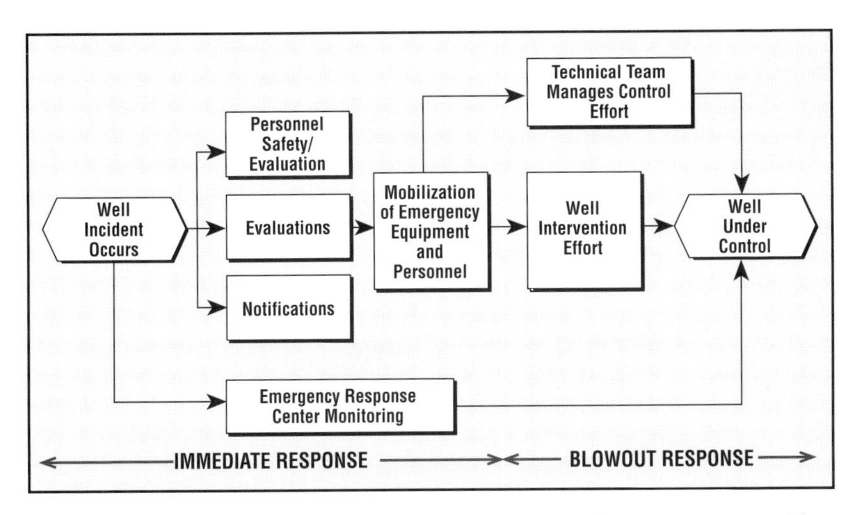

Figure 8.11 A contingency plan helps the response to a blowout run smoothly.

Of course, blowouts are accidental and not an inevitable consequence of drilling, so prevention is the first priority. Drilling companies must have plans for preventing and anticipating blowouts as well as contingency plans for dealing with a blowout (fig. 8.11). Prevention includes fixing minor leaks before they become so big that they require an expensive, complicated process to repair. Operators must train employees how to react in case of a blowout and have breathing equipment readily available. They should also have water available to prevent the well from igniting.

Pipelines

Pipelines carrying both crude oil and petroleum products operate under the Department of Transportation. A pipeline is considered a closed system. However, product sampling and routine station operations expose the petroleum to the air and people to the petroleum. For this reason, OSHA and the EPA are also involved in regulating pipeline companies. Other than these routine hazards, the main environmental concern is with leaks and spills. Pipelines can develop leaks or breaks due to excavation damage or corrosion damage. Heavy rains and flooding can expose lines that have been safely buried. This happened in Houston in October 1994, when flooding exposed lines that had been buried for 30 years, and debris propelled by the water damaged them. The damaged pipelines spilled crude and refined products into the San Jacinto River. In the same month, rains washed out a dam in the Russian Arctic and caused crude to spill from a pipeline. Whenever a leak or spill occurs, fire is also a danger, as is evaporation of the exposed petroleum.

While spills are easy to spot, the main difficulty with leaks in a pipeline is detecting them. Computer systems for detecting leaks have become common. The effectiveness of such a system is measured by how small a leak it can detect and how quickly. The system may include temperature-sensing pipes to detect thawing in permafrost areas, flowmeters to measure flow rate, and transmitters to measure the pressure at pump stations and at several valve sites. One system in Canada polls the signals from these sensors every 20 seconds, stamps the time, and sends them to a computer. The computer software then analyzes the data. It can detect leaks as small as 4 percent.

Figure 8.12 Oil was lightered, or transferred, from the Exxon *Valdez* into another tanker. Only about one-fifth of the tanker's cargo spilled. *(Courtesy of U.S. Coast Guard)*

Spills from Tankers

Exxon's tanker, the *Valdez*, rammed into a reef in 1989 and spilled the largest amount of oil in U.S. history, some 11 million gallons (41.6 million litres) (fig. 8.12). The spill killed plants and animals over 2,000 miles (3,220 kilometres) of shoreline and 500 miles (805 kilometres) out to sea, devastating fishing, hunting, tourism, and recreation in the area. According to the Alaska Oil Spill Commission, a *Valdez*-size spill occurs somewhere in the world once a year; on the average, a million-gallon (3.8 million-litre) spill happens every month.

The *Valdez* spill presented logistical, technical, and scientific challenges never before encountered. Many people and organizations participated in containing it and cleaning it up. Alyeska Pipeline Service Co. (an Alaskan spill response firm) responded immediately, the Coast Guard oversaw the cleanup, Exxon was in charge of directing and paying for the cleanup, and the Alaska Department of Environmental Conservation had jurisdiction over water quality and fisheries. Many volunteers—fishers, conservationists, and residents—donated equipment, expertise, and time to help keep the oil away from salmon hatcheries and parts of the shore.

Since the spill, response plans for remote offshore areas have improved. Several international oil companies have funded the *Marine Spill Response Corporation (MSRC)*. About 70 other smaller firms, mainly cooperatives, have also come into being. These groups are designed to deal with spills from tankers, platforms, rigs, and pipelines off the U.S. mainland, Hawaii, the U.S. Virgin Islands, and Puerto Rico. Alyeska Pipeline Service Co. does the same for Alaska. It keeps an arsenal of boats, booms, and skimming equipment ready for the next spill emergency. In addition, a cottage industry of cleanup consultants has arisen.

Prevention

The Coast Guard, which creates guidelines for offshore safety, reports that human error is the main cause of maritime accidents of all kinds. According to the Coast Guard, fatigue, inadequate pilot-bridge coordination, and inadequate technical knowledge, especially of the use of radar, are the main types of problems. In the offshore oil industry, human error is a particular danger when navigating rough seas in narrow bays, as often happens in the North Sea and off Alaska. Preventing accidents depends in part on training and better use of available technology. Prevention is important because spills are not easy to clean up, especially in bad weather.

The use of double hulls for oil tankers is an engineering solution that may help prevent spills when accidents do occur. Some question the usefulness of double hulls, but the evidence is that they may prevent some spills. In 1996, the *Sea Empress* struck rocks off of South Wales. A single-hulled vessel, it spilled some 77,000 tons (70,000 tonnes) of oil. Earlier, in 1995, a double-hulled vessel, the *Borga*, had struck the same rocks and spilled no oil.

Cleaning Up the Sea

The response to a spill must be fast to work well. Spills are easier to clean up before they reach shore and before the oil weathers.

The first problem is detecting where the spill actually is. Usually this is done visually from ships or aircraft, which means the response team is limited by darkness, fog, and rain.

The first line of defense is to send special response vessels that have extra deck space for helicopters, booms, boom boats, and skimming systems. These vessels also have oil/water separators and storage for the captured oil. A *boom* (fig. 8.13) is a long floating fence that a boom boat tows slowly to enclose the spill or to protect a part of the shore. A boom only works where the waves are not more than two to three feet (60 to 90 centimetres) high and the wind is not over 15–20 knots. Waves on the Outer Continental Shelf, for example, are routinely higher, so the effectiveness of booms is limited to very calm seas.

Figure 8.13 A boom is a flexible air-filled bag that prevents spilled oil from spreading further. *(Courtesy of U.S. Coast Guard)*

Figure 8.14 A skimmer above a storage tank, with a connected boom. The skimmer and boom swing over the side of the boat to enclose and suck up the oil. *(Courtesy of U.S. Coast Guard)*

A *skimmer* is like a huge vacuum cleaner (fig. 8.14). Skimmers work only in an *oil slick*, oil that is floating on the top of the water, and only in calm seas. The crew needs airborne spotters to direct them to the middle of the slick, and they need a place and equipment to offload the captured oil.

Rough seas begin to convert crude and water into an emulsion called a *mousse*. If the response is fast enough to reach the spill before the mousse forms, it can be burned or dispersed using chemicals. Burning may be a good solution in remote areas where logistics prevent other types of quick responses (fig. 8.15). Burning removes 50–90 percent of the oil if it has not yet emulsified because of the action of the waves. As happens in leaks and soil contamination, the volatile fractions evaporate; after this occurs, the oil will not burn. A danger of burning is that it releases toxic gases into the air.

Figure 8.15 Setting fire to the spilled oil can clear away much of it. *(Courtesy of U.S. Coast Guard)*

Figure 8.16 Workers spread absorbent sheets to soak up oil from the beach. *(Courtesy of U.S. Coast Guard)*

Using a *dispersant* to break the oil down into tiny drops requires weather and sea conditions energetic enough to mix the oil and dispersant, but not so rough that an aircraft carrying the application equipment cannot fly. The dispersant must be applied repeatedly and no longer works after the oil has weathered into a mousse. The use of dispersants is controversial. Some are more toxic to marine life than oil.

Other chemicals such as gelling agents, sinking agents, demulsifiers, and burning agents can also be dropped by aircraft.

Cleaning Up the Shore

Waves and tides often bring spilled oil onto shore and then wash it out again to other shores. Its danger to wildlife is due to both its toxicity and the fact that it smothers plants and animals or destroys the ability of fur and feathers to retain heat.

Oil onshore can be removed manually with shovels, buckets, and absorbent materials (fig. 8.16). Some cleanup teams have used high-pressure hot water to wash it out from between rocks or low-pressure cold water to move the oil back into water to be skimmed or collected.

In time, patches of asphalt and tar develop onshore after the volatile factions have evaporated and the oil has weathered. The final effort may be to apply a chemical fertilizer to stimulate the growth of natural oil-degrading bacteria, as described in the section on cleaning contaminated soil. The top layers of soil can be removed with a backhoe and then fertilizer applied, or the fertilizer can be tilled into the soil.

Figure 8.17 Wildlife, such as the birds in this puddle of oil, can suffocate when covered with oil, and topsoil is contaminated. *(Courtesy of U.S. Coast Guard)*

One study, on the cleanup of a marsh in Brittany that had been heavily contaminated by oil from a spill in 1978, showed that the cleaned-up sections recovered more slowly than the sections left alone. The efforts to remove the oil by scraping off the top layer of soil and uprooting plants, combined with the damage done by people trampling the marsh, actually did more harm than good. Such accusations have been leveled at the *Valdez* cleanup (fig. 8.17).

Cleaning Up Shallow Waters

Cleanup in shallow waters can be more problematic because the usual response vessels are too large and have too deep a draft. The traditional way to deal with oil in shallow waters was to let it wash onto shore and then clean it up. Now, however, barges that are trailer-mounted to transport by truck can navigate the water in places like Galveston Bay.

After a spill in Galveston Bay in 1990, the response team used a containment boom and skimming equipment to collect about 30 percent of the oil, and oil-degrading bacteria got 85–100 percent of the rest. Shallow-draft, flat-bottomed boats went up to the marsh areas, carrying the bacteria in dry form. The crew mixed the bacteria with seawater and sprayed the mixture into the marsh through fire hoses with pumps.

Hazards to Cleanup Workers

The workers and volunteers who clean up ocean spills are exposed to a number of dangerous materials and conditions. First, the petroleum itself is hazardous. Inhaling gases, getting chemicals on the skin, and ingesting any petroleum products are potential dangers. In addition, exposure to the environment and weather can cause sunburn, illness from cold or heat, bites and stings, and hearing injuries from equipment noise. The danger of fire and drowning are always present as well. A well-organized cleanup effort will provide protective clothing and gear and a safe place for workers to eat, drink, and receive first aid treatment (fig. 8.18).

Figure 8.18 Face masks and protective clothing are important items of protective gear that cleanup workers may need.

Refining and Petrochemical Production

The environmental concerns in regard to refineries are twofold. First, the refining processes themselves use a lot of water for cooling and other purposes and emit toxic gases and dust. Both water and air emissions must be treated before release from the refinery. Second, increasingly strict regulation of fuel emissions forces refineries to manufacture cleaner-burning gasoline and other petroleum-based fuels. Petrochemical plants have many of the same problems as refineries in regard to controlling water and air pollution caused by the manufacturing processes.

Water Quality

When water is discharged from a refinery or petrochemical plant (*effluent water*), it must be pure enough to meet the demands of local and state water quality boards, the EPA, and other regulatory agencies. Refineries use chemical treatment tanks, holding ponds, and oil-degrading bacteria to remove contaminants such as hydrogen sulfide, ammonia, phenols, and salts from the water (fig. 8.19).

In effluent water storage tanks, hazardous wastes settle on the bottom, so the contents must be separated into water, oil, and solids. The water is recycled through the refinery's treatment plant where any oil is recovered for reuse. Then the water is centrifuged and heated to separate out the solids, which are disposed of in a landfill. Because disposal is costly, refineries try to separate out the most oil and water possible.

Figure 8.19 A bioreactor uses microorganisms to remove impurities from the water. When not fed oil, grease, ammonia, and other contaminants, the bacteria stop working. *(Courtesy of Valero Refining Co.)*

Air Quality

Toxic air emissions that are of particular concern to refineries are *volatile organic compounds*, or *VOCs*. VOCs are the light fractions of petroleum that evaporate at atmospheric temperature and pressure, such as benzene (table 8.3). They cause smog and ozone formation near the ground.

Dust is another problem in refineries. Petroleum coke dust can blow off in air or run off in water. It is a health hazard for workers and fouls machinery, which causes the machines to need more frequent maintenance.

Table 8.3
PETROLEUM CONSTITUENTS

Listed in Decreasing Volatility
1-Pentene
(n) Heptane
(n) Hexane
(n) Pentane
Toluene
Benzene
Ethylbenzene
(o) Xylene
Naphthalene
Phenanthrene
Phenol
Benz (a) Anthracene
Benzo (a) Pyrene

Coke contains hazardous metals such as vanadium, copper, iron, and nickel. The dust is made up of very fine particles that are easily airborne and repel water. For this reason, spraying plain water does not control the dust. A relatively new chemical that can be added to spray water allows better wetting properties; that is, it allows the dust particles to stick to water droplets. This system has reduced dust by 95 percent on roads and 90 percent inside the facility.

Other air pollution controls include floating roofs on storage tanks that eliminate space for vapor buildup, periodic checks for leaks along storage tank walls, vapor collection systems, scrubbers to remove sulfur from combustion gases, and electrostatic precipitators to collect dust.

Refining Clean-Burning Fuels

Since vehicle emissions are among the most significant and widespread air pollution problems, standards for them have become increasingly strict. Lead as an additive to boost octane was phased out in the 1970s, gas volatility in summer temperatures had to be reduced in 1989, and low-sulfur diesel fuel was mandated in 1993. From 1970 to 1990, the emission of particulates dropped 59 percent, VOCs 29 percent, sulfur dioxide 25 percent, and carbon monoxide 41 percent. In 1995 oil companies had to begin introducing *reformulated gasoline*, or *RFG*, in some parts of the United States. RFG is a special gasoline blend that burns with less hydrocarbon and other pollutant emissions. By 2000, the phase-in of RFG should be complete, lowering total air pollution in cities still further.

Starting in 2004, EPA will require trucks and buses to lower emissions. One solution may be the use of compressed natural gas instead of gasoline as a fuel. To burn cleanly, carbon dioxide and hydrogen sulfide must be removed from natural gas. Dow Chemical has developed a process that uses special solvents to remove these two gases. The solvents can be reused until they become contaminated by a hazardous material. When that happens, the contaminant is removed and recycled. Another environmental benefit is that this process is more efficient than previous ones, so it removes more contaminant using smaller pumps and less energy.

Gasification

A trend is for refiners to seek out processes that produce higher-quality, lower-sulfur fuels and gasoline from lower-quality crudes. Forecasts have indicated that crudes of the future will be progressively heavier (have fewer light fractions) and their sulfur content will be higher. Some crudes, particularly some from Mexico and Venezuela, also contain much higher concentrations of metals. At the same time, environmental restrictions are forcing reduced sulfur in fuel oils.

The method usually used to remove sulfur and metals is hydrotreating, but it becomes impractical when the metals content is too high. *Gasification*, or gas cleaning, is a 40-year-old process used for chemicals production that may replace hydrotreating in the future. In gasification, a variety of feedstocks, including intermediate (low-value) refinery streams, petroleum coke, and waste products, are heated to 2,350–2,600°F (1,255–1,427°C). Carbon dioxide, sulfur, steam, and a gas called *synthesis gas*, or *syngas*, are produced. Depending on the feed into the gasifier, the syngas will have different percentages of useful chemicals (table 8.4).

Table 8.4

TYPICAL SYNGAS COMPOSITIONS

Component	Syngas Composition, —Mole%— Heavy Oil Feed	Coke Feed
CO	45.62	47.72
H_2	43.32	30.33
CO_2	8.17	17.88
H_2O	0.27	0.12
CH_4	0.35	0.01
Ar	1.00	0.83
N_2	0.53	1.27
H_2S	0.71	1.76
COS	0.00	0.02
NH_3	0.00	0.00

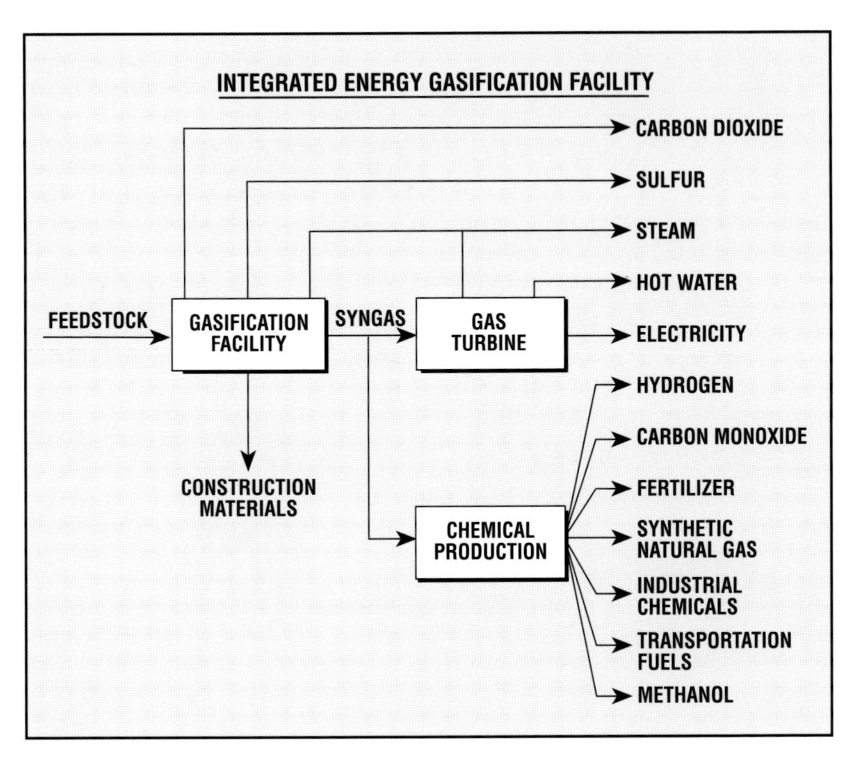

Figure 8.20 A gasification facility converts low-grade crudes to many useful products.

Syngas can be used to produce steam or generate electricity by means of a gas turbine, or it can be further processed to separate it into its components (fig. 8.20). One product of syngas is synthetic natural gas, a clean fuel gas that produces minimal emissions of sulfur dioxide, nitrogen oxides, particulates, VOCs, and carbon monoxide.

Safety

Safety is an important issue to refinery workers. In 1995 refinery explosions and fires caused twelve deaths and numerous injuries. In one explosion, at a Pennzoil refinery, five people were killed and four injured, and the main processing units were down for over two months. This accident stimulated a new emphasis on safety by the union.

9 Petroleum Marketing

Marketing is where the petroleum industry meets the consumer. The marketing department of any organization has two functions: (1) to accurately interpret consumer needs and relay these to its company and (2) to correctly inform the consumer public of the quality, quantity, and kinds of products its company has available. To carry out its functions, the marketing department must manage planning and research, buying and selling, and advertising and public relations so that the right products are in the right place, at the right time, at the right price, and in the right quantity. These five "rights" are the reasons for all marketing functions.

THE CHANGING MARKET

The detail and depth at which marketers pursue the different marketing functions depend, of course, on the nature of the company they are serving and the prevailing market. Political pressures, new technology, new discoveries, and different product uses have all influenced marketers through the years.

Types of Marketers

Basically, two types of marketers operate in the petroleum industry: vertically integrated companies and independent marketers. Naturally, their marketing methods differ.

A *vertically integrated oil company* is one that is large enough to be involved in all aspects of the oil industry. Its marketing practices are integrated within the company's exploration, drilling, production, refining, transporting, and distribution activities. Such a company sells some or all of its automotive products at a service station bearing the company emblem, thereby carrying its petroleum from the reservoir to the retail outlet.

An *independent marketer,* on the other hand, functions at only one of several levels in the marketing chain. An independent buys either the overstock of name-brand refiners or products from independent refineries. The independent company may be a broker that buys and sells mostly on paper, or it may take delivery of the products. An independent company often purchases spontaneously in step with the market of the moment rather than according to a long-range plan. Or it may have a contract with its suppliers for a certain length of time or certain amounts of product. An independent marketer may also work in storage, transportation, and wholesale or retail distribution.

Development of Products Market

Petroleum was first valued in the late 1800s for its potential as a lighting fuel (in the form of kerosene). Since then, the petroleum market has grown astronomically, thanks largely to the automobile. Because of the booming demand for internal-combustion engine fuel, refiners started turning less of the crude oil into kerosene and more into the products that would blend into gasoline.

Until 1912, refiners sold gasoline that was distilled from the original crude (straight-run gasoline) or condensed it from natural gas. With the advent of the refining process called cracking (see Chapter 7), they could produce more gasoline from the crude. This gasoline had a higher *octane rating* than straight-run gasoline from the distilling process. This number represents how efficiently the gasoline burns. The higher the octane rating, the less likely is the gasoline to cause the engine to knock. *Knocking* happens when the gasoline-air mixture in the engine explodes rather than burns evenly. The knock, or ping, is the sound of small explosions as the gasoline is inefficiently burned by the engine. Knocking is harmful, so the octane rating of the gasoline must be matched to the engine's characteristics.

As vehicle engines evolved, they required gasoline with even higher octane ratings to prevent explosions and engine damage. Refiners began adding tetraethyl lead to increase the octane rating, and most gasoline sold from the 1920s through the 1970s contained lead. In the mid-1970s, concerns over the amount of lead emitted in automobile exhaust led to the banning of leaded gasolines in the United States. Now the need is for high-octane unleaded gasolines.

As the octane race intensified, refiners developed new processes: reforming, polymerization, and alkylation. These made it possible to produce a higher percentage of gasoline out of the crude barrel, with an even higher octane rating. In addition, heavy oil cracking units turn residuals into unleaded gasoline with sufficient octane ratings. In response to the requirement for unleaded gasoline, blenders have started adding octane enhancers, such as methyl tertiary butyl ether (MTBE). Varying supplies and prices of such enhancers also contribute to the challenges of marketing gasoline.

Gasoline for automobiles is not the only use for petroleum today. Marketers also manage, buy and sell, and market jet fuels, propane, and petrochemicals.

Natural Gas Industry

In the 1800s, gas was manufactured in Europe and America by heating coal, coke, and other solid fuels and then passing steam over them. This manufactured gas was used for lighting in streets, homes, and businesses but was considered a luxury (fig. 9.1). It had an important effect on the eventual development of the natural gas industry, however, because it was so much cleaner and easier to use than some of the other fuels in common use at the time. By the beginning of the twentieth century, natural gas was available as a fuel for homes and industries, but costs were high because natural gas was difficult to transport from gas wells to urban areas.

The development of more efficient and larger gas pipelines beginning in the late 1920s played a key role in making natural gas an acceptable and affordable fuel in the United States (see Chapter 6). For many years, the United States was by far the main consumer of natural gas in the world, but by the 1970s both environmental concerns and a desire by oil-importing countries to become less dependent on foreign oil led to an increase in use of natural gas. This increase would probably have been even greater if natural gas were not so expensive to transport. Pipelines, which are expensive to build, are the only economical way to move it in its gaseous state, and costs for liquefying it for transport by ship are also high. Still, demand has dramatically increased in Japan, Europe, Latin America, Africa, and the countries of the former Soviet Union.

Figure 9.1 Gaslights were a nineteenth-century luxury. Note the gas table lamp. *(Courtesy of American Petroleum Institute)*

Another of the main reasons demand has increased is that natural gas is plentiful. At the current rate of consumption, the world's gas reserves will last 59 years, as compared to only 32.5 years for oil reserves. In addition, exploration is uncovering more new gas deposits than new oil deposits.

As demand increases, international trade in natural gas will likely continue to grow. The leading exporter is the former Soviet Union, which supplies gas to Europe. The Netherlands and Norway supply other European countries via pipeline, and Canada pipes gas to the United States. Algeria and Indonesia ship gas by both pipeline and as liquid natural gas (LNG)—Algeria to the United States and Indonesia to Japan.

Consumption of natural gas in the United States fell during the two decades after 1970 because of the recession triggered by the 1973 Arab oil embargo and regulation-induced shortages. Proposals first made in 1996 to repeal some of the regulations on production may allow the U.S. gas industry to stabilize and grow in the next decade.

RESEARCH AND PLANNING

The first job of a marketer, whether the marketing department of a vertically integrated company or an independent, is to analyze and try to predict the petroleum market. This is a complex task, often involving seemingly contradictory information and trends. For example, just when the global way of life was committed to using petroleum at an all-time high, the oil industry realized that the worldwide supply of that resource was limited. On the other hand, conservation programs were in their heyday about the time an oil "glut" became apparent—a surplus that resulted from the high cost of imported oil and its effect on domestic drilling. Political factors play an ever-increasing role in conservation and environmental issues. Users of petroleum and its products have been asked, and in some cases forced, to slow down in the past; but the current trend against government regulation of exploration, development, and production, especially in terms of environmental concerns, may have a big effect on marketing in the near future.

Marketers today must prepare for a faster changeover to other energy sources than in the past. Many nations feel that an orderly changeover to other fuels can be accomplished more easily if petroleum is sensibly conserved and therefore used at a slower rate in the coming decades. But predicting this future rate in the face of the expected demand for a continually improving quality of life and the worldwide growth in population is a difficult task for a planner.

Market Assessment

Marketing planners research for three different time periods—the present, the near future, and the more distant future. Policies governing *present* marketing are constantly analyzed for effectiveness. The size of the current market, which brands are selling, and where retail distribution is occurring can all be measured. A planner must take seasonal factors into account—more gasoline sells during the summer vacation months, so prices rise, and heating fuels are needed during the winter months. Because heavier grades of diesel are converted into heating fuels, diesel is not as readily available, thus forcing diesel prices upward in cold months. Research within a company can determine how well its advertising and public relations efforts are working. For example, because gasoline is not a product that consumers like to think much about, creating advertising for it has been difficult. Company marketers watch consumer behavior closely for changing responses to a product or service.

The *near future* is usually defined in the marketing world as the next three years. This phase of research usually involves pretesting of new products, services, methods, or publicity. Marketers distinguish between pretesting and test marketing. *Pretesting* (analysis of a product) is done under simulated market conditions, and *test marketing* is conducted under real conditions of the ever-changing market.

Predicting the *more distant future* requires both science and intuition. Marketers find clues to the future in both the present and the past to predict coming patterns with some accuracy. They can feed demographic indicators such as population numbers, size of families, and educational and financial levels into computers to predict future consumer trends. Marketing research reduces the risk of making decisions for the future; but with the rapid changes in technology, especially in the petroleum industry, petroleum marketers will always need sound judgment.

Supply and Demand

Once the planner assesses the consumer market, the next task is to match marketing requirements with resources—in other words, to balance supply and demand. A marketing planner must be able to project whether the company's supplies will be adequate to meet consumer demand without leading to production surpluses. Demand fluctuates for many reasons, some of which are not easy to discern. Therefore, prices are extremely volatile. Oil prices can rise or drop suddenly when an oil-producing country changes policies or becomes unstable. For example, a decision by President Clinton in 1996 to sell off more of the nation's Strategic Oil Reserves than usual caused world crude prices to drop suddenly, even though the economic impact of his action was much less important than the psychological one.

Traditionally, supply has been able to grow with demand, but a time may be coming in the petroleum market when supply will dictate demand. Along with this challenge, a marketing planner must consider possible sources of crude, refinery capacity, and availability of shipping and pipelining. With massive data input, the planner must be able to predict with some accuracy a balance between input and output of an entire system. In a large oil company, meshing together all activities into an efficient and profitable operation becomes a complex task.

International Influences

Oil prices are ultimately set by supply and demand, but international politics have had a definite impact on prices. Because oil is so lucrative, in most countries the government owns and controls all the petroleum. Therefore, oil is an instrument of internal economic policy and often international policy as well. Perhaps the most dramatic example of this is the Arab oil embargo in 1973, when 13 countries in the Middle East joined together as *OPEC (Organization of Petroleum Exporting Countries)* and forced the price of crude up. Political changes in the Middle East are hugely important because that area possesses 66 percent of the world's known oil reserves (fig. 9.2). A border war in the Middle East can immediately influence the price of oil in Houston. The planner must be flexible enough to take into consideration the unpredictable political implications of the industry.

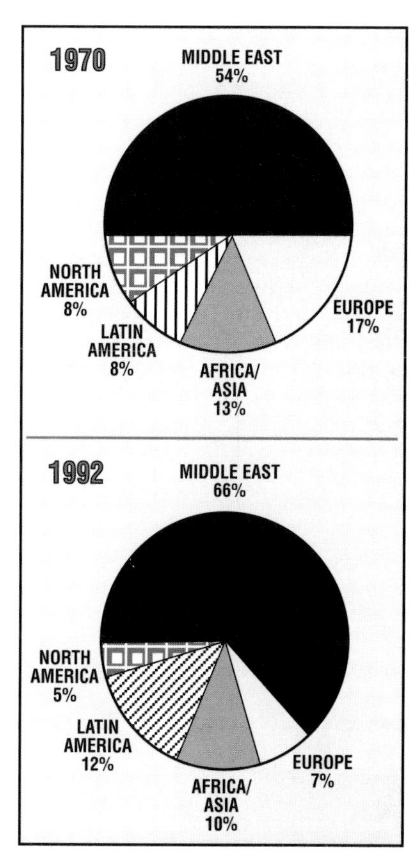

Figure 9.2 Geologists estimate that North America has only about 5 percent of the total oil reserves in the world.

Influence of Technology

Technology also is changing so quickly that a petroleum product that seems irreplaceable today may become obsolete tomorrow. Researchers and planners must keep abreast of technological breakthroughs that, although difficult to predict, may prevent the cost of energy from rising. The exploration and development of two major oil areas discovered in the 1960s—the North Sea and Alaska—cost ten to twenty times as much as developing similar fields in the Middle East. But since the early 1980s, new exploration techniques, such as satellite imaging and better well logging techniques (see Chapter 2), and new drilling technology, such as deepwater platforms (see Chapter 4), have brought down the costs of drilling and development.

Refinery Capacity

Planning for future refinery capability is complicated by the fact that it takes about five years to construct and power up a major new refining plant.

A company must expect to make enough profit over the life of the refinery to justify the amount of capital required to build it. Smaller investment projects at existing refineries may include increased dock facilities, environmental protection facilities, or new storage areas.

New government regulations may require extensive restructuring on the part of some refineries. An example is the effect of the changeover to unleaded gasoline ordered by the Environmental Protection Agency (EPA) beginning in 1984. Most U.S. refineries were able to meet this deadline, but as EPA rules have become even more stringent, many small refineries proved too costly to upgrade and were closed.

Transportation Scheduling

Having a means of transportation ready to pick up and deliver the proper quantities of feedstock and the right selection of products to and from refineries is another task of planners. For example, transportation planners, or schedulers, must accurately estimate how much feedstock a refinery will need in order to take advantage of the fact that the cost per unit of feedstock usually declines if the refinery buys a larger quantity at one time. Schedulers must also select the type of carrier and make sure the trucks or ships are at the proper loading points at the right time, monitor crude and product stocks to have them ready for loading, and choose appropriate routings.

Market fluctuations, weather, vessel engine trouble, or political upsets may affect transportation timing and coordination. Cargoes cannot be shipped before they are available, nor can a planner allow so much feedstock to accumulate at the refinery that lack of storage space becomes a problem. A transportation planner must consider details such as a ship's capacity, pipeline availability, access to inland waterways, and a myriad of other transportation-related factors.

Federal Price Regulations

Government regulations controlling the wellhead prices of crude oil and natural gas greatly affected the end prices of petroleum products from 1971 to 1981. Among the forces that create prices—availability of reserves, entrepreneurial freedom, and market competition—governmental regulation has been a prime contributing factor. The oil industry has long operated under a variety of governmental regulations and tax burdens. The emphasis after the enactment of President Jimmy Carter's National Energy Plan and the *Natural Gas Policy Act (NGPA)* was toward deregulation. Some wellhead prices for oil were allowed to rise to world levels to encourage further exploration and development, but a large portion of the increased industry profits was taken by the government in the form of the *windfall profits tax.* This tax was part of the Energy Security Act of 1980 and continues to be in effect.

Intrastate and Interstate Marketing of Natural Gas

Intrastate gas (gas that does not cross state lines as it is piped from producer to consumer) traditionally has not been federally controlled, whereas the *Federal Energy Regulatory Commission (FERC)* regulates *interstate gas* (gas that travels across state lines). The two markets have developed different strategies and problems as a result.

Most of the gas delivered in intrastate markets is to large industrial firms and electric utilities. This fact, along with minimal regulation and surplus supplies, has created a relatively free market that is very competitive. Profit margins are not large, but the intrastate gas industry is a model of how an unregulated environment can work. The NGPA used this model in creating an elaborate set of rules to phase out natural gas price controls in the interstate market. The immediate result of this deregulation effort was high prices, due to a type of contract known as *take-or-pay*. In take-or-pay contracts, pipeline companies guaranteed producers that they would buy a certain amount of gas. When demand was not as high as expected, the pipelines were stuck with the contracts and prices rose. However, as deregulation has continued, the FERC has taken steps to eliminate take-or-pay contracts.

The interstate market for natural gas is still evolving and faces a number of problems. Should gas sell at prices based on the actual cost of production? Will pipelines be able to renegotiate old contracts based on new market prices? If gas prices are allowed to rise to compete with world market crude, will that action allow oil-producing countries to set the price of U.S. natural gas as well as oil?

Computer Evaluations

Planners use sophisticated computer programs to evaluate the many options available. They enter certain data into the program and then evaluate the practicality of the computer solution in terms of the present, near future, and more distant future. The most economical future solution is not always the most desirable in terms of current conditions.

Computer programs, or *software*, are available to plan almost any aspect of marketing. For example, one type of software may help the marketer plan the best way to transport and store oil or natural gas, by taking into account available routes, volume of the product, final destination, weather conditions, and costs at different points. Another program may simulate buying and selling transactions, or *deals*, using up-to-date prices; forecast changes in risk by using "what-if" models; and track the costs when a deal is made. Yet another type of software may track production, calculate royalties, automate reporting to regulatory agencies, and analyze well performance. The personal computer has changed the way marketers do business, in the same way it has affected almost all other businesses.

BUYING AND SELLING

The methods of buying and selling petroleum products vary depending on the product. A marketer usually specializes in crude oil, natural gas, or petrochemicals, as each requires a thorough knowledge of that aspect of the industry and somewhat different skills.

Crude Oil

A petroleum marketer buying crude plans not only for the immediate needs of the company, but also for its need to build up or deplete stocks for the future. Planners figure the different grades and quantities of crude that the company will have available at any one time. Then they may recommend that the marketing company buy from another company or abroad rather than use its own crude, even when the company is also a producing concern. Or they may recommend that the company trade for different grades of crude than it produces. If an oil company owns storage and transportation facilities, it can be more flexible about holding and selling its petroleum products.

To do this planning, marketers work with researchers to keep up with worldwide statistics on the availability of crude, its prices, and government legislation. Such marketing information, like that shown in Table 9.1, is essential knowledge. Marketers either keep abreast of current news and scientific journals themselves or they employ a consulting service to do so. Commonly used marketing reports are *Platt's Oilgram Price Report* and *Inside FERC.*

Three major methods of buying crude today are the traditional *contract market*, the *spot market* where oil is bought and sold daily, and the *futures market*.

Contract Market

The contract market is the oldest system for trading oil. Suppliers, such as major oil companies, offer contracts to their customers that are in effect for a certain length of time. The contracts specify all terms except price, which is fixed at the time of the sale.

Contract trading is becoming less important to the U.S. oil and gas industries because of fluctuations in prices both seasonally and as a result of political forces. The length of time a contract spans used to be long-term—several years—but now is usually six months to a year and may even be as short as a month. No longer are oil and gas sold from one producer to a pipeline company to a local distribution company to the end user. Instead, buyers and sellers usually trade on the spot market or the futures market or both.

Spot Trading

Spot trading is a one-time transaction in which a product, or commodity, is purchased "on the spot" at the current market price. Trading firms and individuals around the world buy and sell over the phone every day. Usually, large companies with shipping facilities leave a *margin*, or portion, of their shipping capacity to be acquired by spot buying.

Futures Market

Today, most oil and natural gas are bought and sold, or *traded*, on an exchange similar to the stock exchange called the New York Mercantile Exchange (NYMEX). NYMEX is a commodities market. A *commodity* is any material product, as opposed to a service.

Table 9.1

SHORT-TERM WORLD PETROLEUM SUPPLY AND DEMAND, 1991–2Q'93[a]
(millions of barrels per day)

Region	1991	1992 1Q	1992 2Q	1992 3Q	1992 4Q	1992 Year	1991–92 Change	1993 1Q	1993 2Q
Demand									
North America	18.6	18.7	18.5	18.9	19.2	18.8	1.10%	19.2	18.7
Europe	13.4	14.0	12.9	13.2	14.0	13.5	0.70	14.0	13.0
Pacific	6.1	6.7	5.7	5.7	6.4	6.1	0.00	6.8	5.8
Total OECD[b]	38.1	39.4	37.1	37.8	39.6	38.5	1.00	40.0	37.5
China, Asia, and former USSR	16.6	17.1	16.1	15.4	16.1	16.1		16.5	15.8
Eastern Europe	1.2	1.2	1.1	1.0	1.0	1.1		1.1	1.1
Latin America	5.3	5.1	5.4	5.6	5.6	5.4		5.3	5.6
Middle East and Africa	5.6	5.9	5.8	5.9	6.0	5.9		6.0	6.1
Total non-OECD	28.6	29.2	28.3	27.9	28.6	28.5		28.9	28.6
Total demand	**66.7**	**68.6**	**65.4**	**65.7**	**68.2**	**67.0**	**0.60**	**68.9**	**66.1**
Supply									
Total OECD	16.3	16.8	16.2	16.3	16.7	16.6		16.7	16.3
Former USSR	10.4	9.5	9.2	8.8	8.7	9.0		8.5	8.3
China	2.8	2.8	2.8	2.8	2.8	2.8		2.8	2.8
Other non-OECD	10.6	10.7	10.7	10.8	10.9	10.8		10.9	10.9
Total non-OECD	41.4	41.2	40.4	40.2	40.6	40.6		40.3	39.7
OPEC crude production	23.3	24.1	23.6	24.4	25.5[a]	24.4[a]		26.5[a]	
OPEC natural gas liquid production	2.1	2.1	2.1	2.1	2.1[a]	2.1[a]		2.1[a]	
Total OPEC	25.4	26.2	25.7	26.6	27.6[a]	26.5[a]		28.6[a]	
Total supply	**66.8**	**67.3**	**66.1**	**66.8**	**68.2[a]**	**67.1[a]**		**68.9[a]**	
Difference between total demand and total supply	0.1	-1.3	0.7	1.1	0.0[a]	-0.1[a]		0.0[a]	

[a] Estimate
[b] Organization for Economic Cooperation and Development

Source: International Energy Agency

In the futures market, open bidding for contracts determines prices on the NYMEX trading floor. A *futures contract* is a commitment to deliver or receive a specific quantity and grade of oil or product during a designated future month at a preset price. Brokers charge a commission for trading these contracts (fig. 9.3). Actually, however, the crude is delivered to the buyer only rarely because most participants *liquidate*, or resell, their positions before this designated future time arrives.

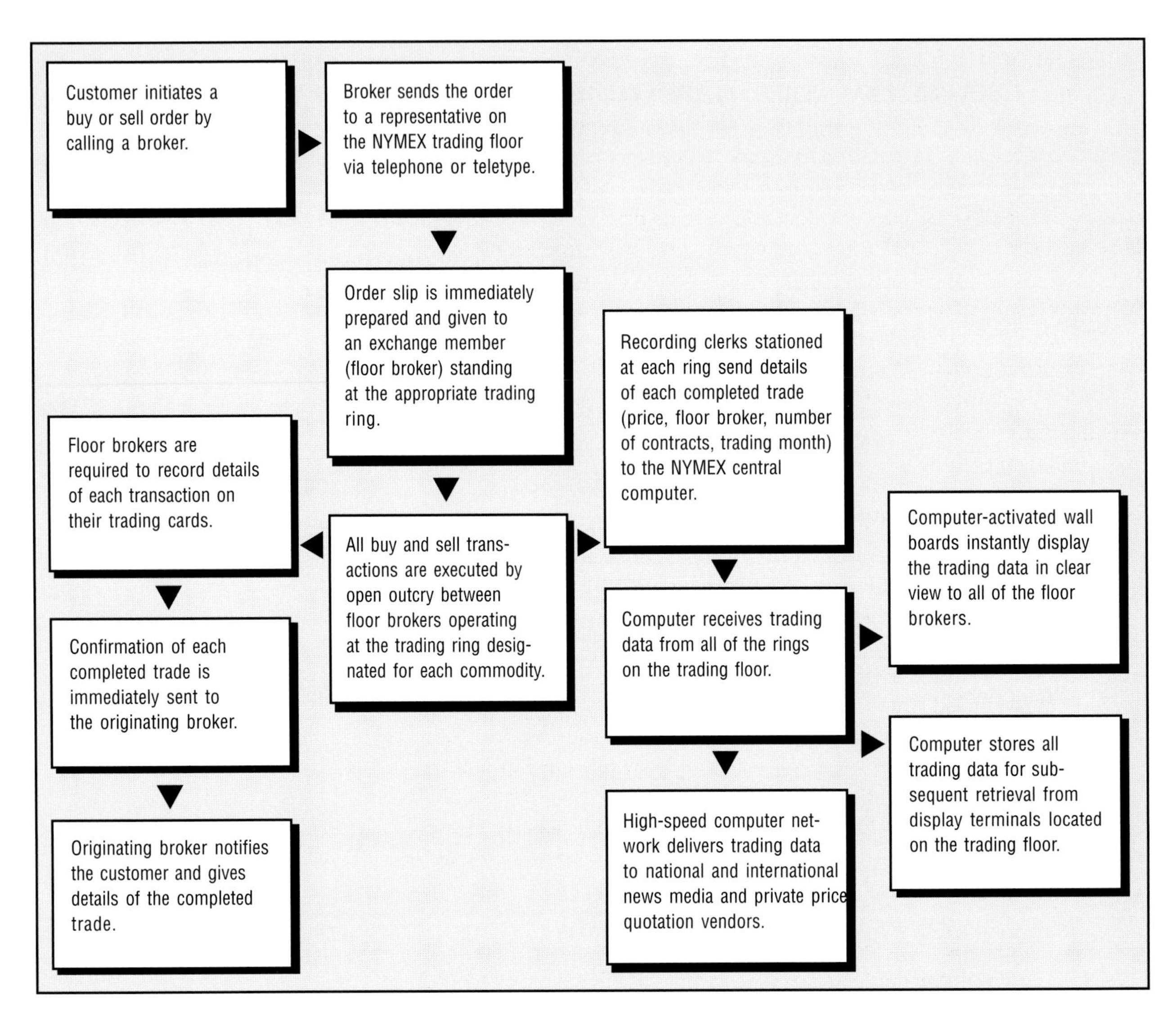

Figure 9.3 A typical futures transaction begins with a customer's buy or sell order by telephone to a broker.

Because of the uncertainty in oil prices in recent years, futures trading has become the marketers' preferred way to buy and sell because it allows them to hedge. *Hedging* is an insurance device for keeping losses down and protecting profits during the production, storage, processing, and marketing of commodities. It works like this: a marketing company can refrain from selling some of its product through contract or spot trading because it believes prices are going up. But if prices drop, the marketer can sell the product on the futures market and lock in a better price for later delivery. That contract may then continue to be exchanged on the futures market among other marketers with similar concerns and expectations. The futures market provides a way to lessen risk in the volatile pricing situation that is the petroleum market.

Trading through NYMEX has helped to "democratize" oil transactions by making them transparent, understandable, and accessible. It has encouraged growth of consumption by allowing oil consumers to manage risk. Through futures trading, the oil industry manages and even exploits oil price uncertainty, and today NYMEX crude oil futures options are the world's most actively traded commodity option.

Petrochemical Sales

Marketing petrochemicals is different in several respects from marketing conventional petroleum products. Few petrochemicals are end products themselves. They are usually feedstocks or intermediate chemicals that are sold to manufacturers, who then turn them into such articles as tires, fabrics, plastics, and many other products. Also, some primary petrochemicals are used as intermediates for the manufacture of other petrochemicals.

The volume of petrochemicals is relatively small compared with those of gasolines and fuel oils, and market development is more of a challenge than for conventional petroleum products. The petrochemical market fluctuates with the demand for products made from them. For this reason, many vertically integrated oil companies have stayed out of petrochemical marketing. They may instead sell their primary feedstocks directly to chemical companies or manufacturers.

Natural Gas Transactions

Before 1985, federal and state regulatory agencies controlled the rate of production and the price of natural gas. A few major pipeline companies bought natural gas at the wellhead and transported it to processing facilities or distribution centers. But in 1978, the Natural Gas Policy Act started deregulation of the price charged by producers for "new" gas—gas discovered since 1977—at the wellhead. This act was designed to achieve deregulation gradually, with the total release of price controls effective January 1, 1985. In addition, starting in 1984, the rulings of the Federal Energy Regulatory Commission (FERC) made the pipeline industry more competitive. FERC forced pipeline companies to release their customers from long-term contracts, to refrain from requiring a minimum gas volume or a minimum dollar contract from their customers, and to transport third-party gas for a reasonable price.

Today, deregulation has created a competitive, fast-paced market for natural gas. Gas is brokered through contracts, over the phone with the help of personal computers. Few contracts exceed six months or a year, and spot trading is more popular every day. The next change is expected to be the integration of the electricity, gas, and oil markets as the electric utilities are deregulated. For example, in the future a gas-fired electric utility might find it more profitable to sell its gas to other buyers and buy its electricity from some other source. Competition is becoming fiercer as industrial and urban customers look for the lowest energy costs.

Products Sales and Distribution

The task of distributing petroleum products from processing plants and refineries to the final consumers is complex and often costly, but it has been streamlined considerably in the past several years. The combined costs of product storage and handling, transportation, and delivery can add up to more than half the total operating costs of a marketing department. The fewer handling or reloading operations done per trip, the lower the total cost will be.

Terminals

Consumer products, such as gasoline and home heating oil, are usually sent directly from the refinery via pipeline, tanker, barge, or railroad to an installation called a *terminal.* A terminal is a storage and loading area close to a large population center and usually owned by a refinery or large oil company.

Figure 9.4 A flash-point indicator is used to verify fuel specifications.

Figure 9.5 A tester checks gravity of the fuel at the terminal.

Personnel at a terminal are responsible for certain quality checks before they can sell products. These tests may include a *flash-point check* (fig. 9.4), to verify that each product vaporizes within the proper temperature range, and a *gravity check* (fig. 9.5), to check the fuel's density. Aviation fuel is sent to a laboratory for testing before it can be sold. These tests have to be repeated at the terminal because products may have altered during transport from the refinery.

From the terminal, products are transported to bulk plants serving smaller communities or directly to service stations, airports, homes, or businesses. Although most large volumes of product are transported by rail, barge, and coastal tanker, truck transport is the most popular form for final distribution. Large industrial consumers, the exception, are usually supplied by water, rail, or pipeline.

Most terminals will have several truck loading devices, commonly called *loading racks* (figs. 9.6–9.9). The rack is equipped to load the truck compartments with several kinds of products at the same time and to pipe vapors off the top of the truck compartments in vapor recovery efforts. Such a terminal often supplies a number of different kinds of consumers.

A refining company may supply its own service stations and other company-brand outlets. The terminal's owning company may contract with a *jobber* (an independent middleman) to act as a wholesale agent for that brand of product in a given trade area. A jobber may also own a retail outlet which resells the product from the terminal. Independent service station owners may purchase their fuels from terminals or from jobbers. Jobbers may purchase propane to deliver to rural homes.

At large terminals, the handling of different types of orders, loading instructions, dispatching, and invoicing has largely been computerized. Automatic invoicing transmits information quickly to a company's central computer, which allows control of credit and recording of sales statistics,

Figure 9.6 A truck operator prepares to load up with several types of products at the terminal.

Figure 9.7 A truck is loaded by attaching the rack hoses.

vehicle statistics, and stock records. A number of physical functions previously performed manually are now automated. During truck loading, the identity of the driver and vehicle, the amount and description of product exchanged, and product destination can be recorded automatically. Some terminals are equipped with visual display units that enable terminal employees to call up computerized files of consumer data and invoices.

When a truck is loaded for final delivery, a planner must design the most efficient route to take. Industrial customers and many service stations can take full truckloads, but farms and domestic customers probably will

Figure 9.8 A truck operator uses a card for automatic invoicing.

Figure 9.9 The terminal meter records type and amount of products purchased.

not need enough fuel to empty a full truck. Planners use computers to plan loads and routes. Efficient routes lower costs, of major importance to all marketing companies.

PETROLEUM PRODUCT CONSUMPTION

Examining how the crude barrel is broken down into its marketable products is important since most of our comforts depend on products and services that use some form of petroleum (fig. 9.10). Besides power for most modes of transportation, petroleum products play an important role in housing (fuel and electricity), clothing (petroleum-based synthetic fabrics), and food (fertilizers, pesticides, cookware, and packaging). Beyond these basic needs, many petroleum-derived products add to our quality of life.

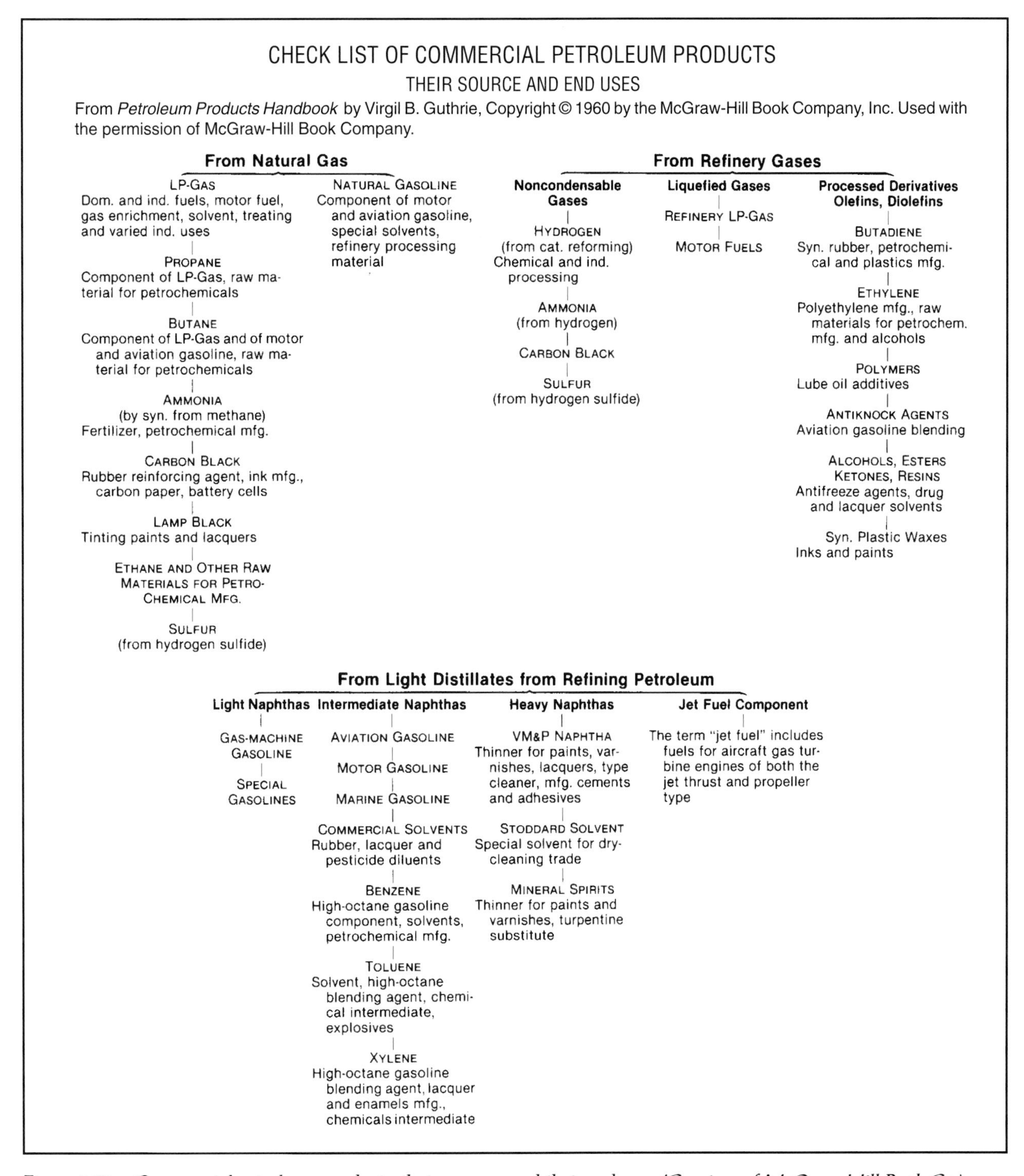

CHECK LIST OF COMMERCIAL PETROLEUM PRODUCTS

THEIR SOURCE AND END USES

From *Petroleum Products Handbook* by Virgil B. Guthrie, Copyright © 1960 by the McGraw-Hill Book Company, Inc. Used with the permission of McGraw-Hill Book Company.

From Natural Gas

LP-GAS
Dom. and ind. fuels, motor fuel, gas enrichment, solvent, treating and varied ind. uses

PROPANE
Component of LP-Gas, raw material for petrochemicals

BUTANE
Component of LP-Gas and of motor and aviation gasoline, raw material for petrochemicals

AMMONIA
(by syn. from methane)
Fertilizer, petrochemical mfg.

CARBON BLACK
Rubber reinforcing agent, ink mfg., carbon paper, battery cells

LAMP BLACK
Tinting paints and lacquers

ETHANE AND OTHER RAW MATERIALS FOR PETRO-CHEMICAL MFG.

SULFUR
(from hydrogen sulfide)

NATURAL GASOLINE
Component of motor and aviation gasoline, special solvents, refinery processing material

From Refinery Gases

Noncondensable Gases

HYDROGEN
(from cat. reforming)
Chemical and ind. processing

AMMONIA
(from hydrogen)

CARBON BLACK

SULFUR
(from hydrogen sulfide)

Liquefied Gases

REFINERY LP-GAS

MOTOR FUELS

Processed Derivatives Olefins, Diolefins

BUTADIENE
Syn. rubber, petrochemical and plastics mfg.

ETHYLENE
Polyethylene mfg., raw materials for petrochem. mfg. and alcohols

POLYMERS
Lube oil additives

ANTIKNOCK AGENTS
Aviation gasoline blending

ALCOHOLS, ESTERS KETONES, RESINS
Antifreeze agents, drug and lacquer solvents

Syn. Plastic Waxes
Inks and paints

From Light Distillates from Refining Petroleum

Light Naphthas

GAS-MACHINE GASOLINE

SPECIAL GASOLINES

Intermediate Naphthas

AVIATION GASOLINE

MOTOR GASOLINE

MARINE GASOLINE

COMMERCIAL SOLVENTS
Rubber, lacquer and pesticide diluents

BENZENE
High-octane gasoline component, solvents, petrochemical mfg.

TOLUENE
Solvent, high-octane blending agent, chemical intermediate, explosives

XYLENE
High-octane gasoline blending agent, lacquer and enamels mfg., chemicals intermediate

Heavy Naphthas

VM&P NAPHTHA
Thinner for paints, varnishes, lacquers, type cleaner, mfg. cements and adhesives

STODDARD SOLVENT
Special solvent for dry-cleaning trade

MINERAL SPIRITS
Thinner for paints and varnishes, turpentine substitute

Jet Fuel Component

The term "jet fuel" includes fuels for aircraft gas turbine engines of both the jet thrust and propeller type

Figure 9.10 Commercial petroleum products, their sources, and their end uses *(Courtesy of McGraw-Hill Book Co.)*

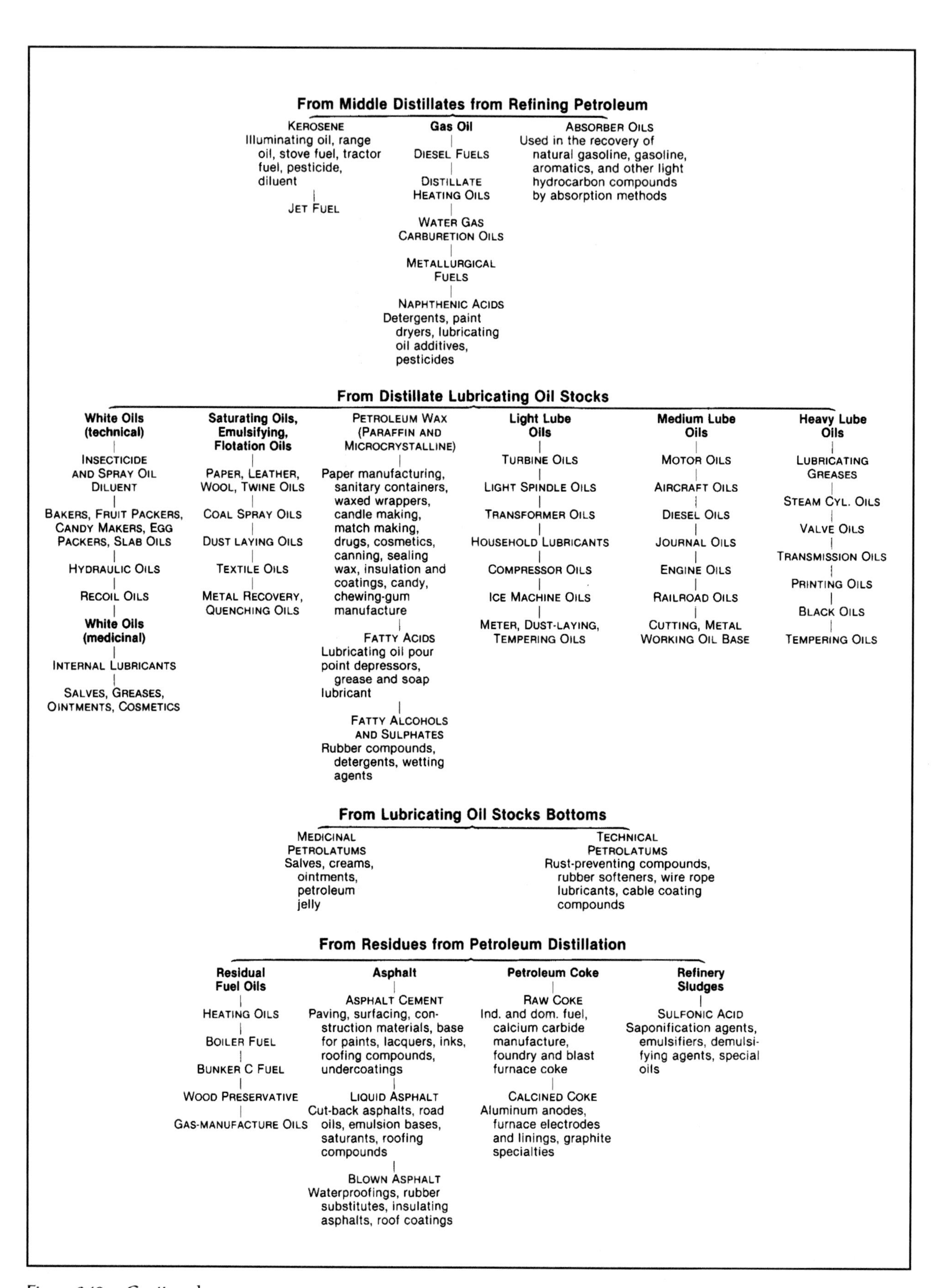

Figure 9.10 Continued

Marketable petroleum products can be divided into broad classes: transportation, electrical generation, industry, home, commerce, and agriculture. These general areas are supplied by different cuts of crude, including gases, liquefied petroleum gas, gasoline, aviation fuels, kerosene, diesel fuel, lubricating oils, waxes, fuel oils, and miscellaneous products.

Automobiles

Consumers of automotive fuels experience the effects of petroleum marketing most directly at the retail outlet—the service station. Several transactions occur in the marketing chain of operations as gasoline moves from the terminal to the station pump. However, deregulation in 1981 eliminated price controls on the sale of petroleum products as well as on the sale of crude oil. Therefore, prices are now largely a function of supply and demand.

To the consumer, the most important characteristic of gasoline is probably its antiknock value, or octane rating. Octane rating is tested and determined under laboratory conditions. A gasoline is marketed according to the average of its octane numbers taken from two calculation methods, called the research method (r) and the motor method (m), thus $(r + m)/2$ = octane number. Acceptable octane ratings for gasolines in most countries range from about 86 to 92.

Aircraft

The need for air-transport fuels has increased dramatically as the size, capacity, and number of aircraft have increased. The demands placed on fuel quality for aircraft are more stringent than those for automotive fuel. Aviation fuel must be free of contaminants (water or dirt), be pumpable at low temperatures, meet regulations for specific gravity and caloric value, and burn cleanly and remain stable when heated to high temperatures.

Two categories of aviation fuel are piston-engine fuel and turbojet fuel. Passenger-carrying and military aircraft have changed over from the piston engines of the past to turbojet engines. However, a need still exists for high-octane aviation fuels for piston engines in older airplanes and smaller private and commercial aircraft (fig. 9.11). Marketing planners try to predict the amount of aviation piston-engine fuel and jet fuel that will be needed each year.

Figure 9.11 Small aircraft, such as this piston-engine plane, use aviation gasoline for fuel.

Figure 9.12 A turbojet commercial airliner is filled with jet fuel.

As the quality of aviation fuel has improved, so also has the design of equipment for its storage and transfer from the refinery to the airplane. Because contamination could cause fatal accidents, aviation fuel movement is monitored very carefully. Laboratory tests are repeated at different stages to guarantee product specifications. Usually, a jobber will buy from a terminal and then store the fuel close to an airport. Different kinds of fuel are color coded for sight identification in loading and servicing.

The speed of refueling has improved greatly to meet scheduled aircraft turnaround time. Large refueling trucks carry the fuel out to an airplane and pump it under controlled pressure into the aircraft's tanks (fig. 9.12). In a hydrant system, underground pipe carries the fuel from storage tanks to aircraft refueling areas.

Generation of Electricity

A large portion of the world's natural gas goes into the production of electricity at generating plants (fig. 9.13). Oil and coal are also fuel sources for generating plants (fig. 9.14). These fuels heat boilers that generate steam; steam turbines drive the generators.

Industry

The industrial sector consumes more than half of the world's total output of petroleum products. A large portion goes to manufacturers who require energy to run their machines and processes. An example is the steel industry, whose plants consume thousands of tons of fuel oil annually. However, with the coming deregulation of the electric industry, major changes are on the horizon. Electric utilities may begin to be competitive suppliers of power to manufacturers. Besides fuel oil and natural gas, petroleum also supplies industry with a wide range of lubricants and process oils.

Products for industry may be marketed directly in bulk to large consumers, such as the armed forces, railways, and major manufacturers, or indirectly through jobbers to smaller consumers. Sales transactions tend to be long and complex.

Figure 9.13 Natural gas enters this electricity-generating plant through pipelines.

Home

The conversion from solid fuel to oil for home heating has been widespread. Particularly since the 1973 rise in crude prices, the use of gas and electricity has intensified. Natural gas and electricity provide the convenience of being piped or wired into the consumer's property, but oil has to be delivered by truck. Natural gas has become the major competitor to oil for heating. Electricity, a secondary energy source, is usually a more expensive alternative for heating but is used extensively for cooling and for operating household appliances, tools, and instruments.

Figure 9.14 Oil is stored in tanks for alternate use as fuel for generating electricity.

Figure 9.15 An LPG tank furnishes heating fuel for this farm home.

Figure 9.16 A scrubber cleans gas as it enters an underground network of pipelines at a city gate.

In rural areas and locations not serviced by natural gas pipelines, liquefied petroleum gas, or LPG (generally propane), is often used for heating. The property owners buy and maintain LPG tanks, and local distributors fill them (fig. 9.15). Owners of recreational vehicles, mobile homes, and remote vacation cabins often use small, portable propane tanks. In many rural locations, LPG dealers may keep small filled tanks on hand for sale or rent.

As of 1996, both natural gas and electricity come under the heading of public utilities. Because gas pipes and electrical wires must be part of a community network, the community contracts as a whole for services. A utility is considered a "public" service and is subject to state regulations, which influence pricing to consumers (fig. 9.16). Deregulation of the utilities is approaching, however. The next several years could see consumers able to choose their power supplier in a competitive market driven by supply and demand.

Commerce

Although gas and electricity are used increasingly for residences, heating oil continues to heat commercial buildings such as hospitals, schools, and factories. Fuel oils for domestic burners are distilled from the deeper cuts of crude after the lighter products have been taken off. The *American Society for Testing and Materials (ASTM)* divides domestic fuel oils into two grades for marketing—*number 1 oil* and *number 2 oil*. Fuel oils are marketed and distributed like gasoline, with jobbers delivering and selling to individual customers.

Agriculture

Petroleum products play a vital role in farming today. The internal-combustion engine was largely responsible for transforming the traditional rural farm into a modern industry. Between 1940 and 1983, agricultural productivity increased by 90 percent, a growth attributed to mechanization, fertilizers, pesticides, and irrigation. The development of diesel fuels and LPG for tractors and other mechanized implements, as well as the development of proper lubricants for those machines, revolutionized farming.

Figure 9.17 Fertilizer is poured into the hopper of a spreader for use on a rancher's grazing land.

Other vital petroleum derivatives used for agriculture are the chemicals used as fertilizers. Farmers also use a range of petrochemicals to protect crops from plant and animal pests. Methods of marketing this wide array of agricultural products vary. Sometimes marketing companies sell directly to the farmer or rancher, sometimes agents or jobbers are used, and sometimes the farmer or rancher buys from a local distributor (fig. 9.17).

ADVERTISING AND PUBLIC RELATIONS

The third aspect of marketing is advertising and public relations. Like planning and trading, it has also evolved over the past 100 years. During the era of allocations, controls, and short supplies, marketers needed little advertising to sell their products. However, as restraints were removed and competition returned to the petroleum marketplace, promotion once again became important to a company's success. These marketing activities may be categorized as institutional advertising, special promotions, or service-oriented public relations.

Institutional Advertising

Advertising today is aggressive as the industry continues to try to overcome its poor image. Major U.S. oil companies traditionally have run newspaper and magazine ads and sponsored radio and television programs. Some advertising and programming is geared toward educating the public on the role of the oil industry in the national and world economies. Another way to get a company's name before the public in a positive light, without actually advertising its products, is to sponsor sports events, the arts, humanities research, and entertainment programs on television. This type of advertising—known as institutional advertising—establishes a company image in the minds of its viewers. It promotes the company, or institution, rather than its products.

The newest outlet for institutional advertising is the Internet. Major oil companies have joined many other businesses in offering Web pages with educational articles, images of the oil industry, and even free software such as a screensaver that shows wildlife pictures.

Special Promotions

Marketing gasoline, the main petroleum product sold directly to the public, is not easy. People cannot usually tell the difference between brands, and buying and consuming it isn't a particularly pleasant sensory experience. One marketing method is the giveaway promotion. An oil company may give its customers stamps with each fillup that are good toward reduced prices at the pump, or pass out lottery-type tickets for a million-dollar grand prize. Other appeals have included giving away road maps, dishes, and other premiums at service stations. The idea behind these types of promotions is to build goodwill and sales volume at the same time.

Service-Oriented Public Relations

After the dramatic increase in the cost of crude oil in 1973, American oil companies began to cut expenses. They closed down unprofitable stations and reduced their marginal services at others. The familiar driveway attendant who checked the water and oil, cleaned the windshield, filled the gas tank, and aired up the tires became practically extinct. Self-service was offered to partly offset the rising price of gasoline (fig. 9.18). Price wars, which had been widely advertised in certain areas to draw customers to the lowest priced stations, were no longer valid promotional devices. Companies no longer solicited credit card accounts but rather discouraged their use by charging higher prices for credit transactions. Many service stations supplemented their gasoline business with sales of groceries, snacks, drinks, and ice for the traveler—a service that has come to be expected by the average American on the road today.

Figure 9.18 Self-service helps the consumer keep the price of gasoline down.

In the 1980s, companies again began to offer added services to promote their products and enhance their public relations. Credit cards are one service that has returned. Many gasoline credit cards can now be used for tires and auto accessories, food and drinks, travelers' insurance, motel accommodations, car rental, and even emergency cash advances. Some companies have long used their credit cards to sell mail-order merchandise of all kinds. To make their stations even more attractive, some companies have installed electronic systems at the pump to process credit card transactions immediately. These systems speed up the fillup for customers and reduce paperwork for the company.

One of the oldest services rendered by the service station is its restroom facility, and the condition of the restrooms has long influenced the traveler's choice of service stations. Not only is a poorly kept station bad for business, it is also a large contributor to bad public relations for its parent company.

Another trend is for major oil companies to join with other companies to offer a combination of services. For example, Citgo has installed gas pumps at 7-Eleven convenience stores, and Exxon has teamed with McDonald's to offer a combination drive-through for gas and burgers.

Good public relations come from customer contact, and that contact is at the retail outlet. To maintain a good overall image, or reputation, dealer-oriented companies conduct marketing seminars for their dealers and jobbers, hold national sales meetings for their sales representatives, and participate in trade shows displaying the latest equipment available for their vendors. The purpose of these activities is to keep up-to-date on market changes and the public's interests, a high priority for marketers.

References

Abbott, John P. "Barging." *Shell News,* February 1985.

Abbott, John P. 'We Just about Cover the Map." *Shell News,* February 1985.

Abel, L. William, "Planning, Training, Equipment All Crucial in H_2S Blowout," Oil & Gas Journal, June 5, 1995.

A Dictionary for the Petroleum Industry, 1st edition, Austin: Petroleum Extension Service, The University of Texas at Austin, 1991.

American Petroleum Institute. *Introduction to Oil & Gas Production.* Dallas: Production Department, American Petroleum Institute, 1983.

Amoco Chemicals Corporation. "Amoco Chemicals Corporation, Texas City Plant." Houston: Amoco Public and Government Affairs Department, 1974.

Amoco Chemicals Corporation. "Chocolate Bayou Plant, Alvin, Texas." Houston: Amoco Public and Government Affairs Department, 1974.

Anderson, Robert O. *Fundamentals of the Petroleum Industry.* Norman: University of Oklahoma Press, 1984.

Australian Institute of Petroleum. "Economic Power Moves from the Atlantic to Pacific." *Petroleum Gazette,* Summer 1984.

Baker, Ron. *A Primer of Oilwell Drilling,* 4th ed. Austin: Petroleum Extension Service, The University of Texas at Austin, 1979.

Baker, Ron. *A Primer of Offshore Operations.* Austin: Petroleum Extension Service, The University of Texas at Austin, 1985.

Baker, Ron. *Oil and Gas: The Production Story.* Austin: Petroleum Extension Service, The University of Texas at Austin, 1983.

Bell, Harold Sill, ed. *Petroleum Transportation Handbook.* New York: McGraw-Hill, 1963.

Berger, Bill D. and Kenneth E. Anderson, *Modern Petroleum.* Tulsa: PennWell Books, 1992.

Berkhout, A. J. "It Is Time to Aim for the Best." *World Oil,* December 1984.

Bland, William F., and Robert L. Davidson, eds. *Petroleum Processing Handbook.* New York: McGraw-Hill, 1967.

British Petroleum Company Limited. *Our Industry Petroleum.* London: British Petroleum Co., 1977.

Chevron Corporation. "Chevron Shipping—Battened Down and Making Way through Rough Seas." *Chevron World,* Summer 1985.

Conoco Inc. "Conoco's First Sealift Headed North." *Conoco World,* July 1985.

Conoco Inc. *World Energy Outlook through 2000.* Wilmington, Del.: Coordinating and Planning Department, Conoco, 1985.

Culbertson, Katherine, "MTBE Prices Take Lead from Gasoline, Gather Momentum from Tight Supplies," *The Oil Daily.* April 12, 1996.

Denny, Sharon. "The Hunt for Offshore Oil Moves into Deeper Waters." *The Oil Daily,* May 6, 1985.

Dobrin, Milton Burnett. *Introduction to Geophysical Prospecting,* 3rd ed. New York: McGraw-Hill, 1976.

Environmental Problems and Solutions, video, Austin: Petroleum Extension Service, The University of Texas at Austin, 1992.

Eubank, Judith. *Land and Leasing.* Austin: Petroleum Extension Service, The University of Texas at Austin, 1984.

Fasullo, Peter, "The Intrastate Pipeline Industry: The Next "Jurassic Park" or a Model for the Future?" Paper presented at the 8th annual *National Gas Transportation Association & Natural Gas Association of Houston Natural Gas Conference,* Houston, February 8, 1995.

Fischer, Joel D., and Constantine D. Fliakos. *British Petroleum: A Basic Report.* New York: Drexel Burnham Lambert, Inc., 1978.

Galloway, W. E.; T. E. Ewing; C. M. Garrett; N. Tyler; and D. G. Bebout. *Atlas of Major Texas Oil Reservoirs.* Austin: Bureau of Economic Geology, The University of Texas at Austin, 1983.

GATX TankTrain. Chicago: General American Transportation Corp., n.d.

Gregston, Terry G., *An Introduction to Federal Environmental Regulations for the Petroleum Industry.* Austin: Petroleum Extension Service, The University of Texas at Austin, 1993.

Guthrie, Virgil B., ed. *Petroleum Products Handbook.* New York: McGraw-Hill, 1960.

Harvie, David M., "International Opportunities in the Industry." Paper presented at the 8th annual *National Gas Transportation Association & Natural Gas Association of Houston Natural Gas Conference*, Houston, February 8, 1995.

Hassell, Greg, "Lighting a Fire Under Gas Sales," *Houston Chronicle,* May 1, 1996.

Hayes, C.; M. M. Thacker; and J. H. Viellenave. "Exploration Techniques Aided by Computer Tiering." *World Oil*, December 1984.

Hebert, H. Josef, "Technology Sending Oil-Drilling Deeper," *Austin American-Statesman*, May 10, 1996.

Heinemann, William H., *Geology Made Simple*. Garden City, NY: Doubleday Publishing Co., 1982.

Hinds, Bill. "Competition Comes to the Gas Industry." *Venture*, American Natural Resources Co., Summer 1985.

Horton, S.T., *The Drill String and Drill Collars*. Austin: Petroleum Extension Service, The University of Texas at Austin, 1995.

Hosmanek, Max. *Pipeline Construction*. Austin: Petroleum Extension Service, The University of Texas at Austin, 1984.

Hunt, Mike, "Drilling Site on a National Seashore Required Extra Environmental Precautions," Bright & Co., San Antonio, on Derrick Equipment's Web page: http://www.derrick equipment.com

International Energy Agency. *World Energy Outlook*. Paris: Organization for Economic Cooperation and Development, 1982.

International Petroleum Encyclopedia. Tulsa, Okla.: PennWell Publishing Company, 1985.

Laitos, Jan G. and Joseph P. Tomaine, *Energy and Natural Resources Law in a Nutshell*. St. Paul, MN: West Publishing Co., 1992.

Levander, Kai; Torsten Heideman; Henrik Segercrantz; and Gustav Lindqvist. *Ice 2000: Ice-Going Ships for the Next 20 Years*. Helsinki, Finland: Wärtsilä Shipbuilding Division, 1985.

Makinen, Eero, and Raimo Roos. "Ice Navigation Capabilities of Lunni-Class Icebreaking Tankers." Paper presented at the meeting of Eastern Canadian Section of SNAME in Montreal, November 22, 1977, and Quebec City, November 23, 1977.

Mangan, Frank. *The Pipeliners*. El Paso: Guynes Press, 1977.

Mathews, Jessica, "Wean Us from Cheap Oil," *Houston Chronicle*, May 12, 1996

Matthews, Charles, "Blueprint for a Rational Energy Policy," *Houston Chronicle*, May 12, 1996.

Megill, Robert E. "Exploration Outlook: Is the Past a Key to the Future?" *World Oil*, June 1984.

Moody, Graham B., ed. *Petroleum Exploration Handbook*. New York: McGraw-Hill, 1961.

Morris, Jeff; Annes McCann Baker; and Richard House. *Practical Petroleum Geology*. Edited by Jodie Leecraft. Austin: Petroleum Extension Service, The University of Texas at Austin, 1985.

Moses, Leslie. "From Lease to Release." *AAPL Guide for Landmen*. Fort Worth, Tex.: American Association of Petroleum Landmen, 1970.

Newton, Jim, "Remediation of Petroleum Contaminated Soils," *Pollution Engineering*, December 1990.

Oil and Gas Pocket Reference. Houston: National Supply Co., 1984.

"Oil on the Move," *Marathon World*, No. 1, 1985.

Oil Pipelines in the United States: Progress and Outlook. Washington, D. C.: Association of Oil Pipe Lines, 1983.

Our Magnificent Earth: A Rand McNally Atlas of Earth Resources. New York: Rand McNally & Co., 1979.

Parshall, Joel S. "The Moving Force." *Marathon World*, No. 1, 1984.

Petrie, Thomas A., editor, *Industry Analysis: The Oil and Gas Industries*. Charlottesville, VA: Association for Investment Management and Research, 1993.

Petroleum Extension Service. *Field Handling of Natural Gas*, 3rd ed. Austin: Petroleum Extension Service, The University of Texas at Austin, 1972.

Petty, O. Scott. *Seismic Reflections*. Houston: Geosource Inc., 1976.

Phillips Petroleum Company. "Pasadena Terminal Operation Makes Fuel Loading Easier." *PhilNews, July* 1985.

Phillips Petroleum Company. "Remote Self-Service Is First." *PhilNews*, September 1985.

Pinkard, Tommie. "Gulf Intracoastal Waterway: Texas' Big Ditch." *Texas Highways*, August 1984.

Pressure Transport, Inc. *Pressure Transport*. Austin: Texas Gas Transport Co., n.d.

Proceedings of the 34th annual *Rocky Mountain Mineral Law Institute*, July 21, 22, and 23, 1988.

Promotional brochures for Xprt™, Xprtise™, and Primo™ marketing software. EDS Gas Services, Houston, Texas.

Royal Dutch/Shell Group of Companies. *The Petroleum Handbook*, 6th ed. New York: Elsevier Science Publishing Co., 1983.

Salpukis, Agis, "Satellite Gives Oil Hunters Better Shot," *Austin American-Statesman*, April 15, 1996.

Schuch, Jon J., "Gas Supply Planning Issues." Paper presented at the 8th annual *National Gas Transportation Association & Natural Gas Association of Houston Natural Gas Conference*, Houston, February 8, 1995.

Seminar handouts, New York Mercantiles Exchange.

Sheppard, Nora, ed. *Introduction to the Oil Pipeline Industry*, 3rd ed. Austin: Petroleum Extension Service, The University of Texas at Austin, 1984.

Simson, Stephen F., and H. Roice Nelson, Jr. "Seismic Stratigraphy Moves toward Interactive Analysis." *World Oil*, December 1984.

Sklarewitz, Norman. "Shell Goes for the Guinness Book of Records with the GATX TankTrain™ System." *Shell News*. Offprint by General American Transportation Corporation, n.d.

STEP Program, American Petroleum Institute brochures and Web page: http://www.api.org/step/

Tank Cars. *The Car and Locomotive Cyclopedia*, 4th ed. New York: Simmons Boardman Publishing Corp., 1980.

Telford, W. M., et al. *Applied Geophysics*. Cambridge and New York: Cambridge University Press, 1976.

The Rail Tank Car Industry. Alexandria, Va.: Railway Progress Institute, n.d.

Thomasson, M. Ray; John Sandy; Robert G. Reeves; James R. Lucas; Bill Ehni; and Benjamin Prokop. "Non-seismic Methods: A Key to Integrated Exploration." *World Oil*, December 1984.

Van Dyke, Kate, *Primer of Oilwell Service, Workover, and Completion*, 1st edition. Austin: Petroleum Extension Service, The University of Texas at Austin, 1997.

Van Dyke, Kate, *The Drawworks and the Compound*, 1st edition. Austin: Petroleum Extension Service, The University of Texas at Austin, 1996.

Wallace, Britann E. "FERC Rule Changes Gas Marketing." *The American Oil & Gas Reporter*, October 1985.

White, D. A. "Assessing Oil and Gas Plays in Facies-Cycle Wedges." *AAPG Bulletin*, Vol. 64, No. 8, 1980.

Williamson, Harold F.; Ralph L. Andreano; Arnold R. Daum; and Gilbert C. Klose. *The American Petroleum Industry: The Age of Energy*, 1899-1959. Evanston: Northwestern University Press, 1963.

Williamson, Harold F., and Arnold R. Daum. *The American Petroleum Industry: The Age of Illumination*, 1859-1899. Evanston: Northwestern University Press, 1959.

Wood, Robert A. "Farm Power." *Marathon World*, No. 3, 1985.

Wood, Robert A. "Charting a Course Downstream." *Marathon World*, No. 1, 1984.

Wood, Robert A. "The Right Stuff." *Marathon World*, No. 1, 1984.

Index

- B -

- C -

- D -

- E -

- F -

- G -

- H -

- I -

- J -

- K -

- L -

- M -

- N -

- O -

- P -

- R -

- S -

- T -

- U -